Rethinking Rehabilitation
Theory and Practice

REHABILITATION SCIENCE IN PRACTICE SERIES

Series Editors

<table>
<tr><td>Marcia J. Scherer, Ph.D.</td><td>Dave Muller, Ph.D.</td></tr>
<tr><td>President
Institute for Matching Person and Technology</td><td>Executive
Suffolk New College</td></tr>
<tr><td>Professor
Physical Medicine & Rehabilitation
University of Rochester Medical Center</td><td>Editor-in-Chief
Disability and Rehabilitation
Founding Editor
Aphasiology</td></tr>
</table>

Published Titles

Assistive Technology Assessment Handbook, *edited by Stefano Federici and Marcia J. Scherer*

Assistive Technology for Blindness and Low Vision, *Roberto Manduchi and Sri Kurniawan*

Computer Access for People with Disabilities: A Human Factors Approach,
  Richard C. Simpson

Computer Systems Experiences of Users with and Without Disabilities: An Evaluation Guide
  for Professionals, *Simone Borsci, Maria Laura Mele, Masaaki Kurosu, and Stefano Federici*

Devices for Mobility and Manipulation for People with Reduced Abilities,
  Teodiano Bastos-Filho, Dinesh Kumar, and Sridhar Poosapadi Arjunan

Multiple Sclerosis Rehabilitation: From Impairment to Participation,
  edited by Marcia Finlayson

Paediatric Rehabilitation Engineering: From Disability to Possibility, *edited by Tom Chau
  and Jillian Fairley*

Quality of Life Technology Handbook, *Richard Schultz*

Rehabilitation Goal Setting: Theory, Practice and Evidence, *edited by Richard J. Siegert
  and William M. M. Levack*

Rethinking Rehabilitation: Theory and Practice, *edited by Kathryn McPherson,
  Barbara E. Gibson, and Alain Leplège*

Forthcoming Titles

Ambient Assisted Living, *Nuno M. Garcia, Joel Jose P. C. Rodrigues, Dirk Christian Elias,
  and Miguel Sales Dias*

Cognitive Rehabilitation, *edited by Charles J. Robinson*

Neurological Rehabilitation: Spasticity and Contractures in Clinical Practice and Research,
  edited by Anand D Pandyan, Hermie J. Hermens, and Bernard A. Conway

Neuroprosthetics: Principles and Applications, *edited by Justin Sanchez*

Physical Rehabilitation, *edited by Charles J. Robinson*

Rehabilitation Engineering, Science, and Technology Handbook, *edited by Charles J. Robinson*

Rethinking Rehabilitation
Theory and Practice

Kathryn McPherson
AUCKLAND UNIVERSITY OF TECHNOLOGY
AUCKLAND, NEW ZEALAND

Barbara E. Gibson
UNIVERSITY OF TORONTO
TORONTO, CANADA

Alain Leplège
UNIVERSITÉ PARIS DIDEROT, FRANCE

CRC Press
Taylor & Francis Group
Boca Raton   London   New York

CRC Press is an imprint of the
Taylor & Francis Group, an **informa** business

CRC Press
Taylor & Francis Group
6000 Broken Sound Parkway NW, Suite 300
Boca Raton, FL 33487-2742

© 2015 by Taylor & Francis Group, LLC
CRC Press is an imprint of Taylor & Francis Group, an Informa business

No claim to original U.S. Government works

Printed on acid-free paper
Version Date: 20141001

International Standard Book Number-13: 978-1-4822-4920-0 (Hardback)

Library of Congress Cataloging-in-Publication Data

Rethinking rehabilitation : theory and practice / [edited by] Kathryn McPherson, Barbara E. Gibson, and Alain Leplège.
    p. ; cm. -- (Rehabilitation science in practice series)
    Includes bibliographical references and index.
    ISBN 978-1-4822-4920-0 (alk. paper)
    I. McPherson, Kathryn, editor. II. Gibson, Barbara E., editor. III. Leplège, Alain, editor. IV. Series: Rehabilitation science in practice series.
    [DNLM: 1. Rehabilitation--methods. 2. Rehabilitation--psychology. 3. Disabled Persons--rehabilitation. WB 320]

RM950
617'.03--dc23   2014037702

Visit the Taylor & Francis Web site at
http://www.taylorandfrancis.com

and the CRC Press Web site at
http://www.crcpress.com

Contents

SECTION I   Rethinking the Past and Re-Envisioning the Future

SECTION II   Philosophy in Action

SECTION III   Rethinking Rehabilitation Delivery, Research, Teaching, and Policy

Foreword

Rethinking Rehabilitation

This unusual book should be essential reading for all people involved professionally in rehabilitation—clinicians, researchers, and managers. I think that each chapter in this book will raise questions that should alter practice. Moreover, it should also interest many other people such as patients, other healthcare professionals, and anyone coming into contact with or interested in rehabilitation; the issues raised will show that there is much more to rehabilitation than therapy. It is an exciting, intellectually challenging field of practice.

This book will particularly help any group of rehabilitation professionals who know that they want to develop and improve their practice and/or their service but do not know what to do. Asking team members to spend time reading and then discussing the content of this book could lead to dramatic changes and improvements in a fairly short time. It would achieve this by encouraging everyone to think about and question current practices, without either appearing to disparage the work of others or to devalue their own work.

Rehabilitation—A Holistic Enterprise?

Rehabilitation takes a holistic view of the patient. Most professionals would subscribe to this statement. In practice it is rarely true.

For a start, should one use the word *patient* in the context of a process that inevitably depends upon full participation by the person involved? The word originally meant "A sufferer; one who suffers patiently" and developed to "One who is under medical treatment" (*Oxford English Dictionary* 2012). All the meanings imply passivity and suffering.

I agree that there may not be a good word—subject? participant? client?—but many professionals and teams still consider the patient as an object to be given therapy, information, etc. As several chapters in this book highlight, this is inappropriate.

Most rehabilitation professionals generally know of and use a biopsychosocial framework or model of illness. Nevertheless, in practice, they often narrow their use of the holistic model to limited aspects.

For example, rehabilitation as a field of endeavor has been blighted by the word *physical* for many years, for example, as in "physically disabled" and "physical medicine." Many people—patients, managers, funding agencies, etc.—still focus on technological interventions. While technology no doubt offers great assistance, it should only be seen as a means toward a more important end.

This dualistic approach to health severely constrains our approach to rehabilitation. This issue is brought up directly or indirectly in several chapters. This book encourages a more holistic approach in two ways. The content of the book draws upon a much broader range of existing knowledge than most other books on rehabilitation. For example, it includes philosophy and sociology, among many other research areas.

Second, this book unashamedly emphasizes the importance of social position and social roles to all people, ill or otherwise. The overriding message is that the person consulting you or your team about the health problem he or she has is, first and foremost, a person. The person is not *disabled* or *a patient* or *suffering* from a problem, or *unfortunate* or any one of a very large number of nouns and adjectives that we use to categorize and describe people who seek rehabilitation services. The person is another human being, just like you.

A further message is that rehabilitation needs to consider and take into account the wide body of knowledge available concerning the motivations and behaviors of people. This text implicitly challenges the profession and its dependence upon models and practices, which often lack both a theoretical and an evidential basis. (No examples will be given but I am sure each reader will easily think of an example; there are many.)

Improving Rehabilitation—The Need for Theories

There is already a substantial body of evidence about what is effective and what is not effective in rehabilitation, at many levels from global organizational level (e.g., specialist services are effective) to specific detail (e.g., exercise does not increase spasticity). However, much practice and much research are entirely pragmatic, lacking any theoretical underpinning.

This book suggests that rehabilitation should join the rest of healthcare in having or developing a body of theories that are relevant to rehabilitation, in order to suggest better ways of helping people and to facilitate more targeted research.

The contributors start the process of theory development in several ways. Some discuss existing theories concerning why people behave as they do; other consider theories underlying attempts to change behavior. And different contributors approach these questions from different perspectives.

The authors are all involved in rehabilitation research and/or practice, and so are aware of the implicit assumptions commonly made. Therefore, they are able to pose questions and challenges that most readers will immediately recognize as relevant. Indeed it is likely that readers may have already thought about some of the questions themselves.

Any readers expecting the answer to the question, "what is (or should be) the theoretical basis of rehabilitation?" will be disappointed. In rehabilitation, there is no "theory of everything" or "Grand Unified Theory," any more than there is in physics. Instead there will be theories appropriate to the situation, just as Newton's theory of gravity is quite sufficient to guide rockets to the outer reaches of the solar system even though it cannot explain other aspects of gravity covered in relativity theories. There will never be a single "theory of rehabilitation."

Instead readers should expect a very interesting and challenging series of chapters. Each chapter contains a mixture of evidence and ideas that will be new to some people and well known to others. However, all chapters use the material to challenge aspects of rehabilitation practice.

Using the Book—When and How

Therefore, this book will be very helpful to rehabilitation professionals who wish to change, develop, and improve their service and their own practice, both individually and more importantly as members of a team. It will help clinicians, managers, and researchers reformulate their problem in ways that lead to solutions.

My advice would be for a group of rehabilitation professionals to agree to read one chapter and then to meet and discuss the contents for an hour or so. The group could be postgraduate staff undergoing further professional training, or teams that want to develop but are not sure what to do, or researchers looking for a new approach to rehabilitation research.

Readers must recognize that the chapters do not and cannot provide a specific correct answer to any question. For most issues, there never will be a single, correct answer. The chapters will generally allow the readers to consider a problem from a different perspective. The emphasis throughout is the perspective of the patient. To that extent, this book is not only encouraging a holistic approach but also a personalized, humanistic approach to clinical practice.

Two Additional Challenges

Finally, I would like to add two challenges that the book does not directly set out, but which arise from the content. I have no answers.

The first concerns the word or term used for the patient. *Patient* is still my preferred word, because it is widely used and understood, but it carries disadvantages. The most obvious is that it perpetuates the biomedical approach to illness, which has many major weaknesses—not least its invalid implication and assumption that the ill person is a passive recipient of interventions that will cure them. There are many other disadvantages.

Closely associated with this is the question of health and illness. It is correctly argued here and elsewhere that someone who has residual losses secondary to tissue damage is not necessarily ill. But rehabilitation is normally provided and funded by health organizations that naturally consider the recipients of their services to be ill.

The principles of rehabilitation are very close to those of education, especially higher education. I initially suggested *student* would be a good word. However, it was pointed out that this also implies an unequal power relationship—if you assume that the patient is the student. *Perhaps the patient should be the teacher, and the professional the student!*

I have no answer—and there probably is none—but discussing terminology would certainly stimulate people to think of rehabilitation quite differently!

The second concerns the attitude and approach of organizations that fund health-care and rehabilitation. Anyone reading this book will understand that it is completely wrong to consider the primary goal of rehabilitation to be independence in personal activities of daily living or even being able to live independently in accommodation.

A similar dilemma is faced by education, especially universities. At one time, universities thought their main goal was to produce well-educated people who were likely to contribute to society in some way, but universities did not consider exactly how any individual student might eventually contribute, or what they might contribute. Now universities judge themselves by the later employment and earning power of their students.

Somehow we will need to persuade funding organizations that the long-term benefits of helping people (re)establish themselves into a socially meaningful and satisfying life are worth the costs of rehabilitation services.

This book should alter your approach to rehabilitation. Read it, disagree with its ideas, debate the questions with colleagues, write something to refute an idea, or in some other way respond to the ideas and challenges it sets out. And enjoy it.

Derick Wade
Professor of Neurological Rehabilitation
Oxford Centre for Enablement
Windmill Road
Oxford, UK

Chapter Précis

Section I: Rethinking the Past and Re-Envisioning the Future

Where rehabilitation has come from and the current state of affairs—what we know and don't know, and where rehabilitation is heading.

1. Rethinking Rehabilitation: Theory, Practice, History—And the Future
   Kathryn McPherson, Barbara E. Gibson, and Alain Leplège
   Purpose: To set the scene and stimulate the reader to think about why "rethinking" the link between the past, theory, philosophy, and practice is actually warranted and worth doing as opposed to focusing on only theory or only practice, or doing research that is disconnected from either. The idea is to provoke interest in the more detailed chapters examining history and theory in readers who may have come to think these are redundant or divorced from the "reality" of clinical practice.
   Key contact: katmcphe@aut.ac.nz

2. Conceptualizing Disability to Inform Rehabilitation: Historical and Epistemological Perspectives
   Alain Leplège, Catherine Barral, and Kathryn McPherson
   Purpose: To provide a solid exploration connecting the past to the present of rehabilitation and questioning how it has contributed to where we currently are and how we might best move forward. Key historical developments in rehabilitation and the key models underpinning what is currently provided—including the ICF, the Disability Creation Process the ICIDH and other models.
   Key contact: alain.leplege@univ-paris-diderot.fr

3. Rethinking Rehabilitation's Assumptions: Challenging "Thinking-as-Usual" and Envisioning a Relevant Future
   Karen Whalley Hammell
   Purpose: Rehabilitation is informed by assumptions that have rarely been examined critically and that are contested by research evidence. These include ideals of "normality," ideologies of "independence" and beliefs about the

relationship between physical function and perceptions of quality of life. Indeed, rehabilitation's assumptions, dominant theories and models of practice reflect specific middle class, urban, Western, ableist perspectives. Informed by the belief that ability is of little use without opportunity, and drawing on insights derived from the experiences of rehabilitation clients, this chapter highlights flawed assumptions that underpin rehabilitation's practices; challenges professional dogma and dominance; and suggests future directions for practice and research that are grounded in both clients' priorities and research evidence.

Key contact: ik.hammell@sasktel.net

4. Rethinking "Normal Development" in Children's Rehabilitation

   Barbara E. Gibson, Gail Teachman, and Yani Hamdani

   Purpose: To question the focus of normal development as the primary organizing concept in children's rehabilitation. Normal development provides the underlying rationale for a host of health and education interventions designed to maximize children's physical and mental capacities in adulthood. The link between rehabilitation therapies and development is accepted almost universally as uncontroversial. However, the notion of child development proceeding in a linear and predictable stepwise fashion toward adulthood is only one way among others of understanding childhood. This chapter situates current dominant understandings of the child and development in their historical origins, asks how well they are serving disabled children, and considers how things could be otherwise.

   Key contact: Barbara.Gibson@utoronto.ca

Section II: Philosophy in Action

Key conceptual and theoretical issues and new directions for reimagining rehabilitation. For example, postmodern, critical, or phenomenological approaches to conceptualizing rehabilitation and its objects of interest.

5. Do Frogs Have Lips?—An Exploration of the Place of "Mind" in Rehabilitation

   Richard J. Siegert and Matthew Maddocks

   Purpose: This chapter argues that the mind/brain/body problem is an important theoretical issue for rehabilitation to consider. Firstly, important stances that ancient and modern philosophers have taken to the mind/body problem will be briefly outlined. Secondly, some key examples of recent advances in psychology and rehabilitation that rely upon an understanding of human functioning and abilities at a cognitive (mental) level of explanation rather than purely neural are explored. Third, the recent work of Denny Borsboom and colleagues will be introduced to illustrate how to actively consider the mind in rehabilitation. This chapter therefore includes a focus on theory that has been explored largely outside the field of rehabilitation but that has relevance for it. It also highlights how advances in psychometrics now permit mathematical modeling of complex relationships between behavior and biology.

   Key contact: rsiegert@aut.ac.nz

6. Rethinking Movement: Postmodern Reflections on a Dominant Rehabilitation Discourse

    David A. Nicholls, Barbara E. Gibson, and Joanna K. Fadyl

    Purpose: To retheorize a core concept in rehabilitation using postmodern theory, both expanding understanding of movement beyond the physical body and problematizing some of the limitations of the narrow conceptualizations that underpin contemporary rehab practice.

    Key contact: david.nicholls@aut.ac.nz

7. Therapeutic Landscape: Rethinking "Place" in Client-Centered Brain Injury Rehabilitation

    Pia Kontos, Karen-Lee Miller, Cheryl Cott, and Angela Colantonio

    Purpose: To explore the interfaces between physical and social spaces and their interactive roles in determining therapeutic care in neurorehabilitation using therapeutic landscape theory as a novel framework.

    Key contact: pia.kontos@uhn.ca

8. Rethinking Social-Relational Perspectives in Rehabilitation: Traumatic Brain Injury as a Case Study

    Jacinta Douglas, Melanie Drummond, Lucy Knox, and Margaret Mealings

    Purpose: To explore and illustrate the potential place and importance of attending to social-relational aspects of individuals' lives in the rehabilitation process reflecting on theoretical perspectives and a range of studies undertaken by the authors.

    Key contact: J.Douglas@latrobe.edu.au

9. Rehabilitation and Recovery of Self-Identity

    Emily J. Thomas, William M. M. Levack, and William J. Taylor

    Purpose: To explore the possible role of rehabilitation in the reconstruction of self-identity after acquired disability. Philosophical, psychological, and cultural conceptualizations of self-identity are discussed as they pertain to rehabilitation and recovery following acquired disability. The authors discuss the distinction between concepts such as self-identity, personhood, personality changes, agency and self-concept. Finally, this chapter looks at emerging interventions designed to influence reconstruction of self-identity after acquired disability.

    Key contact: william.levack@otago.ac.nz and em@doctors.org.uk

Section III: Rethinking Rehabilitation Delivery, Research, Teaching, and Policy

Creative ideas for transforming rehabilitation care, delivery, research, and policy.

10. "This Unfortunate Young Girl...": Rethinking a Necessary Relationship between Disability Studies and Rehabilitation

    Susan Guenther-Mahipaul

    Purpose: Despite being terms that are frequently seen as partners, the disability and rehabilitation movements have in many ways exhibited a lack of partnership, for many reasons. This chapter will take both a personal (experiential) and academic (evidence based) look at how a productive relationship may come about, and some of the barriers to it.

Key contact: guenth@mcmaster.ca

11. Rethinking Measurement in Rehabilitation

   Paula Kersten, Åsa Lundgren-Nilsson, and Charles Sèbiyo Batcho

   Purpose: This chapter discusses the use and misuse of outcome measures in clinical practice. It draws on clinical examples to illustrate key points, discusses the consequences of poor measurement, and then proposes a new future for measurement of outcome.

   Key contact: pkersten@aut.ac.nz

12. *Te Waka Oranga*: Bringing Indigenous Knowledge Forward

   Hinemoa Elder

   Purpose: Rehabilitation (in its individualistic form) fails to meet the needs of many indigenous populations. This chapter will examine why this is, and how progress might be advanced by rethinking mainstream rehabilitation from an indigenous perspective.

   Key contact: hinemoa@xtra.co.nz

13. Whose Behavior Matters? Rethinking Practitioner Behavior and Its Influence on Rehabilitation Outcomes

   Nicola M. Kayes, Suzie Mudge, Felicity A. S. Bright, and Kathryn McPherson

   Purpose: When interventions fail to achieve expected results, the common response is to ascribe failure to the patient. In such cases, the patient is frequently described as "unmotivated," "difficult," "noncompliant," or "not ready," suggesting the practitioner has done all that he or she can do. The purpose of this chapter is to critically discuss how rehabilitation practitioner's thoughts, feelings, and behaviour may directly or indirectly influence outcome. We will draw on existing evidence to demonstrate the role of the rehabilitation practitioner as a potentiating, or limiting, factor in relation to outcome and critically discuss implications for practice.

   Key contact: nkayes@aut.ac.nz

14. Rehab as an Existential, Social Learning Process: A Thought Experiment

   David A. Stone and Christina Papadimitriou

   Purpose: Drawing on a Heideggerian perspective, this chapter presents an example exploring the phenomenological meaning of client-centered care in a particular rehabilitation context. This chapter includes content that aims to challenge thinking about what rehabilitation is, and also proposes a "way" of thinking that might assist further thinking among rehabilitation professionals and researchers.

   Key contact: dastone@niu.edu and cpapadimitriou@niu.edu

Editors

Kathryn McPherson is the Laura Fergusson Chair Professor of Rehabilitation and Director at Person Centred Rehabilitation Centre in the Health and Rehabilitation Research Institute (HRRI), School of Rehabilitation and Occupation Studies, Auckland University of Technology, Auckland, New Zealand. Subsequent to training as a nurse in Australia, Professor McPherson completed midwifery and health visiting training in Edinburgh, Scotland. She earned her psychology honors degree at the Open University (UK) before undertaking her PhD at the University of Edinburgh, graduating in 1998. She joined Auckland University of Technology in Auckland, New Zealand, in late 2004, and along with colleagues, she has developed a successful teaching and research program in rehabilitation. Her research focuses on developing a better understanding of, and response to, what matters most to clients and their family. Recent projects include clinical trials of new approaches to goal setting; living well with a long-term condition; measuring what matters; quality of care; informing rehabilitation by psychological approaches; engagement in rehabilitation; and enhancing understanding of theory in rehabilitation. Professor McPherson uses both qualitative and quantitative approaches recently enjoying the application of participatory designs and a focus on implementation science.

Barbara E. Gibson is an associate professor in the Department of Physical Therapy, University of Toronto, and a senior scientist at the Bloorview Research Institute at the Holland Bloorview Kids Rehabilitation Hospital in Toronto, Canada. She holds Bloorview Children's Hospital Foundation Chair in Childhood Disability Studies. Professor Gibson is a physical therapist and bioethicist whose research examines the sociopolitical dimensions of childhood disability and rehabilitation. Her transdisciplinary research investigates how dominant discourses in health and social practices such as independence, quality of life, and inclusion can be both disabling and enabling. She is particularly focused on postmodern approaches to examine various "assemblages" of bodies, technologies, and places and their effects. Barbara holds cross appointments at the Person Centred Research Centre, Auckland University of Technology, New Zealand, and the CanChild Centre for Childhood Disability Research, McMaster University, Canada. She is an academic fellow at the Centre for Critical Qualitative Health Research, and a member of the Joint Centre for Bioethics at the University of Toronto.

Alain Leplège is a professor in the Department of History and Philosophy of Sciences, Faculty of Life Sciences, Université Paris Diderot, Paris, France. A health service researcher, specialized in outcome measurement, methodology, and epistemology, Dr. Leplège is a psychiatrist by training and earned a PhD in philosophy (Paris 1 Panthéon-Sorbonne) with a postdoc in health service research from Johns Hopkins University. He is the head of the Ville Evrard Mental Health and Disability Research Centre. He is also an adjunct professor at the Person Centred Research Centre, Division of Rehabilitation and Occupation Studies, Health and Rehabilitation Research Institute, Auckland University of Technology, New Zealand. Since 2012, he has been an adjunct professor at the University of the Sunshine Coast, Faculty of Science, Health, Education and Engineering, Sippy Downs, Queensland, Australia.

Contributors

Catherine Barral
EHESP Rennes
Sorbonne Paris Cité
Paris, France

Charles Sèbiyo Batcho
Department of Rehabilitation
Faculty of Medicine Université Laval
Québec, Québec, Canada

Felicity A. S. Bright
Centre for Person Centred Research
Auckland University of Technology
Auckland, New Zealand

Angela Colantonio
Department of Occupational Science and
    Occupational Therapy
University of Toronto
Toronto, Ontario, Canada

Cheryl Cott
Department of Physical Therapy
University of Toronto
Toronto, Ontario, Canada

Jacinta Douglas
Department of Human Communication
    Sciences
La Trobe University
Melbourne, Australia

Melanie Drummond
Department of Human Communication
    Sciences
La Trobe University
Melbourne, Australia

Hinemoa Elder
Te Whare Mātai Aronui
Te Whare Wānanga o Awanuiārangi
Auckland, New Zealand

Joanna K. Fadyl
Centre for Person Centred Research
Health and Rehabilitation Research
    Institute
Auckland University of Technology
Auckland, New Zealand

Susan Guenther-Mahipaul
School of Rehabilitation Science
McMaster University
Hamilton, Ontario, Canada

Yani Hamdani
Dalla Lana School of Public Health
University of Toronto
Toronto, Ontario, Canada

Karen Whalley Hammell
Department of Occupational Science and
    Occupational Therapy
University of British Columbia
Vancouver, British Columbia, Canada

Nicola M. Kayes
Centre for Person Centred Research
School of Rehabilitation and Occupation
 Studies
Auckland University of Technology
Auckland, New Zealand

Paula Kersten
Centre for Person Centred Research
Health and Rehabilitation Research
 Institute
Auckland University of Technology
Auckland, New Zealand

Lucy Knox
Department of Human Communication
 Sciences
La Trobe University
Melbourne, Australia

Pia Kontos
Toronto Rehabilitation Institute–
 University Health Network
Toronto, Ontario, Canada

William M. M. Levack
Rehabilitation Teaching and Research
 Unit
University of Otago
Wellington, New Zealand

Åsa Lundgren-Nilsson
Department of Clinical Neuroscience and
 Rehabilitation
The Sahlgrenska Academy
University of Gothenburg
Gothenburg, Sweden

Matthew Maddocks
Department of Palliative Care, Policy and
 Rehabilitation
School of Medicine
King's College London
London, England

Margaret Mealings
Department of Human Communication
 Sciences
La Trobe University
Melbourne, Australia

Karen-Lee Miller
Toronto Rehabilitation Institute–
 University Health Network
Toronto, Ontario, Canada

Suzie Mudge
Centre for Person Centred Research
Health and Rehabilitation Research
 Institute
Auckland University of Technology
 North Campus
Auckland, New Zealand

David A. Nicholls
School of Rehabilitation and Occupation
 Studies
Auckland University of Technology
Auckland, New Zealand

Christina Papadimitriou
College of Health and Human Services
Northern Illinois University
DeKalb, Illinois

Richard J. Siegert
Centre for Person Centred Research
School of Public Health and Psychosocial
 Studies
Auckland University of Technology
Auckland, New Zealand

David A. Stone
College of Health and Human Services
Northern Illinois University
DeKalb, Illinois

William J. Taylor
Rehabilitation Teaching and Research
Unit
University of Otago
Wellington, New Zealand

Gail Teachman
Graduate Department of Rehabilitation
Sciences
University of Toronto
Toronto, Ontario, Canada

Emily J. Thomas
Rehabilitation Teaching and Research
Unit
University of Otago
Wellington, New Zealand

and

Specialist Registrar in Rehabilitation
Medicine
Wessex Deanery, United Kingdom

I

Rethinking the Past and Re-Envisioning the Future

1

Rethinking Rehabilitation: Theory, Practice, History—And the Future

Kathryn McPherson

Barbara E. Gibson

Alain Leplège

Any fool can know. The point is to understand.

Albert Einstein (Simmons 2003, p. 1)

There is perhaps no better way to begin a book considering thought, theory, and new understandings of our world than to consider Einstein. He ably demonstrated within his own field that true knowledge advancement was augmented by developing and using theory prior to, following, and indeed at times instead of, empirical research. While theorizing "instead of" empirical research is a step too far for many, and the value of theory disconnected from experience or data is disputable for real-world living and practice, our premise is that rethinking the place of theory in rehabilitation, at every level, holds very real power for significant knowledge advance in rehabilitation. Taking inspiration from Einstein, a phrase that typifies our intentions for this book and the challenge we embrace is "Any fool can accept. The point is to question."

Some might interpret this to be advocating *questioning for its own sake* but that is not our intention. However, "unthinking acceptance" is a dangerous approach, despite it being a tempting position to adopt in busy contexts of practice, or when one approach to thinking dominates. Our proposal is that questioning assumptions, challenging accepted practices, and proposing new ways of thinking about how and why we do things, are fundamental to making the most of rehabilitation's potential contribution to the communities we serve. With rapidly changing funding models, and new and expanding rehabilitation populations (including a greater number of survivors from previously fatal illness and/or injury, as well as an aging population), the need to avoid complacency about what we do, and why, has never been more pressing.

What Is Rehabilitation?

There are many formal definitions of rehabilitation; most contemporary approaches share a shift of focus away from treating disease and pathology as core, toward reducing the impairments secondary to illness or injury to minimize disability and maximize independence (World Health Organization 2014). With the introduction of the International Classification of Functioning, Disability, and Health (ICF; World Health Organization 2001) as a framework for rehabilitation practice, an explicit connection was made to rehabilitation's role in enhancing participation. However, interpretations (and realizations) of rehabilitation still vary widely. Some suggest our practice remains very much focused on pathology and/or impairment (Patston 2007; Whalley Hammell 2006) or that it is associated with specific disciplines such as physiotherapy (McPherson 2006). Other arguments (for example, that disability is socially determined [Oliver 1998]) have even raised questions about rehabilitation's relevance, because the archetypal medical model focuses on the individual and their deficits without due consideration of social determinants and contextual factors. Each chapter in this book questions in some way prior definitions and interpretations of what rehabilitation is or is not, what it could, and perhaps even should, be. It might be an interesting exercise to pause and think about your own starting definition and see where you end up after a few chapters.

Other issues that bedevil rehabilitation are that while research has grown in quantity and quality over the last few decades, the field is still criticized for lacking a sound theoretical base (Siegert et al. 2005; Whyte 2008, 2014); being methodologically problematic (Wade et al. 2010); having disappointingly low rates of translation into practice (Teasell 2012; Walker et al. 2013); and problematic and debatable claims that rehabilitation is person-centered (Gzil et al. 2007; Whalley Hammell 2012). These factors prompt our own questioning and efforts to support *rethinking rehabilitation*, particularly among new and emerging generations of practitioners and researchers—those who have the greatest potential to influence the future of rehabilitation. It seems crucial to question issues that are too often taken for granted, such as:

- What should be the purposes of rehabilitation now and in the future?
- Why does rehabilitation look like it does?
- How could rehabilitation be different, better, more effective?

- What are the key outcomes that really matter and that we could and should address?
- Who benefits and who is left out in current practices?
- What are the hidden and/or unintended consequences of rehabilitation practices?

These questions and more are discussed in the chapters that follow, but before we move on to describe these in brief, we consider three points of convergence that underpin the book and run through the chapters:

1. What does history tell us about the past and about what we currently do?
   a. Rethinking the origins of rehabilitation
2. In what ways does or could theory contribute to rehabilitation knowledge advance and better practice?
   a. Rethinking the potential contribution of theory to rehabilitation
3. What are the nagging issues and entrenched problems in rehabilitation and how might we creatively reimagine, revision, and rethink imaginative solutions?
   a. Rethinking what really matters and how to bring about change

Rethinking the Origins of Rehabilitation

Rehabilitation is comparatively new as a formal discipline, consolidating over the latter part of the twentieth century largely in response to the survival of so many who, in previous times, would have died from their injuries or their consequences (see Chapter 2 for a more in-depth consideration of historical influences on current rehabilitation models). During and after World War II (WWII), improved emergency care and antibiotics resulted in many injured soldiers surviving and returning home—but with significant impairments. The result was a set of urgent and overwhelming challenges for health and social services with rehabilitation being a major part of the response. The introduction of dedicated rehabilitation services was necessarily reactive and rapid (Eldar and Jelić 2003; Linker 2011).

Although war did not bring the first experience of impairment and disability to society, it did produce a specific set of demands that challenged the long-standing and relatively passive response, and provision of supports and services for disabled people. Even with the introduction of organizations such as the Veterans Administration in the United States after WWI, services have been described as "custodial not rehabilitative" (Mauk 2011, p. 3). By WWII, society was arguably not in a position to allow, or be seen to allow, such passivity given the size and visibility of the problem. With so many previously fit and healthy men of working age becoming "suddenly" disabled, the potential consequences on economic recovery and productivity greatly concerned the majority of postwar nations. There was also a perceived moral imperative to act, given that these young people had become disabled in the service of their country (Gritzer and Arluke 1985).

The international effort to address "war-disability" was consolidated by the United Nations Rehabilitation Administration collaborative, a group of 44 countries that set out to revise plans for the care of wounded disabled veterans; many countries also introduced legislation around vocational and social rehabilitation (United Nations Relief and

Rehabilitation Administration 1943). Rehabilitation services were broadly seen as a key way to achieve two things that seemed—on the surface—mutually compatible:

1. Reduce dependency, increase function, and help disabled soldiers reintegrate into their lives and communities
2. Reduce the burden on society that the return of so many "dependent" and disabled young men brought

However—are these two aims mutually supportive? Would working toward one necessarily mean working toward the other? Such an assumption would perhaps hold true if the weight apportioned to each driver for action was balanced, and if that balance was explicitly enacted in policy and provision. The former option prioritizes individual human value in function and suggests effort should be concentrated toward outcomes that privilege a meaningful life and being well integrated in the community. The latter aim had a different key stakeholder (society) and a different key outcome (the "reduction of burden," presumably largely economic). Put in these terms, there are numerous tensions and conflicts that might, and did, result. For example, returning soldiers were certainly not supported in their efforts to establish independence and reintegration if the costs of providing that support were perceived to be disproportionate—reducing societal burden was the trump card. What seems on the surface to be simple and straightforward emerges as complex and conflicted. One might imagine that the tension between these two aims has dissipated over time, but a number of chapters in this book identify this conflict as an enduring issue. There remain very real limits to the actions our society undertakes (and the services provided) in relation to enabling disabled people, and what underpins those limits is connected to our history.

War and conflict still play major roles in contemporary rehabilitation. For example, since the Iraq war, the US investment in rehabilitation research has grown exponentially with major efforts to find new solutions for significant injuries and consequent impairments, including traumatic brain injury and multiple amputation. There is no doubt that military research has the potential to impact civilian populations with similar injuries—technological advance in prosthetic and wheelchair design, virtual rehabilitation environments, and smart housing being just a few examples. However, while civilian society could benefit in many ways, it is possible that such a foundation also produces unintended consequences that pervade research and, indeed, practice and policy. Thinking about war as a historical foundation of rehabilitation (and indeed a stimulus to ongoing development of knowledge) is only part of the picture, and thinking about the history of the discipline prompts some fairly challenging questions of relevance. Many of these are addressed in more depth in subsequent chapters:

- Do current dominant definitions and models of rehabilitation address what people and societies want and need today?
- Are there remnants of privileging or valuing some populations over others for rehabilitation input (for example injured previously healthy adults over children with congenital impairments or older people)?

- Is rehabilitation still reactive—responding in largely ad hoc ways where a more coordinated approach to knowledge development could make more of a difference?
- Is rehabilitation doing something society wants and needs?
- How does rehabilitation challenge and/or contribute to negative attitudes toward disability and disabled people?
- What maintains current dominant thinking and action in rehabilitation despite arguments that the status quo is insufficient?

While we spend little time (understandably) thinking about the origins of rehabilitation on a day-to-day basis, it is intriguing to reflect on how this impacts what we *currently do* and the apparent and continued dominance of impairment, independence, and reducing societal burden and limiting liability as key policy and funding imperatives. Although most human life is characterized by interdependence (Diaprose 2013; Gibson 2006; Gibson et al. 2012; Knabb et al. 2012), our largely individualistic models of rehabilitation (a focus on treating the individual) still seem to prize independence above other outcomes that matter. The popularity of, and reliance on, measures such as the Functional Independence Measure or FIM (Skinner and Turner-Stokes 2006) illustrate this point. Despite changes in language brought by the ICF (World Health Organization 2001), an emphasis on independence persists (further discussed in Chapters 3, 9, 12, and 14).

This is not to suggest that independence has no importance or value, just that its dominant positioning among valued outcomes should be questioned rather than assumed or "unthinkingly accepted." Other outcomes are certainly within the lexicon and rhetoric of rehabilitation including autonomy, quality of life, participation, person centeredness, family centeredness, and more. Some are problematic for the same reason as independence (their position or dominance being assumed) and most are absent from the ICF and other models of rehabilitation that underpin practice (or far lower in the order of importance). Intriguingly, health and well-being, issues at the heart of human rights and disability codes (United Nations 1948, 2006), are absent from most definitions and models of rehabilitation. Could it be that the political and economic drivers for rehabilitation have overwhelmed humanistic drivers? Might it be that our response to the economic demands that were so obvious in postwar times (reducing the costs of impairment and dependency) is now so embedded in service provision and in thinking, that the most "human" aspects of rehabilitation care (such as supporting individuals to discover or learn how to "be" and "to live a meaningful life" with impairment [as discussed in Chapter 9]) are marginalized and poorly addressed?

Given the historical drivers of rehabilitation practice (for better and for worse), it seems reasonable to question whether a more organized approach could advance knowledge and practice differently; enter theory. Despite some arguments that theory and theoretical thinking are too abstract and too far removed from real life, we agree with Lewin, who said "There is nothing more practical than a good theory" (Lewin 1952, p. 169).

Rethinking the Potential Contribution of Theory in Rehabilitation

Kerlinger and Lee defined a theory as "a set of interrelated constructs (concepts), definitions and propositions that present a systematic view of phenomena by specifying relations among variables, with the purpose of explaining and predicting the phenomena." (Kerlinger and Lee 2000, p. 11).

In a paper titled "Theory Development and a Science of Rehabilitation," Siegert et al. (2005) made the case that pursuing a greater focus on a theory of rehabilitation may help progress the field. The suggestion was that in much of what we do, we may lack a "systematic view of the phenomena" and that our specification of "relations among variables" was limited and problematic. In the same issue of the journal, the editor invited a set of commentaries and while some welcomed the paper's position that rehabilitation was theoretically undernourished, others were concerned that focusing on theory would deflect energy away from rehabilitation's "core business." However, a central driver behind the paper was to question what rehabilitation's core business was and what it could be; this inquiry requires theory. The paper concluded:

> Progress in scientific research occurs primarily through two processes—theory building and theory testing. We have argued here that rehabilitation has traditionally neglected theory building and that there is much to gain by attending to the dearth of good theory in the field. The criteria for evaluating the merits of a scientific theory are well established and can be applied to any theory of rehabilitation. Hence we do not need to fear a proliferation of "woollyminded theorie." Even the ost rigorous programme of theory testing can at best only refute an inadequate theory. One is then still left with the task of developing a better theory with which to replace it. However, a little more time spent on theory building in rehabilitation may greatly enhance the fruits of our empirical theory testing. (Siegert et al. 2005, p. 1500)

We are not the only ones to think there might be advantage from using theory to drive knowledge and practice in rehabilitation forward. Notably, John Whyte (2008) proposed that:

- There is a need to acknowledge the fact that empirical work alone will not develop the science of rehabilitation.
- Treatment trials are complex, challenging, and expensive and it is critical the investment of resources for such trials produce more than a yes-or-no answer about the benefit of a particular treatment.
- We need to support theoretical development of our field as energetically as we support the gathering of evidence.
- We need to encourage ongoing conversations between theorists and empiricists in the development of our scientific foundations.

Whyte has updated his view on theory in rehabilitation in a special issue of *Archives of Physical Medicine and Rehabilitation* (95(1): Suppl 2014) dedicated to reviewing an attempt to develop a comprehensive taxonomy of rehabilitation interventions. He proposed two broad groups of theory as being relevant to rehabilitation: *treatment theory* being about how to effect change in clinical targets and *engagement theory* being about how changes in a proximal clinical target will influence distal clinical aims' (Whyte 2014).

We agree that these are both reasonable and important areas for theoretical thinking and action—but are they enough? Do these concepts address that key issue of what rehabilitation aims to do and whether it is sufficient for the modern day and the future. If one's starting point is that rehabilitation is solely about influencing "clinical" aims (whether proximal or distal), this might be sufficient. However, rehabilitation is a necessarily multifaceted practice, involving all aspects of human functioning and connected to all aspects of human existence from the physical to the psychological but also the social, relational, and indeed existential. It is contextualized by a wide range of factors that influence the process—many of which are "beyond" the person who is the target of a therapeutic endeavor. It seems logical that we might require, and benefit from, a pluralistic approach and a range of theories to make sense of what rehabilitation could and should be.

Attempting to derive a single "grand unifying theory of rehabilitation" (Siegert et al. 2005; Whyte 2008) seems at best ambitious and at worst futile. However, our proposal is that we could gain much from developing and testing "better," "more explanatory," and "more interconnected" theories. Theories that define, and allow targeted investigation of, the key issues that are troubling to people with impairment and their families, as well as to professionals and indeed society seem fundamental to ensuring what rehabilitation does is worthwhile. There is (to refer again to Einstein) no single theory of physics, but multiple theories from relativity, unified field theory, quantum possibility theory, and string theory, just to mention a few. Theory is simply a set of concepts and the relationship between them. The most simple of these is what Maxwell calls a "barbell theory," that is two concepts (the weights) joined by a proposed relationship (the bar) (Maxwell 1996). The more concepts and relationships, and the more abstraction built into these, the more complex the theory. Theory can be preexisting and applied to better understand a phenomena, or we can theorize relationships from observations and our critical thinking.

Rather than try to cover *all theories* relating to *all aspects* of rehabilitation, we instead propose a framework to facilitate consideration of the many different kinds of theory (and indeed how they might intersect or inform one another) toward providing a mechanism in the face of what can seem complex and fraught. To that end the framework divides theory into three groups or levels: macro, meso or mid, and micro. We do not infer neat or tidy divisions between these, and there may be substantial overlap between them, but we hope they provide a heuristic device for considering the differences between for example philosophical "grand theories" and highly specified treatment theories. The examples noted within the figure are just a few of the many that could be listed (as indicated by the notation of "and more" at the end of the lists). Our aim with this framework is in part to address the many, often confusing, usages of the term *theory* to mean different things. Perhaps more importantly, we wish to encourage rehabilitation practitioners to engage with theory and theorizing, to assist systematic reflection on what we do, to conceptualize, problematize and theorize as freely as possible, so that new thinking about rehabilitation emerges progressively and collectively.

Qualitative research experts have long advocated for continuity between epistemology (one's take on knowledge); an explicit philosophical and theoretical approach; and using that to select one's methodology and methods rather than focus on methods alone (Crotty 1998). A similar pattern of thinking is relevant in quantitative approaches but, until recently, largely absent from discussion, and either implicit or simplistic in publications. The framework above (and expanded in Figure 1.1) suggests how exploration and explication of theory at multiple levels, regardless of preferred research paradigm, could be facilitated. To us, such explication is warranted and would aid clarity, progress, and formation of new questions of importance.

Theory can do more than frame empirical research, helping us question and better understand what we are doing and why, and how things could be otherwise. For example, the idea that disability is not inherently within an individual but is caused at least in part by environmental barriers was a significant theoretical shift in thinking that led to major social changes across the world. These included landmark legislative changes that ensured access and care for disabled people, and the development of the ICF by the WHO (see Chapter 2).

Microlevel Theory

At the microlevel, one could be no more eloquent than Mark Lipsey who said "Finally—a guided fantasy: Imagine a research community in which every report of a treatment effectiveness study included a section labelled 'treatment theory'." (Lipsey 1983, p. 38).

Despite many years having passed since Lipsey's statement, it seems to remain largely a fantasy as few rehabilitation papers (or indeed publications in other health disciplines) include a clear section on treatment theory. Such detail would make clear hypothesized modes of action for why a novel intervention might have worked, which can then be discussed—not merely whether a significant effect was achieved. Likewise utilizing treatment theory should incrementally contribute to the relevance of existing treatment theories and/or lead to their refinement, yielding better explanations of a condition and the rationale for the development of one suite of interventions over another. While treatment theory may seem only relevant to empirical designs, considering mediators of experience and relational factors whether they be "cause and effect" or more indirect seems to have wider applicability. Given the slow rate of change from current research that seems to prevail (Teasell 2012) and the emerging "power" (in the conceptual sense) of alternate approaches such as single case designs (Tate et al. 2013) and participatory approaches (Cook 2008; Seekins and White 2013), there are many good reasons to explicitly consider the microlevel of "research" theory. New methodologies in the future are arguably already required to answer questions that really matter. Finally, there is rarely a section on "measurement theory/science" in papers that present interpretations of measurement data, nor when presenting the development or testing of measures. Does this matter? Well, if the mathematics is flawed, so will be the interpretation (see Chapter 11 for further discussion of this point).

Midlevel Theory

At the midlevel, the approach we propose is to consider contextual theories that explore who people are, and how people function. For example, social and behavioral theories are sometimes most easily connected to specific disciplines, such as psychology or social work. However, they have absolute relevance to everything any of us do in rehabilitation because people (patients/clients, their families, and the professionals working with them) are at the center of rehabilitation practice. Even relatively discrete or unidisciplinary rehabilitation interventions involve, and perhaps depend upon, human interaction, connection, and behavior (Chapters 8 and 13). In acute healthcare, the expertise of the clinician in treating the individual may make the difference between the success or failure of recovery, to survival or death. Most of us know the sense of gratitude for such skill—for the surgical expertise, medical or nursing care that enabled a family member or ourselves to survive an injury or illness. In rehabilitation it is rarely, if ever, the case that what we do alone achieves a good outcome. Our core contributions, across all disciplines, are necessarily done "with" our patients/clients rather than performed by us.

Applying theoretical thinking to *less visible* but possibly *essential components or contexts* of rehabilitation interventions is both important and, fortunately, increasing. For example we, and others, are investigating theoretically derived understandings of and approaches to goal setting (McPherson et al. 2014), engaging patients and families in rehabilitation (Bright et al. 2014; Kayes and McPherson 2012) and team work and collaborative practice (Penney 2013).

Advances in some areas of theoretical thinking suggested at the midlevel theory in Figure 1.1, say neuroscience, may seem distal to one's interests if working in musculoskeletal rehabilitation or vocational rehabilitation. Similarly, learning theory may feel disconnected from the work of physiotherapy. However, explanatory theory in these areas is integrally connected to human performance, functioning, and existence—the very business of rehabilitation. As such, considering them may highlight important factors that are relevant across a wide range of research and practice.

Macrolevel Theory

For the final level of macro theory, we return to the response to our 2005 proposition that there should be a greater focus on theory (Siegert et al. 2005). Each person undoubtedly "fits" with some ways of thinking and different theoretical and philosophical approaches over others. For example, some of us will be most comfortable working within a positivist perspective and others from a postmodern one. So we are not calling for a merging of perspectives or "woollyminded theories." Ongoing conversations between theorists and empiricists are key, with the gains potentially being greatest if those with quite different perspectives were to engage in collaborative discussion. It is at the uncomfortable nexus and interface between different perspectives that we think the opportunity for knowledge advance is at its greatest. However, engaging with people who have different views (particularly at this macrolevel) seems difficult, and indeed this has been our experience on many occasions. Our sense is that rehabilitation research, perhaps even more than other disciplines because of its complexity and inherent

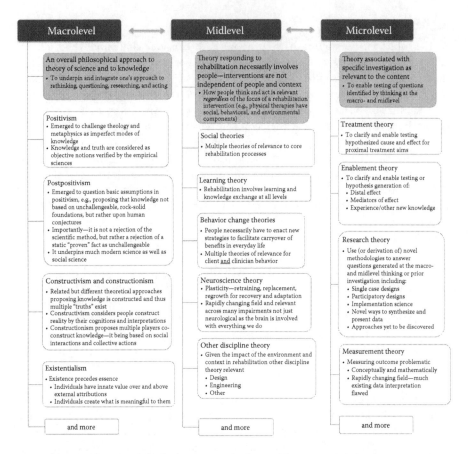

FIGURE 1.1 A framework for facilitating a rethinking of theory in rehabilitation. Introductory reading for the different levels of theory named in the figure: *Macrolevel*—positivism, post-positivism (Crotty 1998; Schick 2000); constructionism (Siebers 2001); constructivism (Crotty 1998); existentialism (Gracey and Ownsworth 2008); critical theory (Kincheloe and McLaren 2011). *Midlevel*—social theory (De Maio 2010); learning theory (Haselgrove and Hogarth 2012); behavior change (McPherson et al. 2014; Shumaker et al. 2009); neuroscience theory (Mishra and Gazzaley 2014; Raskin 2011; Wilson 2009); disability theory (Oliver 1998; Thomas 2012). *Microlevel*—treatment theory (Lipsey 1983); enablement theory (Whyte 2014); research theory (Wade et al. 2010); single case designs (Tate et al. 2013); participatory designs (Seekins and White 2013); implementation science (Straus et al. 2009); measurement theory (as discussed in Chapter 11).

interdisciplinarity, has very real potential to gain from cross fertilization of ideas, and a willingness to engage with others who have a different take on a topic whether you are a positivist struggling to engage with a postmodernist or vice versa. Whyte (2008, 2014) and others have criticized rehabilitation research for largely limiting its focus to *whether something works*, rather than expanding to think about (and test) *how it works* and we would add—*does it really matter*? For us to answer the questions that will enable a sustainable health system that works for clients, we suggest the time has come to shift from simply and strongly defending or rejecting one or other paradigm or way of thinking (McPherson and Kayes 2012).

A number of chapters within this text consider macrolevel theories more commonly explored in relation to disability theory and disability studies, including postmodernism, critical theory, cultural theory, and more (Oliver 1998; Thomas 2012). So as you read this book, if arguments do not sit comfortably with you, we hope that you might engage with why that is the case. It may be that there are some treasured assumptions of your own that are being challenged and, as dialogue is stimulated by the work, our authors might also be challenged. As long as our focus is to advance knowledge and understanding of how to create rehabilitation of the future, we think that is a good thing and that "a little more time spent on theory building in rehabilitation may greatly enhance the fruits of our empirical theory testing" (Siegert et al. 2005, p. 1500).

Rethinking What Really Matters and How to Bring about Change

The intention of this book is to create both a space and a foundation for rethinking rehabilitation. In so doing, we hope it might help you (and us) to creatively envision and envisage imaginative solutions to the most nagging issues and entrenched problems that limit rehabilitation and its contribution.

In this introduction, we have proposed a number of important questions and addressed some of rehabilitation's ideas and practices. Our central purpose is not to just come up with questions—there are likely as many questions as there are people to ask them (or more!). Rather, we suggest the priority is to identify those questions that matter most—those that if answered could and would make the greatest difference for patients/clients, and guide what rehabilitation should look like in the future. Our argument has been that such key issues are more likely to be exposed if we consider the past as a fundamental driver of what happens in the present, and if we consider the extent to which a strong theoretical rationale underpins (or should underpin) our interventions and/or ways of working.

We hope you are persuaded that rethinking rehabilitation might be a worthy pursuit and that your ability to do so will be enhanced by having a perspective that is historically aware and theoretically informed. What follows are a diverse range of chapters that tackle different issues in such a manner, exposing new issues and new opportunities.

Brief Summary of Chapters

This project emerged as a result of a meeting funded by a Canadian Institute for Health Research Networking Grant that brought most, but not all, of the first-named authors together to discuss ideas for the book, for research, for teaching, and for practice. The book is divided into three sections, the first of which pursues ideas we have proposed here in the first chapter to stimulate rethinking of the past and re-envisaging of the future. The second section (Chapters 5 to 9) could broadly be considered an effort to explore "philosophy in action," applying key conceptual or theoretical thinking to new directions for rehabilitation. Concepts considered include theory of mind, postmodernist reflections on movement, as well as spatial, relational, and existential considerations of the self and of rehabilitation. Section 3 (Chapters 10 to 14) examines some key issues that could (and arguably should) impact on rehabilitation care, delivery, research, teaching, and policy. We start with a personal reflection on experiencing disability and rehabilitation, and end with a "thought experiment" about what that experience might be in the future.

Section 1: Rethinking the Past and Re-Envisioning the Future

In Chapter 1, the editors have outlined the purpose of this book and provided a context for the subsequent chapters. We have also proposed a framework to support the consideration of theory at multiple levels and from multiple perspectives and posed important questions for understanding rehabilitation's past, present, and future.

Chapter 2, "Conceptualizing Disability to Inform Rehabilitation: Historical and Epistemological Perspectives" by Leplège et al., explores social and political factors that have influenced how disability and rehabilitation have been conceptualized, and how disabled people have been "categorized" over time. Leplège and colleagues then explore the origins, strengths, and weaknesses of a number of models that have been or are currently used in rehabilitation. They do not aim to provide an exhaustive review of all models, but rather to reflect on how these models might enhance (or limit) our thinking and our practice now and in the future.

Chapter 3, "Rethinking Rehabilitation's Assumptions: Challenging 'Thinking-as-Usual' and Envisioning a Relevant Future" by Karen Whalley Hammell, suggests that rehabilitation is underpinned by a number of assumptions rarely examined critically and contested by research evidence. For example, former rehabilitation clients challenge the premise that simply enhancing physical abilities is sufficient for enhancing quality of life. This chapter draws on data concerning clients' perspectives on the relevance of rehabilitation to their subsequent lives. It is guided by the belief that *ability* is of little use without *opportunity*. Whalley Hammell draws on diverse theoretical perspectives to reflect on the culturally specific ideologies that underpin rehabilitation. She suggests future directions for practice and research that are grounded both in clients' priorities and evidence.

In Chapter 4, "Rethinking 'Normal Development' in Children's Rehabilitation," Gibson et al. question the primacy of normal development as the primary organizing concept in children's rehabilitation. Normal development provides the underlying rationale for a host of health and education interventions designed to maximize children's physical and mental capacities in adulthood. The link between rehabilitation therapies and development is accepted almost universally as uncontroversial. However, the notion of child development proceeding in a linear and predictable step-wise fashion toward adulthood is only one way to understand childhood. The authors situate current dominant understandings of child development in their historical origins, ask how well they are serving disabled children, and consider the implications for children's rehabilitation.

Section 2: Philosophy in Action

In Chapter 5, "Do Frogs Have Lips?—An Exploration of the Place of 'Mind' in Rehabilitation," Siegert and Maddocks argue that the mind/brain/body problem is an important theoretical issue for rehabilitation to consider. The authors consider a range of positions that philosophers have taken to the mind–body connection and explore recent advances in psychology and rehabilitation that rely upon an understanding of human functioning and abilities at a cognitive (mental) level of explanation rather than a purely neural one. Finally, Siegert and Maddocks introduce the work of Denny Borsboom and colleagues to illustrate how the mind can be "actively" considered in rehabilitation.

Chapter 6, "Rethinking Movement: Postmodern Reflections on a Dominant Rehabilitation Discourse" by Nicholls et al., draws on postmodern theory to reconsider a core focus of rehabilitation: movement. It would be fair to say that most of us have probably rarely, if ever, heard the words *movement* and *postmodern* in the same sentence and so this chapter clearly sets out to challenge traditional thinking of this aspect of human functioning. They aim to expand understanding movement beyond the physical body and expose some taken for granted conceptualizations that underpin contemporary rehabilitation practice.

In Chapter 7, "Therapeutic Landscape: Rethinking 'Place' in Client-Centered Brain Injury Rehabilitation," Kontos et al. explore how insights from critical health geography might enhance our ability to deliver client centered rehabilitation. While the ICF includes a focus on the environment as a key context, this chapter suggests it does so insufficiently and in a way that it can seem to be on the "outer" edge rather than a "core" influence we can use to advantage, or ignore to client (and professional) disadvantage. The chapter provides an overview of therapeutic landscape theory and reviews literature that relates these ideas to rehabilitation. The authors draw on qualitative findings from two inpatient rehabilitation hospitals to explore the potentially powerful impacts of sociospatial design on care delivery.

Chapter 8, "Rethinking Social-Relational Perspectives in Rehabilitation: Traumatic Brain Injury as a Case Study" by Douglas et al., defines rehabilitation as a social-relational affair. The authors explore a range of theories that propose human beings as relational and that at least part of how we experience ourselves is in relation to others and hence relevant to rehabilitation interventions and outcomes. The authors draw on empirical data from a number of their own studies to examine how impairment can impact on the relational self in TBI cases, and how cognitive processes (for example, decision making) are socially mediated. These authors make a number of suggestions about how thinking relationally might change how we work in rehabilitation.

In Chapter 9, "Rehabilitation and Recovery of Self-Identity," Thomas et al. pursue the idea that sense of self or self-identify is important and that how we see ourselves is something that may be challenged or changed as a result of changes to our health or abilities. The authors make an argument that this sense of self or self-identify is therefore both a legitimate focus for, and a potentially important outcome from, rehabilitation. Again they draw on their own research, and that of others, to build their case and propose some practical consequences of thinking theoretically about identity.

Section 3: Rethinking Rehabilitation Delivery, Research, Teaching, and Policy

Suzanne Guenther-Mahipaul has titled Chapter 10, "'This Unfortunate Young Girl...': Rethinking a Necessary Relationship between Disability Studies and Rehabilitation." Rehabilitation and disability are at times uncomfortable companions in many ways, for many reasons, and with many consequences. While this is not the only chapter that explores how disability studies could inform rehabilitation, it grapples with the interface between these two disciplines from a personal perspective. Guenther-Mahipaul utilizes her experience to reflect on the dominant models in rehabilitation, as well as the rehabilitation professionals she has worked with as a patient or client and as an occupational therapist colleague. She talks of disability and rehabilitation as both personal *and* political, and of those who have "unintentionally contributed to the marginalization of disabled people" by failing to make that connection.

In Chapter 11, "Rethinking Measurement in Rehabilitation," Kersten and colleagues tackle issues often taken for granted in how we use assessment tools and outcome measures in clinical practice and research. One might imagine from reading the literature that a default position for many is to assume different raw scores on a tool mean something reliably different, and that it makes sense to total or add things up—just because they are numbers. Modern measurement science (actually not so modern, having existed for fifty years) challenges those assumptions and thus some of the research conclusions that we may be relying on. If you use measures, the theoretical thinking and arguments presented here may make you think twice about how you deal with, or interpret, the numbers.

Chapter 12, by Hinemoa Elder, is titled "*Te Waka Oranga*: Bringing Indigenous Knowledge Forward." Elder addresses the challenge raised by rehabilitation's focus on the individual despite clear theoretical and philosophical indicators that human beings are interdependent. Thinking about how rehabilitation currently fits indigenous perspectives is challenging but arguably crucial as disparities in health and disability outcomes prevail. This chapter explores a novel approach to rehabilitation that explicitly involves the collective, both past and present. While the chapter focuses particularly on Māori (the indigenous people of Aotearoa, New Zealand), it points to questions of import for many people and societies across the globe.

In Chapter 13, "Whose Behavior Matters? Rethinking Practitioner Behavior and Its Influence on Rehabilitation Outcomes," Kayes et al. explore a concept related to the notion of "rehabilitation as relational" raised by Douglas et al. in Chapter 8. The authors suggest that despite the long-held belief or assumption that practitioners are a "core ingredient" in rehabilitation, the dearth of research investigating therapist or practitioner behavior has very real consequences. Drawing on a series of theoretical frameworks and empirical data, they suggest we shift our focus within interventions from being about patients or clients alone to identify and operationalize the qualities, skills, and behaviors of clinicians that enhance a therapeutic connection that works.

Chapter 14, the final chapter, is titled "Rehabilitation as an Existential, Social Learning Process: A Thought Experiment." Stone and Papadimitriou invite us to use our imaginations to first open up for question our chosen approaches to rehabilitation and, second, to question the notion that change needs to be incremental. The authors use a thought experiment to question meaning(s) of client-centered care, and ways of rethinking what it could look like or be in the future. They do not propose a single solution, but rather a tool for rethinking, a way to question the essence of what has become rehabilitation.

Conclusion

One of the frustrating things about language and convention is that they can belie what is really intended. We finish this chapter with the word *conclusion* because it is how you indicate the end of a chapter. But by no means do we suggest that this chapter, or the book as a whole, take us to the end of rethinking rehabilitation. We have, however, attempted to make a case for why rethinking rehabilitation's past, present, and future may be warranted, and to illustrate how theoretical ideas beyond those we most naturally align with might be fruitful. We are hopeful that not only will Lipsey's dream come to fruition (that every scientific paper will have a section on treatment theory), but that positivist scientists *and* postpositivist and postmodern thinkers will get together more often to identify new questions that really matter, use differing ways of thinking to advance knowledge, and develop new methods that underpin future rehabilitation approaches that better support the needs of the populations they serve. We hope the chapters in this book illustrate how thinking differently is worthwhile in interesting ways.

Each chapter in this book addresses an issue of importance related to this endeavor and the authors' passion for challenging current thinking is evident. We suspect you will at times find their writing challenging (because the arguments may not sit well with what you currently think or do), and at other times their arguments and thinking will be reassuring (because you have been grappling with the same, or similar, issues). Whatever your response, we hope you will find these reflections better equip you in your own thinking about rehabilitation. We very much hope the book helps you reconsider why things are as they are where you work and live, how they could be different, and what you think are the most important issues for knowledge advance and change in practice. We hope you enjoy what follows and that dipping in and out of the book causes you to think, to rethink what you may have assumed, and to be challenged.

References

Bright, F. A., Kayes, N. M., Worrall, L., and McPherson, K. M. 2014. Engagement in healthcare: A critical review of the concept. *Disability and Rehabilitation* 27: 1–12. [Epub ahead of print].

Cook, W. K. 2008. Integrating research and action: A systematic review of community-based participatory research to address health disparities in environmental and occupational health in the USA. *Journal of Epidemiology and Community Health* 62(8): 668–76.

Crotty, M. 1998. *The Foundations of Social Research: Meaning and Perspective in the Research Process*. St Leonards, Australia: Allen and Unwin.

De Maio, F. 2010. *Health and Social Theory*. Basingstoke: Palgrave Macmillan.

Diaprose, R. 2013. Corporeal interdependence: From vulnerability to dwelling in ethical community. *SubStance* 42(3): 185–204.

Eldar, R., and Jelić, M. 2003. The association of rehabilitation and war. *Disability and Rehabilitation* 25(18): 1019–23.

Gibson, B. E. 2006. Disability, connectivity, and transgressing the autonomous body. *Journal of Medical Humanities* 27(3): 187–96.

Gibson, B. E., Carnevale, F. A., and King, G. A. 2012. "This is my way": Reimagining *Disability and Rehabilitation* 34(22): 1894–99.

Gracey, F., and Ownsworth, T. (eds.) 2008. *The Self and Identity in Rehabilitation: A Special Issue of Neuropsychological Rehabilitation*. Hove: Psychology Press.

Gritzer, G., and Arluke, A. 1985. *The Making of Rehabilitation: A Political Economy of Medical Specialization, 1890–1980*. Berkeley, CA: University of California Press.

Gzil, F., Lefeve, C., Cammelli, M., Pachoud, B., Ravaud, J. F., and Leplege, A. 2007. Why is rehabilitation not yet fully person-centred and should it be more person-centred? *Disability and Rehabilitation* 29(20–21): 1616–24.

Haselgrove, M., and Hogarth, L. 2012. *Clinical Applications of Learning Theory*. Hove: Psychology Press.

Kayes, N. M., and McPherson, K. M. 2012. Human technologies in rehabilitation: "Who" and "How" we are with our clients. *Disability and Rehabilitation* 34(22): 1907–11.

Kerlinger, F. N., and Lee, H. B. 2000. *Foundations of Behavioral Research*. Fort Worth, TX: Harcourt.

Kincheloe, J. L., McLaren, P., and Steinberg, S. R. 2011. Critical pedagogy and qualitative research: Moving to the bricolage. In *The Sage Handbook of Qualitative Research*, eds. N. K. Denzin and Y. S. Lincoln, 4th ed., 163–77. Los Angeles, CA: Sage.

Knabb, J. J., Welsh, R. K., and Alexander, P. 2012. Towards an integrated view of the necessity of human interdependence: Perspectives from theology, philosophy, and psychology. *Journal of Spirituality in Mental Health* 14(3): 166–80.

Lewin, K. 1952. *Field Theory in Social Science: Selected Theoretical Papers by Kurt Lewin*. London: Tavistock.

Linker, B. 2011. *War's Waste: Rehabilitation in World War I America*. Chicago, IL: University of Chicago Press.

Lipsey, M. W. 1983. Theory as method: Small theories of treatment. *New Directions for Program Evaluation* 57: 5–38.

Mauk, K. L. 2011. Overview of Rehabilitation. In *Rehabilitation Nursing: A Contemporary Approach to Practice*. Burlington, MA: Jones and Bartlett Learning.

Maxwell, J. 1996. Conceptual framework: What do you think is going on? In *Qualitative Research Design: An Interactive Approach*, 39–72. Thousands Oaks, CA: Sage Publications.

McPherson, K. M. 2006. Rehabilitation nursing—A final frontier? *International Journal of Nursing Studies* 43(7): 787–9.

McPherson, K. M., and Kayes, N. M. 2012. Qualitative research: Its practical contribution to physiotherapy. *Physical Therapy Reviews* 17(6): 382–9.

McPherson, K. M., Kayes, N. M., and Kersten, P. 2014. MEANING as a Smarter Approach to Goals in Rehabilitation. In *Rehabilitation Goal Setting: Theory, Practice and Evidence*, eds. W. Levack and R. Siegert, 105–120. Boca Raton, FL: CRC Press, Taylor and Francis.

Mishra, J., and Gazzaley, A. 2014. Harnessing the neuroplastic potential of the human brain and the future of cognitive rehabilitation. *Frontiers in Human Neuroscience* 8: 218.

Oliver, M. 1998. Theories in health care and research: Theories of disability in health practice and research. *BMJ* 317(7170): 1446–9.

Patston, P. 2007. Constructive functional diversity: A new paradigm beyond disability and impairment. *Disability and Rehabilitation* 29(20–21): 1625–33.

Penney, P. R. 2013. *Collaborative Practice: A Grounded Theory of Connecting in Community Rehabilitation*. Auckland: Auckland University of Technology.

Raskin, S. A. (ed.) 2011. *Neuroplasticity and Rehabilitation*. New York: Guilford Press.

Schick, T. 2000. *Readings in the Philosophy of Science: From Positivism to Postmodernism*. Mountainview, CA: Mayfield Publishing.

Seekins, T., and White, G. W. 2013. Participatory action research designs in applied disability and rehabilitation science: Protecting against threats to social validity. *Archives of Physical Medicine and Rehabilitation* 94(1 Suppl): S20–9.

Shumaker, S. A., Ockene, J. K., and Riekert, K. A. 2009. *The Handbook of Health Behavior Change*, 3rd ed. New York: Springer.

Siebers, T. 2001. Disability in theory: From social constructionism to the new realism of the body. *American Literary History* 13(4): 737–54.

Siegert, R. J., McPherson, K. M., and Dean, S. G. 2005. Theory development and a science of rehabilitation. *Disability and Rehabilitation* 27(24): 1493–501.

Simmons, G. F. 2003. *Precalculus Mathematics in a Nutshell: Geometry, Algebra, Trigonometry*. Eugene, OR: Resource Publications.

Skinner, A., and Turner-Stokes, L. 2006. The use of standardized outcome measures in rehabilitation centers in the UK. *Clinical Rehabilitation* 20(7): 609–15.

Straus, S. E., Tetroe, J., and Graham, I. D. 2009. *Knowledge Translation in Health Care Moving from Evidence to Practice*. Chichester: Wiley-Blackwell/BMJ.

Tate, R. L., Perdices, M., Rosenkoetter, U., Wakim, D., Godbee, K., Togher, L., and McDonald, S. 2013. Revision of a method quality rating scale for single-case experimental designs and n-of-1 trials: The 15-item Risk of Bias in N-of-1 Trials (RoBiNT) Scale. *Neuropsychological Rehabilitation* 23(5): 619–38.

Teasell, R. 2012. Challenges in the implementation of evidence in stroke rehabilitation. *Topics in Stroke Rehabilitation* 19(2): 93–5.

Thomas, C. 2012. Theorising disability and chronic illness: Where next for perspectives in medical sociology? *Social Theory and Health* 10(3): 209–28.

United Nations. 1948. *The Universal Declaration of Human Rights*. Retrieved March 3, 2014, from http://www.un.org/en/documents/udhr/history.shtml.

United Nations. 2006. *Convention on the Rights of Persons with Disabilities*. Retrieved October 10, 2014, from http://www.un.org/disabilities/convention/conventionfull.shtml.

United Nations Relief and Rehabilitation Administration. 1943. *Agreement for United Nations Relief and Rehabilitation Administration, November 9, 1943*. Retrieved March 24, 2014, from http://www.ibiblio.org/pha/policy/1943/431109a.html.

Wade, D. T., Smeets, R. J., and Verbunt, J. A. 2010. Research in rehabilitation medicine: Methodological challenges. *Journal of Clinical Epidemiology* 63(7): 699–704.

Walker, M. F., Fisher, R. J., Korner-Bitensky, N., McCluskey, A., and Carey, L. M. 2013. From what we know to what we do: Translating stroke rehabilitation research into practice. *International Journal of Stroke* 8(1): 11–17.

Whalley Hammell, K. 2006. *Perspectives on Disability and Rehabilitation. Contesting assumptions; challenging practice.* London: Churchill Livingstone.

Whalley Hammell, K. 2012. Client-centred practice in occupational therapy: Critical reflections. *Scandinavian Journal of Occupational Therapy* 20(3): 174–81.

Whyte, J. 2008. A grand unified theory of rehabilitation (we wish!). The 57th John Stanley Coulter Memorial Lecture. *Archives of Physical Medicine and Rehabilitation* 89(2): 203–9.

Whyte, J. 2014. Contributions of treatment theory and enablement theory to rehabilitation research and practice. *Archives of Physical Medicine and Rehabilitation* 95(1 Suppl): S17–23 e12.

Wilson, B. A. 2009. *Neuropsychological Rehabilitation: Theory, Models, Therapy, and Outcome.* Cambridge: Cambridge University Press.

World Health Organization. 2001. *International Classification of Functioning Disability and Health.* Geneva: World Health Organization.

World Health Organization. 2014. *Rehabilitation.* Retrieved March 1, 2014, from http://www.who.int/topics/rehabilitation/en/.

2

Conceptualizing Disability to Inform Rehabilitation: Historical and Epistemological Perspectives

Alain Leplège

Catherine Barral

Kathryn McPherson

Introduction

In order to *rethink* rehabilitation—it is vital that we *think* about current rehabilitation—what it looks like and why. The dominant models that have emerged to guide development and practice, the frameworks that underpin compensation policies, funding for services, and indeed research, all have historical and political roots. If we better understand these models, their basis or foundation, their strengths, and also their weaknesses, then perhaps we can better understand how to contribute to progress in the future. Our aim in this chapter, therefore, is to discuss issues concerning how past, present, and future understandings of disability and related conceptual models might best inform rehabilitation strategies. The chapter is composed of two main parts. The first part revisits the history of "conceptions of disability." We broadly retrace how disabled people have been described and progressively identified as a specific population within society throughout modern Western history. We mainly focus on the changing social and political perspectives on poverty and assistance, out of which disability has been emerged as a social issue in the twentieth century. This, along with the entrenchment of medical power in Western public policies, leads to administrative categorization of people with impairments in medical terms. The social and political approach to disability, supported by the social movements of disabled people in the twentieth century, challenged this medical dominance, bringing about an ongoing debate on conceptual models of disability. The second part considers the merits of key contemporary disability and disablement models by challenging assumptions and "common" knowledge. We particularly focus on contemporary evolutions of disability models, from the 1960s to today, because in many ways, they reflect or react against prior approaches and underpin much of what is thought of as "modern rehabilitation." Understanding these approaches, and their evolution, may help us better structure and plan future rehabilitation schemes, services, or evaluations.

We conclude the chapter by discussing and suggesting some future lines of reflection.

Evolution of the Conceptions of Disability in Western History

Ability to Work: A Historical Operator of Distinction among Deprived Populations

Throughout the Middle Ages in Western Christian civilization, according to Stone (1986), Castel (1989, 1999, 2003), and Stiker (1997, 1999), the poor who relied on charitable assistance were categorized in two ways, with the ability to work being the dividing line between the two and the cause of a difference in attitudes from society. On one hand, there was what was called "the infirm,"* these being people with impairment, disease, or old age or where circumstances of life (being "insane or impotent," orphans, or widows in charge of a family) rendered them unable to work or maintain their existence by their own means or that of their family circles (Castel 1989, 2003; Stiker 1997). This category of the poor was considered as deserving or worthy; indeed they were seen as being part of the natural order and fulfilling a social and spiritual role as beneficiaries of the alms and obligations of the wealthy members of society. As Christians, the wealthy were required to provide for the poor, either as individuals or by supporting charities. On the other hand, there were the able-bodied poor, reduced to poverty and charitable assistance but, because of conditions relating to the regulations of employment and low wages, were suspected of voluntary idleness and in fact later criminalized

---

* Terms we use currently such as *disability, disabled people,* or *impairments* did not exist before the twentieth century to describe the populations that this chapter explores. We use them to enhance clarity in a modern context. However, we also refer to terms now obsolete and pejorative as they were widely used in history (these are indicated in this chapter by being italicized). Similarly, we anticipate some of the language we use today, even that not yet considered pejorative, may become obsolete in the future.

and stigmatized (Piven and Cloward 1971; Stiker 1997) in a process coined in 1971 by W. Ryan as "blaming the victim" (Castel 1989, p. 18).

According to historians (Castel 1989, 2003; Trattner 1999), the distinction between able-bodied and disabled was made on the basis of the ability to work, and became more clearly established in the middle of the fourteenth century following the Black Death. Repressive measures were taken by monarchs of the time (such as the Statute of Labourers [1349] in England and similar edicts in France, Portugal, and Spain) to prohibit vagrancy and begging that had been generated by dramatic poverty. Any poor with the requisite physical abilities were forced to work, while begging and vagrancy among people with impairments were tolerated and in fact accepted (Geremek 1987; Castel 1989, 2003; Trattner 1999).

In the sixteenth and especially the seventeenth centuries, changes in attitudes toward poverty, begging, and vagrancy meant all "poor people were deemed suspect" (Braddock and Parish 2001, p. 22), including those with impairments. All over Western Europe and in America (Foucault 1973; Trattner 1999), attempts to regulate poverty and to maintain public order resulted in the development of policies that combined repression of, and assistance to, all unemployed people, beggars, and those considered deviant, including the "ill, insane, epileptic, impaired, debauched, vagrants," young and old alike (Foucault 1973; Stiker 1997, p. 114; Fossier 2002). The state's involvement in the administration of assistance, conducted earlier by the church, and in the repression of those considered as deviant and threats to the social order, resulted in the creation of networks of workhouses, almshouses, asylums, and hospitals and the generalization of forced labor (Foucault 1973; Stiker 1999). These centuries, during which assistance itself was a form of repression via what Foucault called "the great confinement" (1973), institutionalized the exclusion, isolation, and marginalization of populations that were seen to deviate from social norms, including people with impairments.

Early Distinctions among Disabled Persons and Early Roots of Rehabilitation

In the eighteenth century, the intellectual revolution of the philosophy of enlightenment challenged religious understandings of the universe, the established social order, and notions of poverty and disability as given or "natural" (Stiker 1997, 1999). In a context of scientific and medical progress, and belief that "people were deemed to be capable of intervening in what had been perceived to be the immutable natural order" (Braddock and Parish 2001, p. 29), the idea took root that with appropriate techniques and teaching methods, disabled people could "be educated" and have access to a previously prohibited intellectual and professional life (Stiker 1997, p. 107). Methods of communication and teaching for the "deaf and mute" were investigated and developed in Spain in the sixteenth century, in England and Switzerland in the seventeenth century, and in France in the eighteenth century (Stiker 1997, p. 106). The widespread interest in this approach across Europe involved the creation of special institutions for those with sensory impairments—not to put people to work, but to educate them. These organizations were forerunners of specialized institutions for all forms of impairment, which became generalized in the twentieth century. Besides being intended as places of solidarity and

stimulation for disabled people, these institutions fostered the invention of new techniques aimed at ameliorating impairments (such as the first embossed dot codes for
people with visual impairment) (Stiker 1997, p. 111; Braddock and Parish 2001, p. 29).

By the end of the eighteenth century, an important and ambitious power rose—that
of the medical profession—that would dominate the two following centuries (Peter
1976), and the biological explanation of disability set in (Stiker 1999). Disability, notably
physical disability, increasingly was identified as biologically rooted but nevertheless
distinct from illness (Stiker 1997, p. 110). Despite what might be seen as progress from
previous epochs, disabled people who were usually poor were still seen as belonging to
the administrative category of the poor, and the repressive institutional answer to the
issue of poverty "had a direct impact on people with disabilities. ... Houses of correction,
workhouses, asylums, and madhouses became more common as the eighteenth century
ended" (Braddock and Parish 2001, p. 29).

Educational and Institutional Developments
in the Nineteenth Century

With the nineteenth century's development in educational and medical approaches to
disability, institutionalization of disabled people in settings specific to their impairment
gained momentum. Braddock and Parish (2001, p. 30) record the following developments: the expansion in Europe and the United States of residential schools for deaf
and blind youth, and the controversy between oralism and sign language; the publication in 1829 of Louis Braille's embossed dot code; J. Itard's educational experiment in
France to educate a "feral child" (a child whose life in the wilderness since birth had
caused muteness and intellectual disability); E. Seguin's "physiological method emphasizing sensory-motor training, intellectual training and moral training or socialisation";
the development in Europe of special schools for persons with intellectual disabilities;
changes in the education of people with speech disorders; and the establishment of
orthopedic institutes and vocational education for physically disabled children and
adults. These developments in a variety of fields and contexts (medical, educational,
and/or social) continued to foster the specialization of institutions aimed at people with
impairment.

From Charity to Entitlement (from Assistance
to Vocational Rehabilitation)

By the early years of the twentieth century, all industrialized countries had gone beyond
charitable contributions to entitlement. During WWI, the pension schemes for wounded
soldiers expanded the notions of collective responsibility and "national debt" (Ewald
1986; Stiker 1997, p. 132) toward those wounded while serving the nation at war (see
Chapter 1). No less important were the needs to offset the wartime labor shortage that
lead countries to establish some form of social insurance for workers. With the introduction of the right to compensation benefits for injuries at work, these early welfare
programs represented a movement from charity to reintegrate all available manpower
in the wheels of production (Ewald 1986; Trattner 1999). Compensation did not simply

consist of "pensioning off" veterans for serving the nation, but was combined with vigorous attempts to enable disabled servicemen to return to work through medical and vocational rehabilitation (Montès 2002).

Complementary to anatomo-pathological medicine, postwar medical rehabilitation developed a new clinical approach, that of focusing on body functions (Frattini 2008). Calling on a multiplicity of medical and paramedical professions, functional rehabilitation triggered the bloom of new medical techniques and the creation of orthopedic and prosthetic centers. Services of occupational training for the recovery of working capacities and the acquisition of skills, as well as services of guidance and work placement, were created (Stiker 1997, 1999). The principle of vocational rehabilitation and its related institutional technologies instituted a new form of relationship between society and its disabled members. The notion that disabled people needed to be *rehabilitated* rather than *assisted* as poor became entrenched and that position gradually framed disability policies in all industrialized countries, with major developments after WWII (Stiker 1997, 1999).

In modern history of disability, the recognition of work- and war-injured people's rights, enshrined in national and international legislation, can be seen as a landmark in Western societies' approaches to disability and a decisive step forward in the struggle of disabled people for recognition as full members of society. These entitlements granted to disabled people (albeit specific categories) triggered another major phenomenon characteristic of the twentieth century—the active mobilization of disabled civilians.

The Rehabilitation Era

In the immediate post-WWI period, disabled civilians rose to claim the same rights to compensation benefits and measures of vocational rehabilitation as those granted to people who were injured at work or war. However, until the 1950s, the creation of occupational training centers for civilian disabled people, often attached to functional rehabilitation centers (Frattini 2008), was almost universally the initiative of, and also funded by, civilian organizations, such as secular charity societies and organizations of disabled persons (Barral 2007).

The establishment of social security programs in the 1930s and 1940s solidified the state's responsibility to provide for civilian disabled people (Stiker 1997, 1999). It also brought with it the development of disability policies based on rehabilitation, which extended to other domains beyond simply vocational (including educational, psychological, and so on). Rehabilitation developed into a system, with a purported relevance to disabled children as well as to adults.

During what might be typified as a medically dominant phase of rehabilitation, disability was conceptualized in terms of deviation from functional and social norms or standards (Becker 1963). Initially rehabilitation aimed to reduce or solve these deviations by developing and maintaining the functional capacities of people with impairment, in order for them to return to the mainstream (Stiker 1997). However, rehabilitation policies still favored the institutionalization rather than the social, professional, and educational reintegration of disabled people (Stiker 1997). This prompted the mobilization of disabled people from the 1960s on, spearheaded in the United Kingdom and

North America, to claim their rights as citizens (UPIAS 1976; DeJong 1979; Oliver 1986; Abberley 1987; Barton 1993).

The Disability Rights Movement

From the late 1960s and early 1970s a nonmedically driven academic interest in disability arose. Key works include Erving Goffman's sociological analyses of the institutional system in terms of "total institution" and the stigmatization process (Goffman 1961). In addition, from the field of social psychology, authors such as Wolfensberger (1972, 1998) developed concepts of the "normalization process" and "social roles valorization," which advocated a shift from institution-based to community-based rehabilitation. Such writers raised awareness about the inadequacies of the then-current system and the inaccurate and arguably unethical conceptualization of disability in terms of deviance.

In the late 1970s and early 1980s, disability began to be more widely conceptualized in radical political terms by disability activists and academics from the United States and Great Britain (Albrecht 1976; UPIAS 1976; Oliver 1983; Barnes 1991). They argued that disability was a social construct and that disabled people were an oppressed minority, discriminated against and excluded from participating in society, as a result of structural and environmental barriers, among which were rehabilitation policies. Disability studies further developed this conceptualization of disability into the "social model" as opposed to the "individual model" of disability (often referred to as the "medical model"); these models were deemed responsible for misleading the understanding of disability (UPIAS 1976; Oliver 1986; Barnes 1991; Barton 1993; Pfeiffer 2002; see also Chapter 10).

Since the 1970s, the Independent Living movement in the United States has engaged in creating "independent living resource centers" as alternatives to institutionalization and to support disabled persons in taking control of their lives. The "paradigm of independent living," to quote DeJong (1979), spread throughout the United States, Canada, and many European countries. Disabled peoples' organizations (DPOs) were set up around the world, and an international disability rights movement, spearheaded by Disabled Peoples' International, was established in 1981. The 1980s saw the involvement of the United Nations (UN) in promoting the rights of disabled people (International Year of the Disabled in 1981, UN Decade of Disabled Persons, and the World Program of action regarding disability in 1982) and further support for community-based rehabilitation (CBR) (Lang 2009). In 1993, the *UN Standard Rules on the Equalization of Opportunities for Persons with Disabilities* provided a guiding document based on the social model of disability for the implementation of a rights-based approach to disability policy. Through ongoing collaboration between the UN and organizations of the disability rights movement, a legally binding international treaty whereby signatory states would be held accountable for the promotion and enforcement of human rights for disabled people was developed. The UN General Assembly adopted the *Convention on the Rights of Persons with Disabilities* in 2006, which came into force in May 2008. However, as Lang (2009) states, the UN Convention is insufficient for ending disability discrimination. Challenges remained (and remain today) in how societies effectively implement these conceptual and legal achievements.

Contemporary Development of Disablement Models and International Classifications

Parallel to the societal evolution of representations and conceptualizations of disability outlined earlier in this chapter, the 1960s was a time that academic interest in disability grew significantly. Papers from the social sciences, public health, and clinical medical fields emerged where ideas were formalized and what would come to be known as "disablement models" were published across health and social science journals.

Generally speaking, a model is a conceptual explanation of a process and its underlying mechanisms (Snyder et al. 2008). A robust model should make explicit its components and their interrelations and enable the translation of the conceptual and explanatory level to the practical level. In other words, it should enable the operationalization of a conceptual structure (see Chapter 1). Some models include a classification system and related measurement instruments, while others do not. Disablement models to date have, despite varying in derivation and approach, largely aimed to articulate the various elements that should be taken into account in order to more fully understand disability. Components within those models might include health status and physical functioning but should arguably go beyond these to include social and contextual factors such as the environment and its diverse relationships and interactions.

Disablement models have different intended uses with some being used as the basic frameworks for public policy, including compensation programs, research, as well as clinical work and rehabilitation services. In this section, we present a chronological overview in the development of disablement models and then focus our discussion on two recent models: the World Health Organization's ICF (WHO 2001) and the Disability Creation Process (Fougeyrollas et al. 1996).

Nagi's Disablement Model

Saad Nagi, an Egyptian sociologist, began his consideration of disability in the 1960s (Nagi 1965), but his work remained unknown until the 1980s. In the 1970s, he shifted his focus from the physical consequences of pathologies to a more dynamic process that mediates changes in function. He defined disability as an "expression of a physical or a mental limitation in a social context, a gap between the individual's capabilities and the demands created by the physical and the social environment," and described disablement through four concepts: active pathology, impairment, functional limitation, and disability (Masala and Petretto 2010).

Nagi's model was aimed at conceptualizing disability and guiding rehabilitation interventions. It did not include a classification scheme (see discussion in The International Classification of Impairments, Disabilities, and Handicaps). This model is interesting in that it implies that the environment can be studied separately from the individual. Nagi (1976) found that correlations among impairments, functional limitations, and social roles were poor, and this initiated research into environmental factors that affect disability, which in turn opened the way to the biopsychosocial model.

Engel's Biopsychosocial Model

In his 1977 landmark paper, George L. Engel identified a need for a new medical model. His idea was to offer a holistic model, one that responded to the criticism of reductionism being levelled at scientific medicine (Engel 1977). He thought of his model as a new framework for clinical medicine and the teaching of psychiatry. Within the model, three dimensions—biological, psychological, and social—were proposed to play a significant role in the functioning of a sick or disabled person, and therefore treatment or care needed to address each of the dimension. Engel also developed methodologies to operationalize his model in clinical observation and medical teaching (Engel 1980). However, despite (or because of) wide acceptance of Engel's proposal, some have criticized it as too general (Dowrick et al. 1996). Indeed others have even called it a myth; McLaren (2002) has suggested that Engel did not in fact develop any such model and simply pleaded for "a more considerate type of medicine" that would focus on a wider range of human functioning than pathology and impairment. Despite criticisms, the language of bio, psycho, and social components of health is now embedded in the lexicon of most health disciplines; there are even dedicated journals that advance the model (for example, www.bpsmedicine.com).

The International Classification of Impairments, Disabilities and Handicaps

Neither Nagi's disablement model nor the biopsychosocial model included a classification system. Developing a classification is considered an important epistemological step forward in the operationalization of a model, especially to underpin measurement (see Chapter 11). In the area of health, the only established classification system available in the late 1970s was the *International Statistical Classification of Diseases and Related Health Problems* (ICD) (WHO 1946). The World Health Organization developed this system in 1946* as the basis for an international classification of diseases to build on an earlier classification of causes of death (developed by Farr in 1855, later by Bertillon in 1893). It is still maintained by WHO, and the system is appropriate for many of the aims of clinical medicine. However, such a classification was inadequate for identifying the consequences of nonlethal diseases, which may be associated with disability, and unable to account for their functional sequelae (Ravaud 1999; WHO 2001). In 1972, the WHO recognized this and decided to complement the ICD with a classification of the consequences of diseases, a decision that resulted in the publication of the *International Classification of Impairments, Disabilities and Handicaps* (ICIDH) (WHO 1980). In the words of Gail Whiteneck (2006), "[The ICIDH] extended the disease-related sequence of aetiology, pathology, and manifestation with the illness-related sequence of disease, impairment, disability, and handicap."

On one hand, among health care professionals, the ICIDH's conceptual segmentation of disability into three distinct dimensions (impairment, disability, and handicap; see Figure 2.1) was seen as providing clarification and as an operational approach to disability. On the other hand, the model was rapidly and roundly criticized by the

---

* http://www.who.int/classifications/icd/en/HistoryOfICD.pdf.

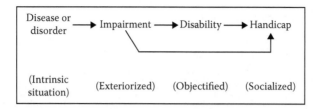

FIGURE 2.1 The International Classification of Impairment, Disability, and Handicap conceptual model. (From World Health Organization, *International Classification of Impairments, Disabilities and Handicaps: A Manual of Classification Relating to the Consequences of Disease.* Geneva: World Health Organization, 1980.)

growing international movement of disabled activists and academics who were developing conceptualizations of disability in sociological and political terms and were opposed to focusing on individual characteristics as the cause of the disability (Albrecht 1976; UPIAS 1976; Barnes 1991).

The Disability Creation Process (Version I)

In 1987, Patrick Fougeyrollas, working from the perspective of the disability movement (Mouvement québecois pour la promotion des droits des personnes handicapées) and acknowledging the epistemologic importance of classification development, began to work on a new model that aimed to take into account the concerns of disability activists by overcoming the opposition between the individual and the social models of disability (Fougeyrollas, 1993, 1995, 2002).

The *Disability Creation Process* (DCP) Fougeyrollas developed aimed "to document and explain the causes and consequences of diseases, trauma, and other effects on the integrity and the development of the person" (http://www.indcp.qc.ca/hdm-dcp/hdm-dcp). For Fougeyrollas, *life habits*—defined as meaningful activities and roles—can be influenced by reinforcing capabilities and compensating for disabilities through rehabilitation, but also by reducing environmental obstacles stemming, for example, from prejudices, lack of assistance or resources, lack of accessibility, and other social factors. He describes life habits as being the result of the interaction between the person and his or her environment. Life habits, he suggested, are indicators of the quality of social participation, which can be measured on a scale ranging from optimal social participation to a completely disabling situation (a measure was developed for that purpose: the Assessment of Life Habits/LIFE-H—http://www.indcp.qc.ca/assessment-tools/introduction).

The DCP model (Figure 2.2) takes into account the concepts of risk factors, organic systems, and capabilities. For this model, risk factors (those factors that might impact the integrity of persons in their context) are related to the causes of impairments in organic systems, but also related to disabilities *and* capabilities (INDCP). This model identifies environmental factors as facilitators or obstacles to a person's social participation, i.e., the realization of his/her life habits. The level of accomplishment of life habits for a person (or population) is said to result from the interaction between personal and environmental factors at a point in time. Social participation or a handicap situation (see Figure 2.2) results from the variation of the interaction of personal and environmental

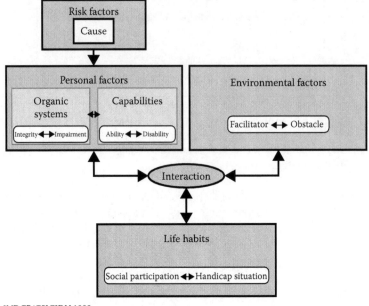

© INDCP/CSICIDH 1999

FIGURE 2.2   The Disability Creation Process (Version 1).

factors. This model has two particularly interesting features: it was the first model that did not causally relate disability to individual impairment, and it also aimed to overcome the opposition between a social and individual approach to disability. A revised version of the DCP, published in 2010, is discussed later in this chapter (see section titled: The Disability Creation Process [Version II]).

The Disablement Process and Other Models

In parallel to Fougeyrollas's work, in 1993 the US National Center for Medical Rehabilitation Research (NCMRR), working in the tradition of Nagi's model, published a model that used a category called "societal limitations," defined as "restrictions attributable to social policy or barriers (structural or attitudinal) which limit fulfilment of roles or deny access to services and opportunities associated with full participation in society" (Nagi 1991; NICHHD and NCMRR 1993, p. 24).

   At that time, some researchers argued that personal factors, such as age, gender, and race and also habits or lifestyle, may influence the disability process and thus should be included in disability theories and models (Whiteneck 2006). Verbrugge and Jette (1994) published a model of the disablement process that took into account such personal and environmental factors. They defined disability as "difficulty doing activities in any domain of life due to a health or physical problem" (p. 1) and suggested that disability is not a personal characteristic, but a gap between environmental demand and personal capability. They distinguished between intrinsic disability (without personal or equipment assistance) and actual disability (with such assistance). They insisted on the importance of measuring both in order to inform research and public health intervention (e.g., compensation policy or prevention).

Concurrently, the US Institute of Medicine published several reports in which the disability process is described as an "interaction of the individual and the environment" (IOM 1991, p. 82), and in which they stated that "the environment plays a critical role in determining whether each stage of disablement occurs and if transitions between the stages occur" (IOM 1997). These reports also introduced the notion of disability risks factors in line with the social determinants of health.

It is notable that while considerations of the social determinants of health have had a long history in the medical literature back to at least the nineteenth century (e.g., Villerme in France, and many studies in Britain as well; see also the development of hygiene and the history of public health; see Fagot-Largeault 1989; Rosen 1993), it has taken so long to explicitly acknowledge their influence in the area of disability and rehabilitation. Interestingly, tackling social determinants of health and disability (see the discussion in The ICF Model) has been recognized as a priority area in the WHO program of work and targeted as a focus for 2014–2019 (WHO 2012). Overall, it seems that these efforts collectively increased the focus on determinants of disability outside the individual impairments and opened the way to the next generation of disablement models.

The ICF Model

As we have seen, if the ICIDH addressed the expectations of rehabilitation professionals, it was far from meeting those of the disability movement. This tension was outlined in the foreword to the fourth reprint of the ICIDH in 1993: "...Concern was expressed that the ICIDH did not state clearly enough the role of social and physical environment in the process of handicap and that it might be construed as encouraging 'the medicalization of disablement' (The term 'disablement' is used here to encompass the full range of impairment, disability, and handicap.)." The WHO subsequently launched in 1992 a revision process of the ICIDH, which ultimately resulted in the publication of the *International Classification of Functioning, Disability and Health* (ICF) (Figure 2.3) in 2001 (Bickenbach et al. 1999; WHO 2001).

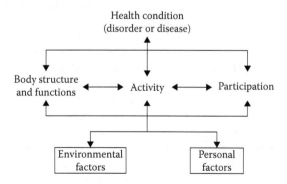

FIGURE 2.3 The ICF Model: interaction between ICF components. (From World Health Organization, *International Classification of Functioning, Disability and Health*. Geneva: World Health Organization, 18, 2001.)

In the WHO's own words, the ICF moved away from being a classification focused on the "consequences of disease" to become a "components of health" classification. "Components of health" classification identifies the constituents of health, whereas "consequences of disease" classification focuses on the impacts of diseases or other health conditions that may follow as a result (WHO 2001, p. 4). In addition, the WHO adopted a neutral position with regards to the etiology of individual health problems. "Impairments may be part or an expression of a health condition, but do not necessarily indicate that a disease is present or that the individual should be regarded as sick" (WHO 2001, p. 13). Last but not least, the ICF explicitly aimed to operationalize the bio-psycho-social model and developed a classification that reflects the biological point of view (the person as a body), the psychological point of view (the person as an individual), and the social point of view (the person as a social being).

Problems with the ICF

Despite the claims described earlier, and the inclusion of environmental factors as an integral part of its conceptual model (undoubtedly the major gain of the revision process of the ICIDH), the ICF has been subjected to numerous and harsh criticisms (e.g., see Roussel 1999).

Undoubtedly, one major change that occurred in the revision from the ICIDH to the ICF was a relabeling of and/or specification of terms such as *handicap* and *disability*, reportedly because of pressure from the disability community who felt the terms were pejorative or misused (Whiteneck 2003). Categories were relabeled to neutral descriptions of capabilities rather than deficits. Finally, the title of the model was changed to the model of functioning, disability, and health. The use of neutral terminology in the ICF is intended to stress the universalism of this tool, acknowledging that every human being can experience a decrement in health or functioning and hence experience disability.

We have been involved in some discussions with those revising the ICF as to whether the term *disability* in the name of the model and in its conceptualization is redundant and inconsistent, given that functioning covers all forms of functioning—with or without disability. Indeed some have suggested that rather than disability or impairment, we should refer to functional diversity (Patston 2007). More generally, the ICF has been reproached for renouncing a biomedical or individual perspective (as operationalized by the ICIDH) on a mere linguistic level, while surreptitiously heading back toward it in its definition of functioning, participation, and health (e.g., Whiteneck 2003). The claim of the ICF to truly be a biopsychosocial model has been challenged by Gzil et al. (2007) who view the ICF as adopting alternatively one *or* the other perspective without integrating one *and* the other, and thus failing to offer a real synthesis between biological, psychological, and social dimensions (see also Solli et al. 2012; Ravenek et al. 2013).

A key question is whether the universal aspects proposed in the ICF are adequately conceptualized and operationalized to allow a full understanding of health, functioning, ability/disability, and rehabilitation. An example where perhaps the operationalization is problematic (beyond consideration of "universal" attributes of health, wellbeing, and disability) is related to the question of which items should belong to the domain of activity (i.e., the functional aspect) and which to the domain of participation (i.e., the social aspect). This has been the topic of much discussion and disagreement during the revision process of the ICIDH. In 2000, in order to close the debate and have the classification officially endorsed by WHO within the planned time frame, WHO representatives, to the surprise of many international experts (including Whiteneck 2006), decided to have

1. A single list of elements for activity and participation domains with four options to use this list.
   a. Distinct sets of activities domains and participation domains (no overlap)
   b. Partial overlap between sets of activities and participation domains
   c. Detailed categories as activities and broad categories as participation, with or without overlap
   d. Use of the same domains for both activities and participation with total overlap of domains
2. Two qualifiers for these two dimensions (capacity and performance).
3. That capacity should be measured in a uniform standard environment (without assistance), while performance should be measured in the real world with the existing barriers and facilitators of the environment of the person (WHO 2001, pp. 234–36).

These decisions seem to have been more political than scientific and have been criticized for lacking an empirical justification (it was expected that empirical data would be collected and conceptual research carried out to differentiate activity and participation). For example, if we revisit the development process of the ICF, participants from different countries acknowledge that WHO hastened the process by drafting and submitting to its collaborating centers successive versions of the classification, without clear consideration or incorporation of data from the field trials carried out on each of them.* The records of the 2000 Madrid meeting of the WHO's network of international collaborating centers also acknowledge the lack of validation of the final version (Barral and Roussel 2002). For many, "the loss of a clear distinction between activity and participation was the step backward" of the revision process (Whiteneck 2006, p. 57). Beyond this, combining different concepts into one is particularly problematic for measurement (see Chapter 11).

Last but not least, the multiplicity of intended uses of the ICF and the attempt to develop a classification that suits them all presents another source of difficulties. In the

---

* Offical Statement of M. Lyazid, the French government's representative at the Madrid International Meeting of WHO's Collaborating Centers (November 15–18, 2000).

 – Correspondence between the heads of WHO's Collaborating Centers and WHO headquarters in charge of the revision of ICIDH (November 2000–February 2001, Archives of the French Collaborating Center).

 – Official response of French officials to Document OMS WP/2000/EIP (12.12.2000) Archives of the French WHO Collaborating Center.

early days of the ICF, it was anticipated that its uses would be broad and various (WHO 2001; Cieza et al. 2008) including

1. A statistical tool for population studies and information management
2. A research tool for outcome measurement and quality of life
3. A clinical tool for measurement planning, assessment, and evaluation
4. An educational tool for curriculum design and awareness raising
5. A social policy tool for all aspects of health and disability policy design implementation and monitoring

Since then, the ICF, it seems, has indeed been used in all of these areas and more besides (Bruyère et al. 2005). However, the question that remains is whether these objectives are compatible, whether a single classification can adequately serve these different aims, or whether this multipurpose classification has led to some confusion and conceptual shortcomings.

As a consequence of these issues, more than ten years after its official adoption by the World Health Assembly, it seems the ICF in its current form is still not fully optimized conceptually, nor as a classification to underpin research in rehabilitation. In particular, the need to accommodate the legitimate conceptual issues raised in response to the ICIDH seems to have led to some blurring of terminology—problematic given the intention that models represent and make clear a complex reality. To that end, aspects of the structure and the concepts of the new classification undermine the capacity of the ICF as a research tool there is not much point in "measuring" anything if the variables to be measured are unclear or loosely defined (Rossignol 2000). Despite its widespread adoption, for the ICF to be considered an optimal tool for framing rehabilitation strategies, revisions are required, with greater clarity of its theoretical foundations and empirical testing of that theory. Further efforts are needed to rework the complexity of environmental factors as represented in the measures, including a focus on evidence-based knowledge of the key social determinants of health and disability, the creation of a taxonomy of personal factors, and greater attention to risk factors. Clear validation of the resulting model is essential and as yet only partly met by the development of "core sets" (Cieza et al. 2004) since their validity fundamentally rests upon being certain of the validity of the underlying conceptual framework.

The Disability Creation Process (Version II)

In order to contribute to these debates and criticisms, Levasseur et al. (2007) compared the conceptual underpinnings of the *Disability Creation Process* (DCP) with those of the ICF. They found more differences than similarities between the DCP and ICF models. Similarities concerned approaches (systematic and universal), objectives (to describe the complex and dynamic process of human functioning), elements of the models' components (most notably the participation component), and a common focus on participation as a key social element of disability. However, substantive differences included the degree of transparency of the underlying theoretical assumptions and principles, particularly regarding how participation is conceptualized and how the measurement tools are produced. Based on the results of this work, Levasseur and his colleagues suggested that occupational therapists (the lead author is an occupational therapist hence perhaps limiting the recommendation to the primary audience for the paper) should consider using the DCP in both research and practice. They note that in each case, empirical studies were needed to clarify the concept of participation.

In 2010, Fougeyrollas published an enhanced version of the DCP, known as Human Model-Disability Creation Development Process II (HDM-DCP2; Fougeyrollas 2010; Fougeyrollas and International Network on the Disability Creation Process 2010). Key changes were made to the risk factors, which were integrated into the three conceptual domains of personal factors, environmental factors, and life habits (the latter being the closest to definitions of participation). Environmental factors were split into societal environment (macro level), community environment (meso level), and personal environment (micro level). Life habits were split into daily activities and social roles. Finally identity factors were expanded and included within personal factors. All these factors are considered to interact with biological or organic systems and the capabilities of the people. In addition, an interesting feature of the model is that it claims to take explicitly into account temporality and to offer a diachronic understanding of disability. This fairly recent model has not yet, to our knowledge, been submitted to a thorough review and, as with the ICF, needs to be empirically validated.

Incorporating Subjective Perspectives and Quality of Life into Disablement Models

Debates about the social versus individual models of disability have somewhat overlooked a perspective that has undergone an important development since the 1990s in the health sciences: the rise of outcome measurement in terms of quality of life.

The question on how to relate the conceptualization of the disablement process (which we have reviewed here) and the operationalization of quality of life as a measurable outcome has not been comprehensively explored nor clarified to date. For most authors, these are two distinct approaches, one focusing on the causes and processes of disablement and the other on its objective and subjective consequences. Gail Whiteneck was among the first (see IOM 1997; Barbotte et al. 2001; Ueda and Okawa 2003) to clearly identify that subjective quality of life is a key missing component of the ICF model. He suggests the structure of the ICF model be revised to situate quality of life on the same level as activity and participation saying, "Only the actual addition of a new quality-of-life domain acknowledges the validity of the subjective perceptions of people with disabilities and that those perceptions are distinct outcomes of the disability process" (Whiteneck 2006, p. 59). Buntinx and Schalock (2010) work along the same line in the area of intellectual disability. They argue that conceiving disability in the context of the person (including subjective perspectives) and environmental interactions enhances our understanding of the disablement process and informs our approaches to diagnosis, classification, assessment, and planning of individual support programs. In this context, they propose that the construct of quality of life can become the link between the general values reflected in social rights and the personal life of the individual. This is because the concept of quality of life captures (or at least aims to capture) the essential dimensions of an individual's life situation and operationalizes these dimensions so they can be considered in an outcome evaluation. In their review of various disability and quality of life models, the authors found that quality of life (QOL) models differ from disablement models in at least five ways:

1. Their content (QOL models tend to be less functionally oriented)
2. The focus of the assessment (the person's status in the case of QOL measures)
3. The evaluation metric (more subjective dimensions in the case of QOL measures)
4. The focus on the point of view of the person
5. The intended purpose (outcome evaluation in the case of QOL measures)

Buntinx and Schalock offer an interesting framework that aims at integrating disability, quality of life, and individualized support models. These perspectives should no doubt be further researched and discussed, and the debate should also take into account the discussions about the limits of the uses and misuses of the term "quality of life" (e.g., Leplège and Hunt 1997; Willems 2010).

Discussion

The debate around disability models in relation to rehabilitation, from the early work of Nagi on, highlights a series of questions related to which factor should be taken into account beyond traditional medical variables (e.g., physiopathology, health status, and physical and/or psychological functioning) and how these factors relate to each other. In this regard, it seems that over time, models have increasingly acknowledged the need to include, in some way, contextual factors. Specifically there is consensus about the necessity to account for the social dimensions of disability. However, the nature and role of personal factors and which of these should be taken into account are much less clear and lacking in consensus. For example, the ICF does not provide any classification of personal factors, although the category is explicitly part of the model. The question of whether and how to include quality of life also remains in need of clarification.

One important difference between models with regard to social factors concerns the direction of causality, and there appear to be three types of models:

1. *Individual models* for which the arrows symbolizing the causal relations among the variables indicate a causal connection from the individual characteristics to the social phenomena of disability (see for example the ICIDH)
2. *Social models* for which the causal relationship goes from the characteristics of the social environment to the disability phenomenon (for example, including exclusion, ostracism, and the social inequality of opportunities)
3. *Interactive models* for which the potential causality is bidirectional examples being the DCP (Quebec Classification) and—although some would argue—the ICF model

Overall, our exploration suggests there remains an unresolved and uncomfortable intersection between rehabilitation and disability where political (and politicized) drivers of thought and development pervade, despite many shared objectives and ideals among those who work in disability and rehabilitation. This situation is understandable given the historical background of disability that we presented in Introduction, and indeed the persistence of systems (including health systems) that may disadvantage disabled people (Chapter 3, 4, and 10). The social model is frequently placed in opposition to the medical model despite the fact that each approach—medical and social—has its history, merits, and drawbacks and neither is completely satisfactory (Thomas 2012). In many papers about disability, the medical model is referred to somewhat rhetorically as a pejorative label (see, for example, Hahn 1985, 1988, 1993) just as disability might be in archetypal medical works.

It is probably noticeable that we have not presented a medical model of disability in this review, although many authors refer to such in the disability literature. We have taken this step because we suggest that the medical model is in many ways a "straw man"—something held up by critics as the basis of clinical rehabilitation practice and research when there is in fact no single medical model. In the medical literature and the history of medicine, there has been neither consensus on nor description of any one medical model that could be opposed to disablement models or opposed by exponents of those disability models. Interestingly, the work of Boorse (1977) represents one of the most serious attempts to propose a model of health that is compatible with contemporary scientific medicine, but it is barely known beyond the circle of philosophers of medicine.

It seems that what proponents of the social model of disability may have in mind when they talk about the medical model would be more accurately called an individualistic model—for example, the ICIDH's apparent or implied causal association between individual characteristics and disability. Calling such an approach the medical model risks ignoring the history of medicine, which from Hippocrates to the modern development of epidemiology has paid much attention to the contextual (environmental, social, etc.) causes and consequences of disease and impairment. The reason for the tension between the social model and the supposedly medical, according to Patrick Fougeyrollas (personal communication), is that the social model is in fact a political model and as such needs the medical model so as to define itself in opposition. It is difficult to see how one type of model could really overcome the other type, or to see the merit of doing so. It is impossible to disentangle the individual from their/our experience of impairment, disability, health, or wellbeing. From our perspective, the real problem is not to resolve the opposition between the medical and the social models by adopting one or the other model. Rather, we think advance is more likely if we overcome the longstanding opposition by using (and if required producing) models that are integrative and adaptive and take temporality into account.

In terms of models, the opposition between the social and so-called medical models may (or may not) be part of the past. The need for integrative models is clear and, we suggest, is the only type of approach and/or instrument that could be used for the implementation of the UN Convention. Of the two arguably integrative approaches we have described here, the DCP is 25 years old and the ICF is aged 13. If one considers that work done thus far has sufficiently advanced critique of thinking about both rehabilitation and disability, it may be that one could refine these models rather than start again, although there are clearly risks in incremental development if there are conceptual problems at the foundation level of a model or framework. Fougeyrollas pursued "refinement" rather than "novel production" in the second version of the DCP, and this appears to be what the WHO and its collaborating centers are heading toward by launching the revision of the ICF, which could take another ten years to be completed. Our sense is that the next iterations of these models and related tools of assessment or measurement would be different, and improved, if disability and rehabilitation researchers—along with practitioners, disabled people, and other stakeholders—were to work more closely together than we have done in the past.

Conclusion

We have explored the progressive history of the conceptualization of disability and the emergence of disablement models in the second half of the twentieth century. These evolutions have not come to a close, and many issues are still pending. Rehabilitation emerged in the twentieth century as a system of practices, contextualized within the framework of medicine. Disability thinking and scholarship have long challenged an individualistic medical perspective of disability and rehabilitation. Indeed, perhaps due to the success of those challenges, many scholars in health and rehabilitation are also now challenging such an individual focus in research, service development, policy, and teaching (see, for example, multiple chapters in this book).

Rethinking rehabilitation appears a matter of political will, given the multiple states that are cosignatories to the UN Convention (United Nations 2006) to frame (or reform) disability policies (including health and rehabilitation). The associated shifts in thinking and scholarly work reveal an evolving epistemology that more fully recognizes the experiences of disabled people as a legitimate repository and source of knowledge and expertise, an interactive and adaptive perspective where there are respective expert roles, and great possibilities for complement in assessing and addressing disability situations. If rehabilitation is to implement the Convention and enhance the participation of persons with disabilities, if it aims to facilitate agency and autonomy, it should arguably include in its definition notions of capability, drawing on the knowledge, will, and actions of the person in their context. A purely social model *or* a purely individual model of rehabilitation may in the near future be considered obsolete in the transformative context in which the person is seen as the origin of action and not only the target of interventions.

This chapter has highlighted a number of the frameworks that we talk about and that to some degree or other underpin decision making in practice and research. We have explored the strengths and weaknesses of these models, but more particularly explored the reasons for those strengths and weaknesses. This is of course not the end point, but a critique of where we are currently and what has brought us here. We would hope that it provides a way of moving forward, for rethinking rehabilitation, for posing new ways of thinking and doing.

Acknowledgments

We thank Jerome Bickenbach, Patrick Fougeyrollas, Rex Billington, Ros and Richard Madden, and Elisabeth Badley who have generously spent some of their time to discuss their contributions to the history that we describe in this chapter and for having provided us with useful information and insights. However, this chapter does not report these conversations (unless explicitly stated) nor reflect their views on the topic. It reflects our interpretations and arguments and we are solely responsible for its contents and any errors in interpretation.

References

Abberley, P. (1987). The concept of oppression and the development of a social theory of disability. *Disability, Handicap and Society* 2(1): 5–19.

Albrecht, G. (1976). *The Sociology of Physical Disability and Rehabilitation.* Pittsburgh, PA: University of Pittsburgh Press.

Barbotte, E., Guillemin, F., and Chau, N. (2001). Prevalence of impairments, disabilities, handicaps, and quality of life in general population: A review of recent literature. *Bulletin of the World Health Organization* 79(11): 1047–55.

Barnes, C. (1991). *Disabled People in Britain and Discrimination: A Case for Anti-Discrimination Legislation.* London: Hurst and Co.

Barral, C. (2007). Disabled Persons' Associations in France. *Scandinavian Journal of Disability Research* 9(3–4): 214–36.

Barral, C., and Roussel, P. (2002). De la CIH à la CIF. Le processus de révision. *Handicap-Revue de Sciences Humaines et Sociales* 94–95: 1–24.

Barton, L. (1993). The struggle for citizenship: The case of disabled people. *Disability, Handicap and Society* 8(3): 235–48.

Becker, H. (1963). *Outsiders.* Glencoe, IL: The Free Press of Glencoe.

Bickenbach, J.E., Chatterji, S., Badley, E.M., and Üstün, T.B. (1999). Models of disablement, universalism and the international classification of impairments, disabilities and handicaps. *Social Science and Medicine* 48: 1173–87.

Boorse, C. (1977). Health as a theoretical concept. *Philosophy of Sciences* 44: 542–73.

Braddock, D.L., and Parish, S.L. (2001). An institutional history of disability. In G.L. Albrecht, K.D. Seelman, and M. Bury (eds.) *Handbook of Disability Studies,* 11–68. London: Sage Publications.

Bruyère, S.M., Van Looy, S.A., and Peterson, D.B. (2005). The International Classification of Functioning, Disability and Health (ICF): Contemporary literature review. *Rehabilitation Psychology* 50: 113–21.

Buntinx, W., and Schalock, R.L. (2010). Models of disability, quality of life, and individualized supports: Implications for professional practice in intellectual disability. *Journal of Policy and Practice in Intellectual Disabilities* 7(4): 283–94.

Castel, R. (1989). Le social aux prises avec l'histoire. *Les Cahiers de la Recherche sur le Travail Social* n°16/89: 9–27.

Castel, R. (1999). *Les métamorphoses de la question sociale: Une chronique du salariat.* Paris: Gallimard.

Castel, R. (2003). *From Manual Workers to Wage Labourers: Transformation of the Social Question.* New Brunswick, NJ: Transaction Publishers.

Cieza, A., Bickenbach, J., and Chatterji, S. (2008). The ICF as a conceptual platform to specify and discuss health and health related concepts. *Gesundheitswesen* 70: 1–10.

Cieza, A., Ewert, T., Ustun, T.B., Chatterji, S., Kostanjsek, N., and Stucki, G. (2004). Development of ICF core sets for patients with chronic conditions. *Journal of Rehabilitation Medicine* 44(Suppl): 9–11.

DeJong, G. (1979). Independent living: From social movement to analytic paradigm. *Archives of Physical Medicine and Rehabilitation* 60: 435–46.

Dowrick, C., May, C., Richardson, M., and Bundred, P. (1996). The biopsychosocial model of general practice: Rhetoric or reality? *British Journal of General Practice* 46(403): 105–7.

Engel, G.L. (1977). The need for a new medical model. *Science* 196: 129–36.

Engel, G.L. (1980). The clinical application of the biopsychosocial model. *The American Journal of Psychiatry* 137(5): 535–44.

Ewald, F. (1986). *L'Etat Providence*. Paris: Grasset.

Fagot-Largeault, A. (1989). *Les causes de la mort, Histoire naturelle et facteurs de risque*. Paris: Librairie Philosophique J. Vrin.

Fossier, A. (2002). Le grand renfermement. *Tracés. Revue de Sciences humaines, 1/2002*. Retrieved from http://traces.revues.org/4130 (accessed April 10, 2014).

Foucault, M. (1973). *Madness and Civilization. A History of Insanity in the Age of Reason* (R. Howard, Trans.). New York: Vintage.

Fougeyrollas, P. (1993). Explanatory models of the consequences of disease and trauma: The handicap creation process. *ICIDH, International Network* 6(2).

Fougeyrollas, P. (1995). Documenting environmental factors for preventing the handicap creation process: Quebec contribution relating to I.C.I.D.H. and social participation of people with functional differences. *Disability and Rehabilitation* 17(3–4): 145–53.

Fougeyrollas, P. (2002). L'évolution conceptuelle internationale dans le champ du handicap: Enjeux socio-politiques et contributions québécoises. *Pistes* 4: 2.

Fougeyrollas, P. (2010). *La funambule, le fil et la toile: Transformations réciproques du handicap*. Laval: Presses de l'Université.

Fougeyrollas, P., Cloutier, R., Bergeron, H., Côté, J., Côté, M., and St-Michel, G. (1996). *Révision de la proposition québécoise de classification: Processus de production du handicap*. Québec: Réseau international sur le processus de production du handicap.

Fougeyrollas, P., and International Network on the Disability Creation Process (INDCP). (2010). *Human Development Model-Disability Creation Process*. Québec, Canada: INDCP. Retrieved from http://www.indcp.qc.ca/hdm-dcp/hdm-dcp.

Frattini, M.-O. (2008). *Dynamique de constitution d'une spécialité médicale fragile: La médecine de rééducation et réadaptation fonctionnelles en France entre médecine et politique*. (Unpublished Masters Dissertation 'Santé, Population, Politiques sociales'). Paris: EHESS. Retrieved from http://www.sofmer.com/download/sofmer/MOF-M2-SPPS-EHESS.pdf.

Geremek, B. (1987). *La potence ou la pitié. L'Europe et les pauvres du Moyen-Age à nos jours*. Paris: Gallimard.

Goffman, E. (1961). *Asylums: Essays on the Social Situation of Mental Patients and Other Inmates*. New York: Anchor Books.

Gzil, F., Lefève, C., Cammelli, M., Pachoud, B., Ravaud, J.F., and Leplège, A. (2007). Why is rehabilitation not yet fully person-centered and should it be more person-centered? *Disability and Rehabilitation* 29: 20–21.

Hahn, H. (1985). Toward a politics of disability: Definitions, disciplines, and policies. *Social Science Journal* 22: 87–105.

Hahn, H. (1988). The politics of physical differences: Disability and discrimination. *Journal of Social Issues* 44: 39–47.

Hahn, H. (1993). The political implications of disability definitions and data. *Journal of Disability Policy Studies* 4: 41–52.

Institute of Medicine. (1991). *Disability in America. Toward a National Agenda for Prevention*. Washington, DC: National Academy Press.

Institute of Medicine. (1997). *Enabling America: Assessing the Role of Rehabilitation Science and Engineering*. Washington, DC: National Academy Press.

Lang, R. (2009). The United Nations Convention on the right and dignities for persons with disability: A panacea for ending disability discrimination? *Alter-European Journal of Disability Research* 3(3): 266–86.

Leplège, A., and Hunt, S. (1997). The problem of quality of life in medicine. *Journal of the American Medical Association (JAMA)* 278(1): 47–50.

Levasseur, M., Desrosiers, J., and St-Cyr, T.D. (2007). Comparing the disability creation process and international classification of functioning, disability and health models. *Canadian Journal of Occupational Therapy* 74(3 Suppl): 233–42.

Masala, C., and Petretto, D.R. (2010). Models of disability. In J.H. Stone, and M. Blouin (eds.) *International Encyclopedia of Rehabilitation*. State University of New York College of Buffalo, Buffalo, New York: Center for International Rehabilitation Research Information and Exchange (CIRRIE). Retrieved from http://cirrie.buffalo.edu/encyclopedia/en/article/135/ (accessed October 24, 2014).

McLaren, N. (2002). The myth of the biopsychosocial model. *Australian and New Zealand Journal of Psychiatry* 36(5): 701–3.

Montès, J.-F. (2002). L'office national des anciens combattants et victimes de guerre. Création et actions durant l'entre-deux-guerres. *Guerres mondiales et conflits contemporains* 2002/1, n°205: 71–83.

Nagi, S.Z. (1965). Some conceptual issues in disability and rehabilitation. In M.B. Sussman (ed.) *Sociology and Rehabilitation*, 100–13. Washington, DC: American Sociological Association.

Nagi, S.Z. (1976). An epidemiology of disability among adults in the United States. *The Milbank Memorial Fund Quarterly* 54: 439–67.

Nagi, S.Z. (1991). Disability concepts revised: Implications for prevention. In A.M. Pope, and A.R. Tarlov (eds.) *Disability in America: Toward a National Agenda for Prevention*. Washington, DC: National Academy Press.

National Institute of Child Health and Human Development and National Center for Medical Rehabilitation Research. (1993). *Research Plan for the National Center for Medical Rehabilitation Research*. Washington, DC: US Government Printing Office.

Oliver, M. (1983). The politics of disability. Paper given at The Annual General Meeting of The Disability Alliance. Retrieved from http://disability-studies.leeds.ac.uk/files/library/Oliver-dis-alliance.pdf (accessed October 24, 2014).

Oliver, M. (1986). Social policy and disability: Some theoretical issues. *Disability, Handicap and Society* 1(1): 5–17.

Patston, P. (2007). Constructive functional diversity: A new paradigm beyond disability and impairment. *Disability and Rehabilitation* 29(20–21): 1625–33.

Peter, J.-P. (1976). Le grand rêve de l'ordre médical en 1770 et aujourd'hui. *Autrement* n°4, *Gouverner la misère*, 183–92.

Pfeiffer, D. (2002). The philosophical foundations of disability studies. *Disability Studies Quarterly* 22(2): 3–23.

Piven, F., and Cloward, R.A. (1971). *Regulating the Poor. The Functions of Public Welfare.* New York: Vintage.

Ravaud, J.F. (1999). Modèle individuel, modèle médical, modèle social: La question du sujet. *Handicap-Revue de Sciences Humaines et Sociales* 81: 64–75.

Ravenek, M.J., Skarakis-Doyle, E., Spaulding, S.J., Jenkins, M.E., and Doyle, P.C. (2013). Enhancing the conceptual clarity and utility of the international classification of functioning, disability and health: The potential of a new graphic representation. *Disability and Rehabilitation* 35(12): 1015–25.

Rosen, G. (1993). *A History of Public Health.* Baltimore: Johns Hopkins University Press.

Rossignol, C. (2000). ICIDH2. Projet Béta2: Analyse textuelle et formelle. Rapport d'expertise. Laboratoire Langage et parole. CNRS.

Roussel, P. (1999). CIH-1/CIH-2: Rénovation complète ou ravalement de façade? *Handicap-Revue de Sciences Humaines et Sociales* 81: 7–19.

Snyder, A.R., Parsons, J.T., Tamara, C., McLeod, V., Curtis Bay, R., Michener, L.A., and Sauers, E.L. (2008). Using disablement models and clinical outcomes assessment to enable evidence-based athletic training practice, part i: Disablement models. *Journal of Athletic Training* 43(4): 428–36.

Solli, H.M., and Barbosa da Silva, A. (2012). The holistic claims of the biopsychosocial conception of WHO's International Classification of Functioning, Disability, and Health (ICF): A conceptual analysis on the basis of a pluralistic-holistic ontology and multidimensional view of the human being. *The Journal of Medicine and Philosophy* 37(3): 277–94.

Stiker, H.J. (1997). *Corps infirmes et sociétés*, 2nd ed. Paris: Dunod.

Stiker, H.J. (1999). *A History of Disability* (W. Sayers, Trans.). Ann Arbor, MI: University of Michigan Press.

Stone, D.A. (1986). *The Disabled State.* Philadelphia, PA: Temple University Press.

Thomas, C. (2012). Theorising disability and chronic illness: Where next for perspectives in medical sociology? *Social Theory and Health* 10(3): 209–28.

Trattner, W.I. (1999). *From Poor Law to Welfare State. A History of Social Welfare in America*, 6th ed. New York: The Free Press.

Ueda, S., and Okawa, Y. (2003). The subjective dimension of functioning and disability: What is it and what is it for? *Disability and Rehabilitation* 25(11–12): 596–601.

United Nations. (2006). *Convention on the Rights of Disabled Persons.* New York: United Nations.

UPIAS and Disability Alliance. (1976). *The Union of the Physically Impaired Against Segregation and The Disability Alliance Discuss Fundamental Principles of Disability.* London: The Disability Alliance, Portland Place. Union of the Physically Impaired Against Segregation, Ealing.

Verbrugge, L.M., and Jette, A.M. (1994). The disablement process. *Social Science and Medicine* 438: 1–14.

Whiteneck, G. (2003). The ICF: One step forward, one step back, and a few steps to go to a complete model. Presentation to Conference on State of the Science: Outcome Research in Post-Acute Care, April 24–25, 2003, Washington, DC. Organized by Boston University. Retrieved from http://www.bu.edu/cre/webcast/testwhiteneck.doc.

Whiteneck, G. (2006). Conceptual models of disability: Past, present, and future. In M.J. Field, A.M. Jette, and L. Martin (eds.) *Workshop on Disability in America: A New Look*, 50–66. Washington, DC: National Academies Press.

Willems, D. (2010). Quality of life: Measure or listen. A reflection for disability studies. *Medische Antropologie* 22(2): 289–91.

Wolfensberger, W. (1972). *The Principle of Normalization in Human Service*. Toronto: National Institute on Mental Retardation.

Wolfensberger, W. (1998). *A Brief Introduction to Social Role Valorization: A High-Order Concept for Addressing the Plight of Societally Devalued People, and for Structuring Human Services*, 3rd ed. Syracuse, NY: Training Institute for Human Service Planning, Leadership and Change Agentry, Syracuse University.

World Health Organization. (1946). *International Statistical Classification of Diseases and Related Health Problems (ICD)*. Geneva: World Health Organization.

World Health Organization. (1980). *International Classification of Impairments, Disabilities and Handicaps: A Manual of Classification Relating to the Consequences of Disease*. Geneva: World Health Organization.

World Health Organization. (2001). *International Classification of Functioning, Disability and Health*. Geneva: World Health Organization.

World Health Organization. (2012). *Social Determinants of Health*. Geneva: World Health Organization.

3

Rethinking Rehabilitation's Assumptions: Challenging "Thinking-as-Usual" and Envisioning a Relevant Future

Karen
Whalley Hammell

This chapter offers a few of the reasons why I believe we need to rethink rehabilitation's assumptions and practices, and outlines some ideas for how rehabilitation practices and research might be made more useful and more relevant.

Challenging Assumptions

Assumptions are those ideas that we assume to be "right" or "common sense," and that are taken for granted because they often exist outside our conscious awareness (Eakin et al. 1996; Hammell 2009a). The rehabilitation literature reveals several common assumptions, such as the nature of the therapy professions (e.g., helpful, needs-based, client-centered) and the nature of their goals (e.g., striving for normality, increasing physical function and physical independence to enhance quality of life) (Hammell 2010a). The assumption that these goals are legitimate and beneficial appears to the rehabilitation professions to be self-evident, and it is precisely the common-sense nature of rehabilitation's assumptions that assures their uncritical acceptance and reproduction.

As professionals, however, integrity requires us to challenge the assumptions that are inherent in our field, to question, to confront dogma, to unmask conventional and accepted ideas, to take nothing for granted, and to be unwilling "to let half-truths or received ideas" steer us along (Said 1996, p. 23). Integrity also requires us critically to interrogate whether our assumptions—and the practices they inform—are in any way adequate, relevant, or evidence-based. In fact, because a "scientific" discipline is one that assures "a culture of healthy scepticism: a readiness to doubt claims and assumptions about the 'rightness' of any particular theory or intervention" (Brechin and Sidell 2000, p. 12), the critical appraisal of rehabilitation's assumptions and practices ought to be a constant professional endeavor.

Challenging Assumptions: Rehabilitation's Goals

A significant body of research demonstrates that the assumptions of those of us who are able-bodied bear little relationship to the realities of life for disabled people (DeLisa 2002).* This is illustrated by the following three examples.

Challenging Assumptions about "Normality" as a Goal

The rehabilitation professions often assume that they have a mandate to restore a disabled person to "normality," and failing this, to restore the person to a state that is as close to normality as possible (Gibson and Teachman 2012; Hammell 2010a; Kielhofner 2004). They have largely failed to acknowledge that the norms toward which their clients are exhorted to strive are neither scientific nor objective realities, but rather the creation of cultural values and judgments (Baylies 2002; Hammell 2004a); and that this is why medical statisticians are careful to refer to "deviations from *the assumption of normality*" (Bland 1991, p. 179, emphasis added). Development of statistics in the nineteenth century created the concepts of the norm, the normal bell curve, and consequently the concept of deviance from *assumed* norms (Davis 2002). Because the criteria for "normality" are the product of cultural and social forces rather than of science, medical historians can demonstrate how constructions of normality and deviance have changed

---

* Critical disability theorists employ the term *disabled people* to denote people who are disabled (disadvantaged) by social and political responses to their impairments. Their perspective renders the euphemism "people with disabilities" completely meaningless.

through time. Norms are not, therefore, value-free, unbiased scientific truths, but rather the products of ideology and of value judgments that determine, for example, culturally preferred forms of appearance, behavior, and ability.

Despite its constructed nature, the ideology of normality is central to the International Classification of Functioning, Disability, and Health (ICF, World Health Organization 2001), which has been embraced by the rehabilitation professions as if it is somehow "correct" (Hammell 2004a). Further, "It is rarely acknowledged that the purpose of the ICF is *not* to enable the assessment of human needs, but to *classify* deviations from assumed norms in every area of human life" (Hammell 2010a, p. 43); yet this purpose is clearly stated in its title.

Critical disability theorists contend that normalizing goals reflect an effort to induce social conformity rather than to enhance function (Sandahl 2003; Stiker 1999) and that physical normality is sometimes pursued by rehabilitation professionals to a degree that constitutes abuse (Priestley 2003; Swain et al. 2003). Moreover, far from reflecting deviations from the norm, impairments are a normal component of human diversity, existing throughout history and within every society (Hammell 2006).

The rehabilitation professions have largely failed to contest the assumption that normality is an appropriate or even a kind goal for people who have impairments, and have accordingly proceeded to create their own norms (goals) such as normal posture, normal walking, normal speech, and normal handwriting that constitute therapists' favored forms of physical functioning (Gibson and Teachman 2012; Hammell 2010a). This is not a disinterested endeavor. Jarman et al. (2002, p. 561) suggest that each rehabilitation profession "must foster society's ongoing acceptance of its norms (e.g., 'normal' gait, 'normal' speech) in order to perpetuate its viability and gain access to new economic niches." The ideology of normality may thus be promulgated, not because it serves clients' best interests, but because it contributes to reinforcing professional power (Hammell 2006). (See Chapter 4 for a discussion of how "normal development" shapes children's rehabilitation.)

Challenging Assumptions about Physical Function and Quality of Life

Rehabilitation professionals frequently claim that the principle aim of the rehabilitation enterprise is to enhance quality of life (e.g., Hampton and Qin-Hilliard 2004; van Leeuwen et al. 2012). However, those forms of assessment and outcome measures usually chosen by therapists indicate that quality of life is assumed to be enhanced by simply enhancing individuals' physical abilities! Indeed, the assumption that enhancement of physical function and physical independence somehow translates into enhanced quality of life pervades rehabilitation (Gibson et al. 2009; Hammell 2004b; Secrest and Thomas 1999). This is not an assumption that enjoys evidence-based support.

A significant quantity of research evidence demonstrates that perceptions of quality of life among disabled people correlate with neither their degree of physical impairment nor level of functional ability (e.g., Hammell 2004c; Secrest and Thomas 1999). And of course, as rehabilitation professionals, we are only too aware that increased physical function for many clients is often, anyway, a futile goal.

Similarly, rehabilitation professionals often assume that depression and anxiety correlate with the level and severity of physical impairment (Hammell 2010b), such that those with high, complete spinal cord injury are expected to be more distressed than those with low, incomplete lesions. But again, this assumption is contested by a significant quantity of research evidence demonstrating that psychological distress correlates with neither the severity of physical impairment nor degree of functional ability (e.g., Krause et al. 2000; Post et al. 1998).

Reynolds and Prior (2003) summarized the major difficulties presented by multiple sclerosis, and this provided a useful glimpse at the sort of issues faced by someone living with a significant injury or illness. These included mobility impairment, incontinence, fatigue, and pain: the sort of issues that are readily addressed by a focus on physical functions. But having multiple sclerosis also entailed "many social and psychological difficulties, including loss of roles and occupations, unwanted changes in relationships with family and friends, encounters with social discrimination and stigma, and barriers to accessing public places, and, therefore, problems with carrying out 'normal' tasks of daily living, such as shopping" (Reynolds and Prior 2003, p. 1231). Clearly, if rehabilitation only addresses the physical consequences of disease and impairment, it is woefully inadequate.

Of course, I'm not arguing that efforts to enhance physical function are unimportant, but it is apparent that this is an insufficient response to the challenges faced by many, or most, rehabilitation clients. Moreover, research evidence shows that following serious injury or illness, the ability to return to a meaningful life is of greater concern than the achievement of physical abilities (Doolittle 1991; Hammell 2007a; Kagan and Duchan 2004), and surely, this should not be surprising? In my own life, for example, although I appreciate being able to walk and to brush my own teeth, these are not the things that make my life meaningful or rewarding, and these are not the qualities that make me valuable to the people who love me.

Challenging Assumptions about Physical Independence and Quality of Life

There is a frequently stated assumption that the purpose of rehabilitation is to teach self-care skills to enable the highest level of physical independence (Swain et al. 2003), as if this is a goal to which all people aspire, or ought to aspire, irrespective of culture, role demands, or personal values (Hammell 2006). Although dominant Western culture extols individualism and applauds independence, the majority of the world's people value interdependence, reciprocity, and sense of belonging (Hammell and Iwama 2012; Iwama 2006). In fact, in some cultures, striving for independence is viewed as a sign of immaturity (Ng et al. 2003). And research into rehabilitation outcomes demonstrates that disabled people often choose to live interdependently, preferring to use their time and energies in those activities that have personal value and meaning, rather than in performing self-care tasks without assistance (Holcomb 2000; Marks 1999).

Although the rehabilitation professions portray independence as a universally valued goal, this assumption lacks evidence-based support and has been exposed as an ideology that reflects specific ableist* Western values and that serves to demean those who are dependent or who value interdependence (Hammell 2006). Cross-cultural research

* *Ableism* refers to discriminatory policies and practices arising from the assumption that everyone's physical abilities conform to the valued "norms" of the dominant culture.

demonstrates that interdependence is an integral dimension of being human (Reindal 1999; Ryff and Singer 1998), and even within Western culture, research evidence shows that the ability to contribute to others is associated with lower levels of depression, higher self-esteem, and fewer health problems (Schwartz and Sendor 1999; Stewart and Bhagwanjee 1999): findings that are *not* associated with physical independence.

Critiquing "Thinking-as-Usual"

These brief examples—and there are more where these came from—suggest that efforts to rethink and re-examine rehabilitation's assumptions and theories are overdue. Of course, taking a skeptical approach to the assumption that normality, physical function, and physical independence are adequate responses to disability and legitimate preoccupations for rehabilitation challenges both traditional rehabilitation dogma and professional dominance (Hammell 2004b). And this endeavor, also, is overdue!

Professional integrity requires that rehabilitation professions reflect critically on such questions as, Should therapists be enthusiastic proponents of physical independence or should they encourage disabled people to focus on other valuable abilities? Should therapists be exhorting disabled people to conform to the norms of the dominant population or, rather, assisting disabled people to attain the rights and opportunities currently reserved for those deemed normal (Hammell 2010a)?

Theorists within occupational therapy have recently stated the obvious: that the profession's assumptions, dominant theories, and models of practice reflect a specific Western perspective that privileges well-educated, white, middle class, middle-aged, Judeo-Christian, heterosexual, urban values and ways of thinking (Awaad 2003; Hammell 2009a,b, 2010a, 2011; Iwama 2006; Iwama et al. 2011; Nelson 2007, 2009); and that these ethnocentric perspectives are promoted as if they are universal rather than culturally specific (Gerlach 2012; Hammell 2009a; Hammell and Iwama 2012; Iwama 2003; Jull and Giles 2012). Further, it has been observed that the assumptions underpinning occupational therapy are class-bound, ableist, and lacking in supportive evidence (Hammell 2009a; Hammell and Iwama 2012), and that the dissemination of these assumptions in a global context constitutes theoretical imperialism—or intellectual colonialism (Gerlach 2012; Hammell 2009a, 2011). Although occupational therapy is the rehabilitation profession with which I am most familiar, this is not, unfortunately, a problem unique to occupational therapy.

I tend to think that if the rehabilitation professions can get *so* many assumptions *so* wrong, then perhaps we ought to be paying more attention to what rehabilitation users—past and present—have to say about the rehabilitation endeavor. How shall we know if our assessments are assessing anything relevant (bearing in mind that therapists treat what they measure), whether our interventions are aiming at the right target, or what constitutes a relevant outcome worth measuring, if we don't know what former and present clients value and need in their daily lives?

What Is "Rehabilitation"?

I ought to make clear that I define rehabilitation not as the treatment of body parts but as a process of enhancing well-being and of attaining—or regaining—a meaningful life in the context of disease, illness, or impairment. Someone who has torn a cruciate ligament, for example, may derive enormous benefit from a standardized program of exercises in physical therapy, but I regard this as treatment, not rehabilitation (Hammell 2004b). Rehabilitation, I suggest, is not concerned with treating the aftermath of Colles' fractures or Achilles' tears, but with learning to live well, in the context of one's own environment, with potentially life-altering problems such as stroke, multiple sclerosis, traumatic brain injury, spinal cord injury, severe burns, severe arthritis, and serious and persistent mental illness. Thus, I suggest that rehabilitation could usefully be regarded as a *process of enhancing engagement in living.* (For a related discussion, see Chapter 14.)

What Is Important?

Having highlighted the impotence of an approach to rehabilitation focused primarily on physical abilities, I want, also, to state that a very considerable quantity of quality qualitative research has identified the factors that *are* important to people, across cultures, as they adjust their lives to incorporate their changed or changing bodies and abilities. It is no coincidence that these factors are those already identified by philosophers, who have been studying the constituents of human well-being for millennia (see Hammell 2004d, 2006; Lavine 1984; Ryff and Singer 1998). These factors may be summarized as follows:

- Experiencing meaning and purpose through engagement in personally valued roles and occupations.
- Having a positive sense of self-worth.
- Having the ability and opportunity to make choices and exert control over one's life.
- Having the ability and opportunity to belong and contribute within reciprocal relationships (e.g., families, friends, assistants, communities).
- Envisioning continuity within "disrupted biographies," such that one's future life after injury or illness is experienced as continuous with life before (including, for example, one's capabilities, roles, interests, and sense of identity). This may be summarized as "hope" (see Chapter 9).

Traditionally, the rehabilitation professions have had little interest in the social sciences and humanities, aspiring to align themselves with the biological sciences and especially with medicine, presumably in an effort to assert professional worth and status (Davis 2004; Stalker and Jones 1998). This has equipped therapists with knowledge of little more than the anatomy and physiology of the body: an inadequate basis from which to understand the needs and experience of living with impairments (Hammell 2006). Theorists, scholars, and researchers within the social sciences and humanities have already engaged with many issues of relevance to disability (e.g., Hammell 2006; McRuer and Wilkerson 2003; Meade and Serlin 2006; Shakespeare 1998; Snyder et al. 2002; Stiker 1999). The alignment of qualitative rehabilitation research with philosophy, cited above, is an example of this congruence.

How Might "Health" Be Defined within Rehabilitation?

I should also explain how I define the concept of *health*. I am persuaded that the definition of health articulated by people living with cancer has considerable utility for rehabilitation. They defined health "based upon their ability to maintain a sense of integrity as productive, able, and valued individuals within their social spheres, despite their physical condition" (Kagawa-Singer 1993, p. 295). This melds with the observation that although therapists view recovery in terms of improved functional abilities, people following stroke, for example, view recovery as a return to valued activities and meaningful lives (Doolittle 1991); and it fits with evidence that people often define themselves as healthy, despite the presence of impairments, severe injuries, and chronic illnesses (Hammell 2006). Thus, disability and health are not mutually incompatible, nor are disability and ill health necessarily congruent.

Experiences of Rehabilitation

As a basis for rethinking rehabilitation, it seems appropriate to examine research that has explored the experience of rehabilitation from clients' perspectives and its relevance to their subsequent lives. I believe we need these perspectives to inform modes of intervention and models of service delivery, and to teach us what is important for us to assess and to measure (Hammell 2004a). For example, I have already alluded to therapists' assumption that normality is a relevant goal for rehabilitation. But what do clients think? Reflecting on her own experience of rehabilitation, "Kate" observed: "what concerns me most of all is this focus on trying to make me 'normal.' I get that from all the therapists... I want them to say, 'What sort of things would help you to lead a full life in the context of your impairment?'" (cited in French 2004, p. 103).

Many years ago, Carpenter (1994, p. 626) noted that there was "a dearth of studies" ascertaining how people perceive the rehabilitation they receive, and that clinicians have therefore been able "to deceive themselves that the traditional rehabilitation programme constitutes the most effective and optimal method of service delivery." Although two decades have passed since she made her astute observations, and despite the rehabilitation professions' claims to operate in a client-centered, needs-led manner (e.g., AOTA 2002; CAOT 1983; COT 2005; CSP 1996), a surprisingly small body of research has explored clients' experiences of rehabilitation or their perceptions of its relevance (Hammell 2007a, 2013a). (See also Guenther-Mahipaul's Chapter 10 for a personal reflection on her experiences with rehabilitation.) Despite the small number of studies, consistent themes have emerged, with the rehabilitation process frequently depicted as oppressive, boring, meaningless, and irrelevant to users' lives, inappropriate to their roles and values, incongruent with their goals, and reinforcing their sense of powerlessness and dependency (Abberley 1995; Corring and Cook 1999; Dalley 1999; Fleming et al. 2012; French 2004; Johnson 1993). Moreover, therapy has been described by clients as being problem-centered rather than client-centered (Johnson 1993), and as focused on the implementation of standard treatment regimens rather than upon the client as a person (Jorgensen 2000; Lucke 1999).

Three specific problems with traditional rehabilitation programs are identifiable in the research literature: a preoccupation with enhancing physical function (and neglect of every other dimension of living with a life-disrupting injury or illness), a lack of individuality, and a failure to prepare clients for the real world.

I realize, of course, that rehabilitation, by definition, often occurs during an unwelcome phase in someone's life that has been disrupted by illness or injury, and that it can be a time of profound uncertainty and struggle. However, it does not appear that the passage of time softens the more negative appraisals of the rehabilitation experience, nor does it serve to persuade former clients of the relevance, usefulness, or excellence of their rehabilitation programs (Hammell 2007a). Neither does it appear that those rehabilitation services experienced in recent years are appraised more favorably than those of previous decades (Hammell 2007a). In reality, there is little difference between the reports of current rehabilitation inpatients and those of people who have been living in the community for many years, although the latter have considerably more to say than the former about the lack of preparation they received for living in the real world (Hammell 2007a).

The assumption—held by rehabilitation professionals—that rehabilitation is consistently relevant and useful would appear to lack evidence-based support. In a meta-synthesis of qualitative research exploring experiences of rehabilitation following spinal cord injury, rehabilitation services were often described as homogenizing, inflexible, regulated, restrictive environments that were "beset with policies and procedures" such that clients described themselves as being imprisoned (Hammell 2007a; see also Barnes and Mercer 2003; Swain and French 2010; Ylvisaker and Feeney 2000).

Sadly, any assumption that the rehabilitation professionals are consistently perceived by clients to be client-centered, kind, and helpful also lacks evidence-based support, with therapists frequently described by clients as coercive; domineering; manipulative; hierarchical; cold; distant; disempowering; accountable to employers rather than to clients; indifferent to clients as human beings; willing gate-keepers to the equipment, resources, and services clients need; and sharing a pessimistic, deflating ethos (Abberley 2004; Barbara and Curtin 2008; Cant 1997; Corring and Cook 1999; Hammell 2007b, 2013b; Kemp 2002; Swain et al. 2003; Twigg 2000).

A surprisingly small body of research has examined those factors perceived by current and former clients to contribute positively to their rehabilitation. This research highlights the importance of peer mentors, who enable future life possibilities to be envisioned and who foster hope (Hammell 2007a; Sand et al. 2006). It also identifies the imperative for rehabilitation to enable clients to meet the needs of the real world and to assist them to connect their future lives to their past interests and valued roles (Cott 2004; Hammell 2007a). Notably, clients do not seem to mention the physical environment of rehabilitation: the benefits—or the lack—of state-of-the-art buildings, large gymnasia, extensive sporting facilities, high-tech gadgetry, or well-equipped therapy areas. However, every study identifies the importance of rehabilitation professionals who strive to tailor their programs to meet the real-world needs, interests, and priorities of their individual clients and who enable their clients to incorporate their physical changes into their ongoing lives. Indeed, "while the health-care professions foster and reward academic and technical excellence and social conformity, what patients appear to value most in their therapists is a sense of being valued as a human being" (Hammell 2007a, p. 271; see also Corring and Cook 1999; Fadyl et al. 2011; Mangset et al. 2008; Meade et al. 2011). Further, research consistently demonstrates that clients view the interpersonal qualities of their therapists as more important than their technical skills, valuing interventions that are responsive to their needs rather than their "conditions" (French 2004).

Professionals who are felt to facilitate the rehabilitation process are contrasted with those who are perceived as thwarting the process, those who are slavish adherents to institutional rules and regulations, or proponents of standard treatment templates.

I have therefore suggested that if financial resources were allocated according to the priorities of rehabilitation clients, these would be targeted not at elegant buildings or cool equipment but at the recruitment and retention of caring, client-orientated staff (Hammell 2007a). (See also Kontos et al.'s nuanced discussion in Chapter 7 of "therapeutic landscapes" and important aspects of rehabilitation environments.)

Rethinking Assessment: Cataloging Deficits or Identifying Abilities and Resources?

If interventions are to be responsive to clients' priorities and needs, it is self-evident that assessments must be able to appraise these needs, and that the effectiveness and usefulness of rehabilitation will subsequently be determined by outcome measures that are sensitive to clients' priorities. Because the rehabilitation professions have a history of deciding for themselves what should be assessed and how the impact of their interventions ought to be measured, these practices also require rethinking.

The assessment, in rehabilitation, is akin to a map for a navigator. It explicates "where we are now" and, because therapists treat what they measure, it provides the basis not only for "where we are going" (goals and objectives) but "how we plan to get there" (selected interventions). The nature of the assessment process determines whether therapists can identify clients' priorities, abilities, and resources and the needs of their specific situations, roles, and environments. The relevance—or irrelevance—of the assessment will determine whether "where we are going" is somewhere the client actually wishes to go. And it is therefore puzzling that clients' priorities, abilities, and resources are not assessed as consistently as their deficits and deviations. The assessment process can potentially bring the meaning and context of the life disruption caused by an injury or illness into sharp focus, for both client and therapist, and it can provide the foundation for interventions that are relevant to the real-world needs of clients. Note, however, that I use the word *can* rather than *does*.

Although it has long been recognized that the use of a set protocol of assessments for diagnostically defined categories of clients is inappropriate within client-centered rehabilitation practice (Law et al. 1995), therapist-centered approaches to assessment remain ubiquitous and are exemplified by the use of standardized, diagnosis-specific assessments designed to identify predetermined deficiencies and deficits that are important to the therapist or simple to assess (Hammell 2006). Thus, the rehabilitation professions frequently compile detailed assessments describing impairments and the degree of assistance required to accomplish various activities of daily living: dimensions with little or no bearing on quality of life (Whiteneck 1994).* Critics within medicine and rehabilitation have pointed out that focusing assessments on therapists' priorities is a modus operandi that benefits the rehabilitation professions by reinforcing their professional dominance (Basnett 2001; French 1994). Moreover, "in focusing on deficits and adaptations to these deficits, we focus less on things that really matter to the individual" (Secrest and Thomas 1999, p. 246).

A very considerable amount of time, effort, financial resources, and journal space has been expended in the development, testing, and promotion of a countless number of rehabilitation assessments with only rare attempts to determine whether the factors being assessed are important to clients' lives, or whether *only* those things being assessed are important to clients' lives. Prestige and economic benefits accrue to the developers of popular assessments, and influential members of the professions may have vested interests in promoting the use of specific tests on which they have expended research time, funding, and reputations.

A recent audit among occupational therapists who work with people with Parkinson's disease observed the "need for more client-centred engagement when identifying goals" (COT 2012, p. 7) but also complained that "only" about 30% of occupational therapists used formal standard assessments. Given that the occupational therapists were seeing people who had been living with this disease for an average of six years, it is intriguing to ponder why the lack of formal standard assessment was lamented. In reality, perhaps therapists were asking clients to explain their current problems and self-defined needs in an effort both to foster client-centered engagement and ensure the relevance and usefulness of professional endeavors. A formal, standardized assessment of clients' deficits might have contributed very little to the therapists' understanding of their clients' needs. Moreover, using precious time in compiling a standardized catalog of deficits and dysfunctions—*many of which are not amenable to therapists' interventions*—would surely have appeared to clients as unkind and irrelevant, at best.

Rethinking Outcome "Measurement"

Despite the oft-cited assumption—that one of the principle aims of rehabilitation is to enhance quality of life (e.g., Hampton and Qin-Hilliard 2004; van Leeuwen et al. 2012)—rehabilitation practitioners have rarely measured outcomes in terms of their clients' active participation in their community, engagement in personally meaningful or productive activities, or quality of life (Whiteneck 1994; Hammell 2006). Critics contend that such measurements as range of motion, exercise capacity, or functional independence are surrogates for what rehabilitation really ought to assess: the effect of interventions on clients' lives (Guyatt et al. 1997). Clearly, an assessment of physical function is an insufficient gauge of the usefulness or value of rehabilitation services (Hammell 2004b) and conveys little about the impact of rehabilitation on clients' lives. It is not original to observe that outcome measurement tends to focus on those aspects of intervention that are most easily measured rather than those of greatest relevance to

---

* Because there are few exemplars in the literature of rigorous skeptical interrogation of rehabilitation practices, several of my citations are old. I do not believe these critiques are outdated.

disabled people (Magasi 2008). As Ray and Mayan (2001) observed, a "clinically significant" outcome may be significant only to a clinician.

Bizarrely, therapists often appraise the "success" of their interventions by whether a
client is able to perform a specific activity, irrespective of whether the client wishes to
do so, or could do so in their own environment or in a span of time they consider to be
reasonable. More than two decades ago, critics queried whether the purpose of judging
outcomes by measuring those skills prioritized by therapists is simply a means to justify
customary rehabilitation practice (Eisenberg and Saltz 1991). This is a question that continues to merit serious introspection.

Kagan and Duchan (2004) observe that by not paying sufficient attention to the views
of clients, evaluation tools may be missing outcomes critical to success in living long-
term with impairments. Moreover, despite the espoused "client-centered" approach to
rehabilitation practice, there is little evidence to suggest that outcome measures developed from a user perspective are employed consistently to inform patterns of service
delivery (Hammell 2006; McPherson et al. 2001). The issue of *who* evaluates outcomes is
rarely discussed, yet it is a clear conflict of interest when therapy services are evaluated
by service providers, and when success is defined according to therapists' satisfaction
with the results of those outcomes they have chosen to measure (Abberley 1995). Clearly,
the impact and outcome of rehabilitation cannot be derived from the viewpoints of service providers, because they are not the ones who have to live with the outcomes.

It has been proposed that reducing the gap between traditional outcome measures
and the needs of disabled people requires researchers to ensure that their research is
guided by those outcomes that are valued most highly by disabled people (DeLisa 2002).
And this requires us to rethink rehabilitation research.

Rethinking Rehabilitation Research
(to Enable Us to Think Differently)

Although the rehabilitation professions proclaim themselves to be client-centered,
when they undertake research they tend to demonstrate a perceived entitlement to dis-
pense with this ethical stance and become, instead, researcher-centered—deciding on a
research question and dictating every aspect of the research process (Hammell et al. 2012).
If our theories and our practices are to be informed, both by relevant evidence, and by clients' perspectives—congruent with our frequently espoused client-centered principles—
then our research practices have to change (Hammell 2013b). Further, I support the
premise that if researchers and disabled people pool their expertise, there is the potential
to achieve better research (Hammell et al. 2012; Priestley et al. 2010a,b; White et al. 2004).

Although the imperative for rehabilitation to be evidence-based continues to ener-
gize the professions' interest in research, critics contend that much of the rehabilitation evidence base is flawed because it has been developed from research agendas that
did not consider the issues that matter to disabled people (Basnett 2001; Glasby and
Beresford 2006). For many years, critical disability theorists have bemoaned the quality
and outcome of much disability research, the theoretical models employed, the questions posed, the modes of analysis, the locations of publication, and the absence of tangible results (Barnartt and Altman 2001; Barnes et al. 1999). They have also noted that

many researchers use disability primarily as a means of advancing their own careers (e.g., Barnes et al. 1999).

In addition, people who have been participants in rehabilitation research complain that they have invested time and energy as subjects in studies that had little relevance, and that the findings seemed to end up in a desk drawer, with no apparent commitment from the researchers to translate research findings into action (Abma 2005; Barnitt and Partridge 1999; Hammell et al. 2012). Critical disability theorists and patients' rights advocates have called for researchers to enhance both the relevance and usefulness of their research by working in partnership with people living with impairments, to identify appropriate research initiatives and relevant research outcomes (Abma 2005, 2006; Stone and Priestley 1996; White 2002; White et al. 2001). Cleary, this would fit with the rehabilitation professions' self-proclaimed client-centered approach to practices (Hammell 2013b). Moreover, because the impetus for changes that benefit disabled people tends, almost without exception, to derive from disabled people themselves rather than from researchers, therapists, or policy-makers (Hammell 2004c, 2006; Richardson 1997), research findings are more likely to be acted upon when disabled people have access to information that *they* can use to influence policy and practice (Shakespeare 1996).

Rethinking Rehabilitation

I want to suggest some different ways of looking at the rehabilitation enterprise. In view of the increasing incidence, prevalence, and longevity of many chronic conditions, I suggest that some of our rehabilitation models are outdated, being designed for the remediation of acute orthopedic and sports injuries rather than for managing chronic and deteriorating conditions in the context of clients' own environments and lives.

I think it is useful to view the adult onset of illness or impairment as a significant life disruption: not just a dysfunctional body, but a disruption of daily occupations, valued roles, and future hopes. Sociologists have suggested that this sudden disruption should be viewed in terms of its consequences—its impact on valued activities and roles—and its significance, that is, its impact on one's sense of competence and self-worth (Bury 1991). I have suggested that rehabilitation could assist people to change for themselves the meaning of both the consequences and significance of impairment, through helping them to shift their focus from inabilities to abilities and thus to view themselves as competent and capable, able, and valuable (Hammell 2004d). To do so, however, requires that rehabilitation professionals shift from solely measuring deficits and modifying inabilities to also identifying and fostering abilities and resources (Siegert et al. 2007). This would be a useful strategy, also, for people with congenital impairments.

Moreover, existing research evidence suggests that the work of rehabilitation could usefully address two issues: enabling people to engage in personally valued roles and occupations that enhance their perceptions of being valuable, capable, and competent (e.g., Hammell 2004e; Rebeiro 2004; Vrkljan and Miller-Polgar 2001); and encouraging the (re)building of a sense of self that is not dependent upon physique or physical abilities but upon other abilities and competencies (Hammell 2006; Kleiber and Hutchinson

1999). How might this be accomplished? I believe that those engaged in physical rehabilitation could usefully learn a new definition of "recovery" from practitioners in the
field of mental health.

Rethinking "Recovery"

Within the mental health literature, there has been a recent shift in emphasis from illness management to recovery. Recovery is defined in this context not as a cure but as
a process of changing one's attitudes, values, and goals, and of living a hopeful, satisfying, meaningful, purposeful, and contributing life within the limitations caused by
one's disease or impairment (Anthony 1993; Rebeiro Gruhl 2005). This recovery model
focuses not on inabilities, deficits, or dysfunctions but on capabilities and the mobilization of resources.

I referred earlier to evidence that following serious injury or illness, return to a
meaningful and satisfying life is of greater concern than the achievement of physical
abilities. I also drew attention to those elements that are consistently found to contribute positively to the experience of a life worth living: engagement in valued roles and
meaningful occupations, a positive sense of self-worth, the ability to enact choices, the
opportunity to belong and contribute within valued social relationships, and a sense of
hope (Evans 2011; Hammell 2004d, 2006; Lavine 1984; Ryff and Singer 1998). Indeed,
Ryff and Singer's (1998, p. 2) observation, that "human well-being is ultimately an issue
of engagement in living" could serve as a useful mantra for those who seek to engage in
relevant and meaningful rehabilitation practices.

People with spinal cord injuries often say that they began to assume responsibility,
take control of their disrupted lives, and rebuild meaningful lives when they stopped
focusing on what they could *not* do and began to focus on what they *could* do (Carpenter
1994; Hammell 2007c). Perhaps if our assessments and interventions focused on abilities, strengths, and resources rather than solely on deficits (Siegert et al. 2007), we might
enable clients to begin this process sooner. Because perceptions of quality of life after
spinal cord injury, for example, are more closely associated with mental health problems
(anxiety, depression, hopelessness, helplessness, pain catastrophizing) than with physical abilities (degree of physical function or of physical independence) (Hammell 2010b),
a move toward an emphasis on "recovery" that addresses hope, engagement in purposeful activities, and community integration might be worthwhile. (See Chapter 14 for a
related discussion from an alternative viewpoint.)

The literature provides support for this shift, with evidence that those people with
serious neurological impairments who believe that they lack the ability to bring about
their desired outcomes in life—who think negatively and lack hope—place themselves
at higher risk for pain, depression, and chronic anxiety, and for reduced life satisfaction
(Middleton et al. 2007). Further support for this approach is found in evidence that perceptions of positive quality of life are associated with quality relationships and participation in social, voluntary, leisure, employment, and community activities (Hammell
2004c, 2007c, 2010b; Kemp 2004); such participation may also reduce the incidence of
secondary medical problems (Elliott et al. 2006) and even the risk of mortality (Krause
1991; Krause and Kjorsvig 1992).

Moreover, because research suggests that dissatisfaction with life after spinal cord injury, for example, may arise, not from the physical effects of the injury, per se, but from social disadvantage (Hammell 2004c), I believe that interventions ought to focus not solely on modifying individuals but on modifying environments.

Rethinking the Preoccupation with Modifying People

If rehabilitation professionals are committed to reducing such problems as pain, spasticity, and pressure sores (factors correlated with poor quality of life among disabled people), they ought, surely, to be committed to reducing such problems as nursing home admissions, occupational deprivation, inadequate access to health care, and poverty (factors correlated with poor quality of life among disabled people) (Hammell 2004c). It is obviously inadequate to teach wheelchair skills, for example, without also addressing the policy barriers and inaccessible transportation systems that prevent those people who use wheelchairs from equitably participating in their communities and activities of choice.

Amundson (1992, p. 114) observed that "unlike ill people, disabled people are not (typically) globally incapacitated *except insofar as an environment helps to make them so.*" If the rehabilitation professions acknowledge that social, physical, economic, legal, and political environments create or exacerbate the problems confronted by disabled people and reduce the quality of their lives (and this is demonstrated by a significant evidence base), then interventions must be targeted not solely at individuals—teaching them to live in a world designed and organized to meet the needs of the dominant population—but at influencing public policies and community planning (Hammell 2004c; Siegert et al. 2007).

Disabled people throughout the world are treated as second-class citizens who experience disproportionate levels of poverty; discriminatory health-care services; inequitable access to housing, education, and employment opportunities; and inaccessible transport and buildings, even in developed countries (Hammell 2013c; McVeigh 2011). Addressing these inequities would seem to fit with the ICF's espoused interest in issues of participation. This would also fit the research priorities identified by disabled people concerning equipment, policies, community access, and participation (Abma 2005; Weaver et al. 2001). And it fits with the assertion that ability is of little use without opportunity (Hammell 2006).

Rethinking the Politics of Rehabilitation

It is apparent that when rehabilitation professionals strive to change a dimension of the physical, cultural, social, political, legal, or economic environment to counteract discrimination and to equalize opportunities, this is a political act: It is important to understand "that when rehabilitation professionals strive instead to change dimensions of individuals so that they can better fit within segregated environments *this is also a political act.* Acquiescing to the inequities of the status quo may be politically conservative, but it *is* political" (Hammell 2006, p. 142).

It would be disingenuous to pretend that the rehabilitation professions are not already engaged in political activism. They usually have both professional associations and trade unions continually advocating on their own behalf in overtly political activities (Hammell 2007b). Indeed, politicians, policy-makers, and the media frequently hear from rehabilitation professionals when their own employment, income, and status appear threatened but only rarely when their clients' resources and services are imperiled (Hammell 2007b). It has been noted that occupational therapists have an inauspicious record in the struggle to change policies that might benefit disabled people, reserving advocacy and action in political and institutional arenas for issues pertaining to their own professional self-interests (Jongbloed and Crichton 1990; Schriner 2001). Regrettably, it does not appear as if the other rehabilitation professions are doing any better. Davis (2004) observed that disability professionals are preoccupied not with the struggle to equalize opportunities for disabled people but in enhancing their own career opportunities and power. Indeed, it is difficult to counter critics' long-held contention: that the rehabilitation profession's espoused commitment to client centeredness is little more than a rhetorical device (McKnight 1981).

Conclusion

More than a decade ago, it was noted that "the time is ripe for a thorough evaluation of rehabilitation theory and practice, as well as the possibility for a *transformative rehabilitation practice* that focuses on the societal conditions that create disadvantage for people whose individual characteristics are outside the societal norm" (Schriner 2001, p. 653). This chapter has supported the premise that rehabilitation's assumptions and practices ought to be evaluated and rethought. I have suggested that rehabilitation could usefully be viewed as the process of enhancing engagement in living—of assisting people to live well with their impairments in the context of their own environments—and have also suggested that rehabilitation professionals ought to take seriously the idea that ability is of little use without opportunity. I have proposed that assessments and outcome measures should be focused on issues of relevance and importance to clients (rather than on cataloging their deviations from culturally valued norms), and have proposed that if we are to generate evidence on which to base our practices, this evidence needs to be grounded in clients' priorities and perspectives.

But is change likely to occur? To challenge the assumptions and entrenched practices of the rehabilitation professions is to challenge the ideologies perpetuated by many of those who wield the most influence within these professions: people who have invested heavily in rehabilitation's conventional practices and who have significant vested interests in maintaining their own power and the status quo. It is also important to recognize that the rehabilitation business is circumscribed by significant structural forces that include specific institutional policies, processes, procedures, practices, and priorities. Townsend (1998) astutely observed that many models and modes of service delivery, and the policies that govern these, appear designed for the benefit of the organization (first) and professionals (second) rather than service users.

However, clients perceive their therapists not as hapless victims of oppressive institutional structures but as willing collaborators in enforcing institutional policies that disable and disempower their clients (e.g., Barnes and Mercer 2003; French and Swain 2001). Thus, "it is *our* collusion with disabling policies, decision-making processes, organizational practices, financial priorities and government-funded services that we can and must address" (Hammell 2007b, p. 265).

Many years ago it was observed that "when core assumptions, which are considered essential to the underpinning of occupational therapy, are perceived by some to be under threat, considerable efforts are made to enforce professional conformity" (Mocellin 1995, p. 502). These observations resonate with the experiences of those of us who have sought to challenge the status quo of this particular profession in recent years (Hammell 2012). Regrettably, it appears unlikely that these machinations of power are unique to occupational therapy.

Change is possible; it is necessary and it is overdue. But change will only occur if we have the courage to challenge professional dominance, vested interests, and the unsupported assumptions and practices this status quo promotes, and if we insist on the sort of client-centered practices that match our existing professional rhetoric and that enhance the likelihood that we will meet the needs of those who seek our help.

Acknowledgment

I am grateful to Nicole Thomson both for her encouragement and her helpful and insightful critique of an earlier draft of this chapter.

References

Abberley, P. (1995) Disabling ideology in health and welfare: The case of occupational therapy. *Disability and Society* 10(2), 221–232.

Abberley, P. (2004) A critique of professional support and intervention. In: J. Swain, S. French, C. Barnes and C. Thomas (Eds.). *Disabling Barriers—Enabling Environments* (2nd edn). London, Sage, pp. 239–244.

Abma, T.A. (2005) Patient participation in health research: Research with and for people with spinal cord injuries. *Qualitative Health Research* 15(10), 1310–1328.

Abma, T.A. (2006) Patients as partners in a health research agenda setting. The feasibility of a participatory methodology. *Evaluation and the Health Care Professions* 29(4), 424–439.

American Occupational Therapy Association (AOTA). (2002) Occupational therapy practice framework: Domain and process. *American Journal of Occupational Therapy* 56, 609–639.

Amundson, R. (1992) Disability, handicap and the environment. *Journal of Social Philosophy* 23(1), 105–118.

Anthony, W.A. (1993) Recovery from mental illness: The guiding vision of the mental health services system in the 1990s. *Psychosocial Rehabilitation Journal* 16(4), 11–23.

Awaad, T. (2003) Culture, cultural competency and occupational therapy: A review of the literature. *British Journal of Occupational Therapy* 66, 356–362.

Barbara, A. and Curtin, M. (2008) Gatekeepers or advocates? Occupational therapists and equipment funding schemes. *Australian Occupational Therapy Journal* 55(1), 57–60.

Barnartt, S. and Altman, B. (2001) Exploring theories and expanding methodologies: Where we are and where we need to go. *Research in Social Science and Disability* 2, 1–7.

Barnes, C. and Mercer, G. (2003) *Disability*. Cambridge, Polity.

Barnes, C., Mercer, G. and Shakespeare, T. (1999) *Exploring Disability: A Sociological Introduction*. Cambridge, Polity.

Barnitt, R. and Partridge, C. (1999) The legacy of being a research subject: Follow-up studies of participants in therapy research. *Physiotherapy Research International* 4(4), 250–261.

Basnett, I. (2001) Health care professionals and their attitudes toward and decisions affecting disabled people. In: G.L. Albrecht, K.D. Seelman and M. Bury (Eds.). *Handbook of Disability Studies*. London, Sage, pp. 450–467.

Baylies, C. (2002) Disability and the notion of human development: Questions of rights and capabilities. *Disability and Society* 17(7), 725–739.

Bland, M. (1991) *An Introduction to Medical Statistics*. Oxford, Oxford University Press.

Brechin, A. and Sidell, M. (2000) Ways of knowing. In: R. Gomm and C. Davies (Eds.). *Using Evidence in Health and Social Care*. London, Sage, pp. 3–25.

Bury, M. (1991) The sociology of chronic illness: A review of research and prospects. *Sociology of Health and Illness* 13(4), 451–468.

Canadian Association of Occupational Therapists (CAOT) and the Health Services Directorate. (1983) *Guidelines for the Client-Centred Practice of Occupational Therapy*. Ottawa, Health Services Directorate.

Cant, R. (1997) Rehabilitation following a stroke: A participant perspective. *Disability and Rehabilitation* 19(7), 297–304.

Carpenter, C. (1994) The experience of spinal cord injury: The individual's perspective—Implications for rehabilitation practice. *Physical Therapy* 74(7), 614–629.

Chartered Society of Physiotherapy (CSP). (1996) Rules of professional conduct. *Physiotherapy* 81, 460.

College of Occupational Therapists (COT). (2005) *Code of Ethics and Professional Conduct for Occupational Therapists*. London, COT.

College of Occupational Therapists (COT). (2012) A quarter of Parkinson's OTs get no CPD access. *Occupational Therapy News* June, 7.

Corring, D. and Cook, J. (1999) Client-centred care means that I am a valued human being. *Canadian Journal of Occupational Therapy* 66(2), 71–82.

Cott, C.A. (2004) Client-centred rehabilitation: Client perspectives. *Disability and Rehabilitation* 26(24), 1411–1422.

Dalley, J. (1999) Evaluation of clinical practice: Is a client-centred approach compatible with professional issues? *Physiotherapy* 85(9), 491–497.

Davis, K. (2004) The crafting of good clients. In: J. Swain, S. French, C. Barnes and C. Thomas (Eds.). *Disabling Barriers—Enabling Environments* (2nd edn). London, Sage, pp. 203–205.

Davis, L.J. (2002) Bodies of difference: Politics, disability and representation. In: S.L. Snyder, B.J. Brueggemann and R. Garland-Thomson (Eds.). *Disability Studies. Enabling the Humanities*. New York, The Modern Language Association of America, pp. 100–106.

DeLisa, J.A. (2002) Quality of life for individuals with SCI: Let's keep up the good work. *Journal of Spinal Cord Medicine* 25(1), 1.

Doolittle, N.D. (1991) Clinical ethnography of lacunar stroke: Implications for acute care. *Journal of Neuroscience Nursing* 23, 235–239.

Eakin, J., Robertson, A., Poland, B., Coburn, D. and Edwards, R. (1996) Towards a critical social science perspective on health promotion research. *Health Promotion International* 11(2), 157–165.

Eisenberg, M. and Saltz, C. (1991) Quality of life among aging spinal cord injured persons: Long term rehabilitation outcomes. *Paraplegia* 29, 514–520.

Elliott, T.R., Bush, B.A. and Chen, Y. (2006) Social problem-solving abilities predict pressure sore occurrence in the first 3 years of spinal cord injury. *Rehabilitation Psychology* 51, 69–77.

Evans, J.J. (2011) Positive psychology and brain injury rehabilitation. *Brain Impairment* 12(2), 117–127.

Fadyl, J.K., McPherson, K.M. and Kayes, N.M. (2011) Perspectives on quality of care for people who experience disability. *BMJ Quality and Safety* 20, 87–95.

Fleming, J., Sampson, J., Cornwell, P., Turner, B. and Griffin, J. (2012) Brain injury rehabilitation: The lived experience of inpatients and their family caregivers. *Scandinavian Journal of Occupational Therapy* 19, 184–193.

French, S. (1994) What is disability? In: S. French (Ed.). *On Equal Terms. Working with Disabled People*. Oxford, Butterworth-Heinemann, pp. 3–16.

French, S. (2004) Enabling relationships in therapy practice. In: J. Swain, J. Clark, K. Parry, S. French and F. Reynolds (Eds.). *Enabling Relationships in Health and Social Care*. Oxford, Butterworth-Heinemann, pp. 95–107.

French, S. and Swain, J. (2001) The relationship between disabled people and health and welfare professionals. In: G.L. Albrecht, K.D. Seelman and M. Bury (Eds.). *Handbook of Disability Studies*. London, Sage, pp. 734–753.

Gerlach, A.J. (2012) A critical reflection on the concept of cultural safety. *Canadian Journal of Occupational Therapy* 79, 151–158.

Gibson, B.E., Darrah, J., Cameron, D. et al. (2009) Revisiting therapy assumptions in children's rehabilitation: Clinical and research implications. *Disability and Rehabilitation* 31(17), 1446–1453.

Gibson, B.E. and Teachman, G. (2012) Critical approaches in physical therapy research: Investigating the symbolic value of walking. *Physiotherapy Theory and Practice* 28(6), 474–484.

Glasby, J. and Beresford, P. (2006) Who knows best? Evidence-based practice and the service user contribution. *Critical Social Policy* 26, 268–284.

Guyatt, G.H., Naylor, C.D., Juniper, E., Heyland, D.K., Jaeschke, R. and Cook, D.J. (1997) Users' guide to the medical literature XII. How to use articles about health-related quality of life. *Journal of the American Medical Association* 277(15), 1232–1237.

Hammell, K.W. (2004a) Deviating from the norm: A sceptical interrogation of the classificatory practices of the ICF. *British Journal of Occupational Therapy* 67(9), 408–411.

Hammell, K.W. (2004b) The rehabilitation process. In: M. Stokes (Ed.). *Physical Management in Neurological Rehabilitation* (2nd edn). Edinburgh, Elsevier, pp. 379–392.

Hammell, K.W. (2004c) Exploring quality of life following high spinal cord injury: A review and critique. *Spinal Cord* 42(9), 491–502.

Hammell, K.W. (2004d) Dimensions of meaning in the occupations of daily life. *Canadian Journal of Occupational Therapy* 71(5), 296–305.

Hammell, K.W. (2004e) Quality of life among people with high spinal cord injury living in the community. *Spinal Cord* 42(11), 607–620.

Hammell, K.W. (2006) *Perspectives on Disability and Rehabilitation: Contesting Assumptions, Challenging Practice.* Edinburgh, Elsevier.

Hammell, K.W. (2007a) The experience of rehabilitation following spinal cord injury: A meta-synthesis of qualitative findings. *Spinal Cord* 45(4), 260–274.

Hammell, K.W. (2007b) Client-centred practice: Ethical obligation or professional obfuscation? *British Journal of Occupational Therapy* 70(6), 264–266.

Hammell, K.W. (2007c) Quality of life after spinal cord injury: A meta-synthesis of qualitative findings. *Spinal Cord* 45(2), 124–139.

Hammell, K.W. (2009a) Sacred texts: A sceptical exploration of the assumptions underpinning theories of occupation. *Canadian Journal of Occupational Therapy* 76(1), 6–13.

Hammell, K.W. (2009b) Self-care, productivity and leisure, or dimensions of occupational experience? Rethinking occupational "categories." *Canadian Journal of Occupational Therapy* 76(2), 107–114.

Hammell, K.W. (2010a) Contesting assumptions in occupational therapy. In: M. Curtin, M. Molineux and J. Supyk-Mellson (Eds.). *Occupational Therapy and Physical Dysfunction: Enabling Occupation* (6th edn). Edinburgh, Elsevier, pp. 39–54.

Hammell, K.W. (2010b) Spinal cord injury rehabilitation research: Patient priorities, current deficiencies and potential directions. *Disability and Rehabilitation* 32(14), 1209–1218.

Hammell, K.W. (2011) Resisting theoretical imperialism in the disciplines of occupational science and occupational therapy. *British Journal of Occupational Therapy* 74(1), 27–33.

Hammell, K.W. (2012) Exposing the emperor: Meditations on credulity and occupational therapy. *Occupational Therapy Now* 14(6), 13–17.

Hammell, K.W. (2013a) Client-centred occupational therapy in Canada: Refocusing on core values. *Canadian Journal of Occupational Therapy* 80(3), 141–149.

Hammell, K.W. (2013b) Client-centred practice in occupational therapy: Critical reflections. *Scandinavian Journal of Occupational Therapy* 20(3), 174–181.

Hammell, K.W. (2013c) Occupation, well-being and culture: Theory and cultural humility. *Canadian Journal of Occupational Therapy* 80(4), 224–234.

Hammell, K.W. and Iwama, M.K. (2012) Wellbeing and occupational rights: An imperative for critical occupational therapy. *Scandinavian Journal of Occupational Therapy* 19(5), 385–394.

Hammell, K.W., Miller, W.C., Forwell, S.J., Forman, B.E. and Jacobsen, B.A. (2012) Sharing the agenda: Pondering the politics and practices of occupational therapy research. *Scandinavian Journal of Occupational Therapy* 19(3), 297–304.

Hampton, N.Z. and Qin-Hilliard, D.B. (2004) Dimensions of quality of life for Chinese adults with spinal cord injury: A qualitative study. *Disability and Rehabilitation* 26(4), 203–212.

Holcomb, L.O. (2000) Community reintegration and chronic spinal cord injury. *SCI Nursing* 17(2), 52–58.

Iwama, M. (2003) Toward culturally relevant epistemologies in occupational therapy. *American Journal of Occupational Therapy* 57(5), 582–588.

Iwama, M. (2006) *The Kawa Model. Culturally Relevant Occupational Therapy*. Edinburgh, Elsevier.

Iwama, M., Thomson, N.A. and Macdonald, R.M. (2011) Situated meaning: A matter of cultural safety, inclusion, and occupational therapy. In: F. Kronenberg, N. Pollard and D. Sakellariou (Eds.). *Occupational Therapies without Borders*. Edinburgh, Elsevier, pp. 85–92.

Jarman, M., Lamp, S., Mitchell, D., Nepveux, D., Nowell, N. and Snyder, S. (2002) Theorising disability as political subjectivity: Work by the UIC Disability Collective on political subjectivities. *Disability and Society* 17(5), 555–569.

Johnson, R. (1993) "Attitudes don't just hang in the air...": Disabled people's perceptions of physiotherapists. *Physiotherapy* 79(9), 619–627.

Jongbloed, L. and Crichton, A. (1990) A new definition of disability: Implications for rehabilitation practice and social policy. *Canadian Journal of Occupational Therapy* 57, 32–38.

Jorgensen, P. (2000) Concepts of body and health in physiotherapy: The meaning of the social/cultural aspects of life. *Physiotherapy Theory and Practice* 16, 105–115.

Jull, J.E. and Giles, A.R. (2012) Health equity, Aboriginal peoples and occupational therapy. *Canadian Journal of Occupational Therapy* 79, 70–76.

Kagan, A. and Duchan, J.F. (2004) Consumers' views of what makes therapy worthwhile. In: J.F. Duchan and S. Byng (Eds.). *Challenging Aphasia Therapies: Broadening the Discourse and Extending the Boundaries*. Hove, Psychology Press, pp. 158–172.

Kagawa-Singer, M. (1993) Redefining health: Living with cancer. *Social Science and Medicine* 37, 295–304.

Kemp, B.J. (2004) Quality of life, coping and depression. In: B.J. Kemp and L.M. Mosqueda (Eds.). *Aging with a Disability*. Baltimore, Johns Hopkins University Press, pp. 48–67.

Kemp, L. (2002) Why are some people's needs unmet? *Disability and Society* 17(2), 205–218.

Kielhofner, G. (2004) *Conceptual Foundations of Occupational Therapy* (3rd edn). Philadelphia, PA, FA Davis.

Kleiber, D.A. and Hutchinson, S.L. (1999) Heroic masculinity in the recovery from spinal cord injury. In: A. Sparkes and M. Silvennoinen (Eds.). *Talking Bodies: Men's Narratives of the Body and Sport*. Jyväskylä, Finland, SoPhi, University of Jyväskylä, pp. 135–155.

Krause, J.S. (1991) Survival following spinal cord injury: A fifteen-year prospective study. *Rehabilitation Psychology* 36, 89–98.

Krause, J.S., Coker, J., Charlifue, S. and Whiteneck, G. (2000) Health outcomes among American Indians with spinal cord injury. *Archives of Physical Medicine and Rehabilitation* 81(7), 924–931.

Krause, J.S. and Kjorsvig, J.M. (1992) Mortality after spinal cord injury: A four year prospective study. *Archives of Physical Medicine and Rehabilitation* 73, 558–563.

Lavine, T.Z. (1984) *From Socrates to Sartre: The Philosophic Quest*. New York, Bantam.

Law, M., Baptiste, S. and Mills, J. (1995) Client-centred practice: What does it mean and does it make a difference? *Canadian Journal of Occupational Therapy* 62, 250–257.

Lucke, K.T. (1999) Outcomes of nurse caring as perceived by individuals with spinal cord injury during rehabilitation. *Rehabilitation Nursing* 24, 247–253.

Magasi, S. (2008) Disability studies in practice: A work in progress. *Topics in Stroke Rehabilitation* 15(6), 611–617.

Mangset, M., Dahl, T.E., Førde, R. and Wyller, T.B. (2008) "We're just sick people, nothing else": ... factors contributing to elderly stroke patients' satisfaction with rehabilitation. *Clinical Rehabilitation* 22, 825–835.

Marks, D. (1999) *Disability. Controversial Debates and Psychosocial Perspectives.* London, Routledge.

McKnight, J. (1981) Professionalised service and disabling help. In: A. Brechin, P. Liddiard and J. Swain (Eds.). *Handicap in a Social World.* Sevenoaks, Hodder and Stoughton, pp. 24–33.

McPherson, K.M., Brander, P., Taylor, W.J. and McNaughton, H.K. (2001) Living with arthritis—What is important? *Disability and Rehabilitation* 23(16), 706–721.

McRuer, R. and Wilkerson, A.L. (Eds.). (2003) *Desiring Disability: Queer Theory Meets Disability Studies.* Durham, NC, Duke University Press.

McVeigh, K. (2011) Disabled total 1 billion, but are still seen as second-class citizens. *Guardian Weekly* June 17, 10.

Meade, M.A., Carr, L., Ellenbogen, P. and Barrett, K. (2011) Perceptions of provider education and attitude by individuals with spinal cord injury: Implications for health care disparities. *Topics in Spinal Cord Injury Rehabilitation* 17(2), 25–37.

Meade, T. and Serlin, D. (Eds.). (2006) Disability and history. *Radical History Review* 94, 1–267.

Middleton, J., Tran, Y. and Craig, A. (2007) Relationship between quality of life and self-efficacy in persons with spinal cord injuries. *Archives of Physical Medicine and Rehabilitation* 88, 1643–1648.

Mocellin, G. (1995) Occupational therapy: A critical overview, part 1. *British Journal of Occupational Therapy* 58, 502–506.

Nelson, A. (2007) Seeing white: A critical exploration of occupational therapy with Indigenous Australian people. *Occupational Therapy International* 14, 237–255.

Nelson, A. (2009) Learning from the past, looking to the future: Exploring our place with Indigenous Australians. *Australian Occupational Therapy Journal* 56, 97–102.

Ng, A.K., Ho, D.Y.F., Wong, S.S. and Smith, I. (2003) In search of the good life: A cultural odyssey in the East and West. *Genetic, Social, and General Psychology Monographs* 129(4), 317–363.

Post, M., de Witte, L., van Asbek, F., van Dijk, A. and Schrijvers, A. (1998) Predictors of health status and life satisfaction in spinal cord injury. *Archives of Physical Medicine and Rehabilitation* 78(4), 395–402.

Priestley, M. (2003) *Disability. A Life Course Approach.* Cambridge, Polity.

Priestley, M., Waddington, L. and Bessozi, C. (2010a) New priorities for disability research in Europe: Towards a user-led agenda. *ALTER: European Journal of Disability Research* 4, 239–255.

Priestley, M., Waddington, L. and Bessozi, C. (2010b) Towards an agenda for disability research in Europe: Learning from disabled people's organisations. *Disability and Society* 25, 731–746.

Ray, L.D. and Mayan, M. (2001) Who decides what counts as evidence? In: J. Morse, J. Swanson and A. Kuzel (Eds.). *The Nature of Qualitative Evidence*. Thousand Oaks, CA, Sage, pp. 50–73.

Rebeiro, K. (2004) How qualitative research can inform and challenge occupational therapy practice. In: K.W. Hammell and C. Carpenter (Eds.). *Qualitative Research in Evidence-Based Rehabilitation*. Edinburgh, Churchill Livingstone, pp. 89–102.

Rebeiro Gruhl, K.L. (2005) Reflections on…The recovery paradigm: Should occupational therapists be interested? *Canadian Journal of Occupational Therapy* 72, 96–102.

Reindal, S.M. (1999) Independence, dependence, interdependence: Some reflections on the subject and personal autonomy. *Disability and Society* 14(3), 353–367.

Reynolds, F. and Prior, S. (2003) "Sticking jewels in your life": Exploring women's strategies for negotiating an acceptable quality of life with multiple sclerosis. *Qualitative Health Research* 13(9), 1225–1251.

Richardson, M. (1997) Addressing barriers: Disabled rights and the implications for nursing of the social construct of disability. *Journal of Advanced Nursing* 25, 1269–1275.

Ryff, C.D. and Singer, B. (1998) The contours of positive human health. *Psychological Inquiry* 9, 1–28.

Said, E.W. (1996) *Representations of the Intellectual*. New York, Random House.

Sand, Å., Karlberg, I. and Kreuter, M. (2006) Spinal cord injured persons' conceptions of hospital care, rehabilitation, and a new life situation. *Scandinavian Journal of Occupational Therapy* 13, 183–192.

Sandahl, C. (2003) Queering the crip or cripping the queer? In: R. McRuer and A.L. Wilkerson (Eds.). *Desiring Disability: Queer Theory Meets Disability Studies*. Durham, NC, Duke University Press, pp. 25–56.

Schriner, K. (2001) A disability studies perspective on employment issues and policies for disabled people. In: G.L. Albrecht, K.D. Seelman and M. Bury (Eds.). *Handbook of Disability Studies*. London, Sage, pp. 642–662.

Schwartz, C.E. and Sendor, M. (1999) Helping others helps oneself: Response shift effects in peer support. *Social Science and Medicine* 48, 1563–1575.

Secrest, J.A. and Thomas, S.P. (1999) Continuity and discontinuity: The quality of life following stroke. *Rehabilitation Nursing* 24(6), 240–246.

Shakespeare, T. (1996) Rules of engagement: Doing disability research. *Disability and Society* 11(1), 115–119.

Shakespeare, T. (1998) *The Disability Reader. Social Science Perspectives*. London, Cassell.

Siegert, R.J., Ward, T., Levack, W.M.M. and McPherson, K.M. (2007) A good lives model of clinical and community rehabilitation. *Disability and Rehabilitation* 29(20–21), 1604–1615.

Snyder, S.L., Brueggemann, B.J. and Garland-Thomson, R. (Eds.). (2002) *Disability Studies. Enabling the Humanities*. New York, The Modern Language Association of America.

Stalker, K. and Jones, C. (1998) Normalization and critical disability theory. In: D. Jones, S. Blair, T. Hartery and R. Jones (Eds.). *Sociology and Occupational Therapy: An Integrated Approach*. Edinburgh, Churchill Livingstone, pp. 171–183.

Stewart, R. and Bhagwanjee, A. (1999) Promoting group empowerment and self-reliance through participatory research: A case study of people with physical disability. *Disability and Rehabilitation* 21(7), 338–345.

Stiker, H.-J. (1999) *A History of Disability* (Trans. W. Sayers). Ann Arbor, MI, University of Michigan Press.

Stone, E. and Priestley, M. (1996) Parasites, pawns and partners: Disability research and the role of non-disabled researchers. *British Journal of Sociology* 47(4), 699–716.

Swain, J. and French, S. (2010) Occupational therapy: A disability perspective. In: M. Curtin, M. Molineux and J. Supyk-Mellson (Eds.). *Occupational Therapy and Physical Dysfunction. Enabling Occupation* (6th edn). Edinburgh, Churchill Livingstone Elsevier, pp. 27–37.

Swain, J., French, S. and Cameron, C. (2003) *Controversial Issues in a Disabling Society.* Buckingham, Open University Press.

Townsend, E. (1998) *Good Intentions Overruled. A Critique of Empowerment in the Routine Organization of Mental Health Services.* Toronto, University of Toronto Press.

Twigg, J. (2000) *Bathing—The Body and Community Care.* London, Routledge.

van Leeuwen, C.M.C., Kraaijeveld, S., Lindeman, E. and Post, M.W.M. (2012) Associations between psychological factors and quality of life ratings in persons with spinal cord injury: A systematic review. *Spinal Cord* 50, 174–187.

Vrkljan, B. and Miller-Polgar, J. (2001) Meaning of occupational engagement in life-threatening illness: A qualitative pilot project. *Canadian Journal of Occupational Therapy* 68, 237–246.

Weaver, F., Guihan, M., Pape, T. et al. (2001) Creating a research agenda in SCI based on provider and consumer input. *SCI Psychosocial Process* 14(2), 77–88.

White, G.W. (2002) Consumer participation in disability research: The golden rule as a guide for ethical practice. *Rehabilitation Psychology* 47(4), 438–446.

White, G.W., Nary, D.E. and Froehlich, A.K. (2001) Consumers as collaborators in research and action. *Journal of Prevention and Intervention* 21, 15–34.

White, G.W., Suchowierska, M.A. and Campbell, M. (2004) Developing and systematically implementing participatory action research. *Archives of Physical Medicine and Rehabilitation* 85(4 Suppl. 2), S3–S12.

Whiteneck, G.G. (1994) Measuring what matters: Key rehabilitation outcomes. *Archives of Physical Medicine and Rehabilitation* 75(10), 1073–1976.

World Health Organization. (2001) *International Classification of Functioning, Disability and Health.* Geneva, World Health Organization.

Ylvisaker, M. and Feeney, T. (2000) Reflections on Dobermanns, poodles, and social rehabilitation for difficult-to-serve individuals with traumatic brain injury. *Aphasiology* 14(4), 407–431.

4

Rethinking "Normal Development" in Children's Rehabilitation

Barbara E. Gibson

Gail Teachman

Yani Hamdani

In this chapter we question and rethink "normal development" as the primary organizing concept in children's rehabilitation. Normal development understands childhood as a predictable trajectory from infancy to adulthood characterized by a series of developmental stages that cover all aspects of personhood including physical, intellectual, emotional, and social. This "ages and stages" framing provides the underlying rationale for a host of health and education interventions designed to maximize children and youth's physical and mental capacities as they "progress" toward adulthood. In the traditional rehabilitation context, therapeutic goals and treatments are designed to assist disabled children to achieve developmental milestones and approximate developmental norms and trajectories. This link between rehabilitation therapies and development is accepted almost universally as uncontroversial (Goodley and Runswick-Cole 2010) and underpins dozens of standardized scales used to assess motor, language, and social development and guide interventions (Effgen 2005, p. 42). Pick up any text on children's rehabilitation and you will find tables, charts, and descriptions of the stages of development, accompanied by statements regarding how an understanding of developmental progression is essential for the design of treatments aimed at ameliorating disability. For example, *Pediatric Therapy* (Porr and Rainville 1999, p. 70) suggests that "knowledge of 'normal' growth and development is essential to therapy practice. Coupled with a knowledge of disease and disability, this enables the therapist to understand the impact of disability on an individual child and his or her... capability and dysfunction. *Obviously*, this underlies any successful intervention (p. 70, emphasis added)."

However "obvious" it may seem, development is only one way of understanding childhood and approaching children's rehabilitation. In what follows, we situate current dominant understandings of the child and development in their historical origins, ask how well they are serving disabled children, and consider how things could be otherwise. Our project of examining the "taken for grantedness" of "normal" development and the effects it produces overlaps with other chapters in this volume that critically examine some of the deeply entrenched notions in rehabilitation research and practice (see Chapters 3, 6, and 14). In all of this work, we engage in "critical" scholarship that examines claims to truth; how they emerged; how they shape rehabilitation; how we divide, classify, and judge the recipients of care; and how we may be contributing to unintentional harms. The goal is not to suggest that we necessarily discard entrenched notions like development, but that we remain ever vigilant in recognizing that they are not natural but contingent human inventions that have myriad effects and are open to revision.

Developmentalism

Child development is a relatively recent idea that came to the fore in the late nineteenth and early twentieth centuries (Rose 1999, pp. 144–54). During this period, observations of young children were thought to inform evolutionary theory and help distinguish the emergence of characteristics that separate humans from animals. The development of the human embryo was seen in parallel to human physical evolution and the cultural evolution from primitive to civilized (Rose 1999, p. 145). Methods to measure, document, and influence children's physical and social development began to emerge. Key proponents of developmentalism like Piaget and Gesell constructed the standardized developmental stages and the normal development curves that are still widely used today to guide healthcare and education policies and practices for children and youth (James and James 2004; Rose 1999).

The emerging interest in development *as* evolution, and the systematic measurement of children's bodies and abilities, cemented a place for the science of establishing the normal child. Identifying and charting developmental milestones and age-linked performance norms provided new ways of thinking about children that were taken up by health and education professionals and the broader society—including parents. The idea of a standardized path for children's growth and development was used to underpin a wide variety of education, social, and health policies and practices that formally and informally regulated children's lives. During the late nineteenth and early twentieth centuries, children were increasingly viewed as adults-in-the-making whose progress could be tracked. The widespread measurement of markers such as height and weight were used to develop population norms against which every child could be assessed. Deviations from these norms were pathologized and thus viewed as requiring intervention. Through a series of checklists, scales, norms, and percentiles, professionals as agents of state education and health-care apparatuses could evaluate what children should be doing by a certain age, and pronounce who was normal or abnormal, advanced or slow, disabled or nondisabled. "At risk" children could be identified and acted upon. By the early twentieth century, childhood and adolescence emerged as an impermanent time of *transition*, whereby ensuring successful outcomes for children became inextricably linked with the achievement of adulthood, and more specifically a productive adulthood in the neoliberal tradition of creating *contributing* citizens (James and James 2004, pp. 143–44). Rather than subsiding, these ideas have gained further traction and are aligned with moral imperatives for life-long individual self-improvement, personal growth, change, and physical optimization that pervade contemporary life.

Developmental measures are not merely aids to assess or categorize children but are in themselves, as Rose (1999, p. 153) notes, "revolutions in consciousness." They, in short, construct the ways we think about what is and is not a child in ways that are today deeply ingrained and taken for granted such that other ways of thinking about children are difficult to imagine. Discussing the monitoring of children's growth, James and James (2004, p. 145) noted

> Although the height and weight chart claims to depict a child's unique and individual development, this uniqueness only exists in the context of a "generalised" child, derived epidemiologically from the population as a whole. Against this, the "normality" or "abnormality" of each individual child is measured and in this, "age"—that is, time passing—is critical, for it provides the context within which "successful" or "pathological" height/weight trajectories are charted. Thus the health of the child's body is "delineated not by the absolute categories of physiology and pathology, but by the characteristics of the normal population (Armstrong 1995, p. 397)," shared and common characteristics that become standardized as "normality."

A critique of developmentalism has emerged from within developmental psychology (Burman 2001, 2008, 2012, 2013; Morss 1995; Motzkau 2009; Walkerdine 1993), the social studies of childhood (James and James 2004; James et al. 1998; Matthews 2007; Mayall 1996), and disability studies (Goodley and Runswick-Cole 2010, 2011, 2013; Priestley 2003). Critics note that development is not the necessary way to understand childhood. Critical work has exposed the apparent factuality of development, which is unproblematically presented as a scientifically determined "truth." Normal development is not "natural" in these critiques but culturally produced and grounded in disparate assumptions that derive from the term's historical roots and misrecognized as scientific facts. As a result, development has evolved into a powerful instrument of categorization that allows for value-loaded judgments regarding what is best for children and youth under the guise of putatively neutral scientific and clinical descriptions.

Within disability studies, Goodley and Runswick-Cole (2010) suggest that play has been coopted to act as the agent of normal development. They state that the pervasive idea that "play is a child's work" takes for granted that the play is, or should be, purpose driven and always oriented to physical, social, and cognitive development and growth. Such a view of play obscures any notion of play for play's sake, or rather play (and the recreational pursuits of youth) as fun. For disabled children, Goodley and Runswick-Cole suggest that this dominant understanding of play constructs children and youth with physical or cognitive differences as "nonplaying objects" who require professional therapeutic interventions. In this way, play is a site of intervention for disabled children and pleasure in play is incidental or a means to an end, used to entice disabled children and youth to engage in therapy-driven activities. The lives of disabled children are thus pervaded with formal therapy, reinforced through informal play activities that are geared toward normalizing their bodies and abilities. Despite therapeutic goals to the contrary, children thus risk developing a sense of their bodies and selves as problems-to-be-fixed that differentiates them from their peers and exacerbates their exclusion.

Rethinking Children's Rehabilitation

Rethinking children's rehabilitation invites us to consider the multiple effects of developmental discourses on disabled children and their families. Rehabilitation and healthcare are rooted in dominant contemporary understandings about disability, normality, and what constitutes a good quality of life. These beliefs organize programs, policies, and practices at a fundamental level and are mostly unquestioned. Rather, they operate as tacit background understandings that determine a vast range of policies and practices ranging from what programs get funded to how goals are discussed with families and to the type and range of interventions offered (Gibson and Teachman 2012). Development is embedded in how children's rehabilitation is *structured*, organized, and delivered; institutional missions; the organization of professional teams; and referral/discharge criteria. It mediates how goals are constructed and pursued, what counts as success ("positive outcomes") or failure, what is measured and how.

Measures of child health and development in a very real way construct how we view children as certain kinds of persons who are in transition—potentialities who can be shaped into productive, self-reliant adult citizens. Quality of life and health status measures typically suggest better outcomes and improved quality of life the higher up on the development chain a child scores, i.e., walking scores higher than wheeling. These measures, which mimic the hierarchical structuring of normal development charts, determine how children and youth are viewed as progressing, regressing, or plateauing. Yet, these scores may bear little or no relationship to young people's self-assessments (Young et al. 2013) or may not be applicable for large groups of disabled children with significant physical or cognitive impairments. Measures in the burgeoning field of "transitions" in the health and education sectors rely on the notion of childhood and youth as precursors to adulthood. Transitions "checklists" set out key social developmental milestones that signal readiness for adulthood, for example, post-secondary education, career, and independent living goals (Gall et al. 2006; Transition Developmental Checklist). Wyn and Woodman (2006, p. 497) have noted that both the concepts of development and youth transitions assume a "linear progression from one identifiable status to another" with youth reduced to a stage in transforming children into capable (working) adults. Transitions work is grounded in universal norms for a set of behaviors and abilities that are assumed to correspond with integration into adult society, and which is further conflated with good "quality of life." The assumptions built into measures of transitions are consistent with neoliberal notions that the entrance to adulthood is marked by achievement of residential and financial independence, as well as the attainment of emotional self-reliance, cognitive self-sufficiency, and behavioral self-control (Arnett and Taber 1994). Enabling children to be successful *as children/youth* is almost unthinkable because the ideas of maximizing potential and personal development is so deeply entrenched in the collective social conscious.

Children's rehabilitation does not occur in a social vacuum but is embedded in dominant social and medical discourses that understand disability in particular ways. Bodies that fall outside the limits set for normal bodies and that cannot be fixed through surgery, therapy, and/or medications risk bearing the stigma of pathology. By internalizing these ideas at an early age, children may struggle to develop positive self and body images and may experience a sense of failure when they cannot achieve or maintain developmental milestones (Bottos 2003; Piggot et al. 2002). Through having their bodies singled out as in need of fixing, children learn to distinguish between normal and stigmatized bodies and internalize these negative valuations. Multiple medical, educational, and social encounters reproduce and reinforce what eventually becomes a tacit understanding of social difference, a positioning outside of an accepted norm (Priestley 1998).

We are not suggesting that the current reliance on developmentalism in rehabilitation is necessarily or uniformly bad. What we wish to explore, however, are the potential harms that can result from misplaced applications of developmental thinking in rehabilitation practices. All conceptual commitments exert effects, which may be positive and negative, intended and unintended, and as ethical practitioners we have the responsibility to understand these effects in all their subtleties. Developmental norms always and already construct disabled children as failed children, as those who require interventions in the hopes that they can approximate the normal as closely as possible. Children with the various impairments that result in referrals to rehabilitation professionals are viewed as in need of fixing; and because most will only achieve limited successes in approximating developmental norms, they can be harmed in a number of ways. In the following, we discuss research that suggests children internalize these messages to self-evaluate themselves as deficient, and describe the extraordinary efforts undertaken by parents and children to normalize children's lives and identities. As we discuss, these efforts may be accompanied by feelings of guilt, anxiety, and despair. We question if there are no other ways to support children with life-long disabilities and their families, ways to help them live well with their impairments, or different narratives of disability that do not mark children as deficient.

Research with disabled children illuminates the many ways children are affected by normative ideas about disability. Walking is an example par excellence of a rehabilitation goal that is a taken-for-granted good in rehabilitation. Intense ambulation training is a major rehabilitation intervention for disabled children and is built into the system such that there may be very little individual decision making about treatment. Notwithstanding intense therapy efforts, a number of children will not achieve functional walking and/or will use wheelchairs for some or all of their mobility needs (Bottos et al. 2001b). This approach, however, has been questioned within rehabilitation (Bottos et al. 2001a; Mulderij 2000; Wiart and Darrah 2002). Mulderij (2000), for example, has argued that children have a "right" to explore their worlds regardless of their disabilities and should thus be encouraged to use their own creative alternative forms of movement (crawling, rolling, carried, mobility devices).

In our research with children with cerebral palsy, we found that children held conflicting and ambivalent notions about the value of walking and walking therapies (Gibson and Teachman 2012; Gibson et al. 2012b). They weighed complex factors related to energy expenditure, the activity, the environment, and their personal preferences when making mobility choices (walk, crawl, or wheel). Moreover, they resisted negative views of disability through direct expression of pride in different aspects of their wheelchairs (speed, color, functionality) and in identification as wheelchair users. Nevertheless, children over the age of 11 years clearly conveyed the personal importance of being identified among their peers as "someone who can walk." These findings help demonstrate how children are socialized to divide the world into walkers and nonwalkers and internalize which group is valorized or stigmatized. They learn a dominant message, reinforced by years of rehabilitation, that nonwalking and nonwalkers are problems to be fixed. As others have commented, by internalizing these ideas at an early age, children may struggle to develop positive self and body images and may experience a sense of failure when they cannot achieve or maintain functional walking (Bottos 2003; Piggot et al. 2002).

Similarly, parents are embedded in discourses of normal development and struggle to make sense of how these imperatives apply to their disabled children. The high value placed on having a "perfect baby" in Western cultures results in parents of disabled children grieving over the "normal" child they did not have (Piggot et al. 2002). Mothers in particular are subjected to competing responsibilities, as both advocates for care and defenders of their children's worth, thus placing a mother in a paradox of saying to her child "I love you as you are" and "I would do anything to change you" (Landsman 2003). Goodley and Runswick-Cole (2011) have noted that the rise in popularity of parenting manuals and advice columns in the popular media means that parents are fully aware of the developmental milestones set for their children and will pursue these diligently. McKeever and Miller (2004) have shown that mothers expect themselves and are expected by others to play a pivotal role in enhancing their child's future value and productivity as an adult and are socially rewarded as "good mothers" for these activities. The (in)ability of their children to reach such milestones constructs both the mother's image of her child and the image of herself as (un)successful mother.

In our interviews exploring the value of walking with parents of children with cerebral palsy (Gibson and Teachman 2012; Gibson et al. 2012b), parents were committed to doing everything possible to improve their children's abilities, but were also plagued by doubts about the wisdom of doing so. They told stories of anxiety and doubt related to their decisions to forgo, stop, or decrease the intensity of an intervention. All parents discussed the efforts needed to enable their children to "experience normal activities" and positioned their children as normal within the interviews (e.g., "To me, he is normal. He walks different than others, he might think a bit differently but he's normal."). Absent from their accounts was any questioning of the normal/abnormal distinction that rewards conformity to arbitrary norms of bodily forms and abilities (Thomas 2007, pp. 67–68). Rather, they reconfigured their internalized understandings to fit with their children's abilities and claim a reworked normality. By claiming that their children are normal or the same because they do the same things as other children do, parents resist negative disability discourses and work to create a positive space for their children. However, their strategies also reproduce and sustain ingrained social values that assign preferential status to nondisabled bodies—ideas that are reinforced by rehabilitation and transmitted to children.

This nascent research begins to reveal how the pursuit of normal developmental milestones may perpetrate an ongoing symbolic violence against disabled children and their parents. Bourdieu described symbolic violence as "a gentle violence, usually imperceptible and invisible even to its victims, exerted for the most part through the purely symbolic channels of communication and cognition (more precisely, misrecognition), recognition, or even feeling" (Bourdieu 2001, pp. 1–2, parentheses in the original). The visibility of bodily differences not only affects the way others perceive and act toward disabled children; it also shapes how children and their parents come to internalize these meanings. Persons come to "know their place" in the world where the meanings attached to their bodies can appear as uncontestable and "true," rather than open to revision (Edwards and Imrie 2003).

Implications for Rehabilitation Practice

Beliefs about how rehabilitation can contribute toward achievement of a good life are understood as particular to evidence and expertise in rehabilitation; yet, they map onto broader neoliberal beliefs about individuals' responsibilities to manage the work of becoming "successful" and achieving a good, happy, or productive life, e.g., using imperatives toward self-improvement, being the best you can be, and reaching for your potential (Burman 2013). We need to continually ask if and how these goals "fit" rehabilitation, or if we are unreflectively imposing them because they are the "way of the world" and thus must be best for everyone. Is the job of rehabilitation to help people to "fit in" and to what? How can we imagine different ways of living well? Promoting acceptance of diversity and difference are given short shrift when compared to the amount of time, energy, and money spent on achieving or approximating normal— normal bodies, normal behaviors, normal movement patterns, normal roles, and normal activities. How do we partner with children/youth and their parents to acknowledge their fears of the unknown and the challenges of living a trajectory that is the road not taken? In what ways are transition services acting to reproduce beliefs and practices about normal childhood? How could they better assist children with forging a path that fits best with their abilities and desires?

In thinking about other ways we might understand childhood and support disabled children, we can first consider the great variation in children's bodies, abilities, and rates of development, and ask if normalizing those differences is the right goal. We must also be aware that trying to approximate a typical developmental trajectory may harm some children and families in ways that are not immediately apparent. Some of the physical and social end points of child development will be unattainable for many disabled children who internalize developmental roles and expectations over time. This can have profound effects when they cannot achieve these norms, including anxiety, distress, depression, feelings of worthlessness, or "nothing to offer" (Gibson et al. 2007, 2014). Walkerdine (1993, p. 466) has argued that the dominant narrative of development is part of a bigger European patriarchal story that subsumes other stories. She suggests that, "(a) move away from a universal developmentalism must be a move away from a pathologization of Otherness... (T)he way to deal with this is to produce new narratives which tell of change and transformation in the very specific conditions in which they are produced."

In rehabilitation, focusing on normal development closes off other possibilities that may better suit young people's needs and abilities. What would these alternatives look like? First of all, we would suggest exploring ways of "living well in the present" (Davies 1997) should be considered alongside development goals and the pursuit of typical adult milestones. This might include attention to achieving or maintaining relationships, personally meaningful projects, "or just plain enjoyment of life" (Hammell 2004). These life choices do not figure in the current transitions rhetoric and we suspect many professionals working in transitions programs might struggle with supporting them. Nevertheless, we suggest that only focusing on approximating normative life trajectories closes off other possibilities that may in some cases better suit young people's needs and abilities. Recreation activities may take on a greater significance, as may hobbies, caring for others or managing one's own care needs. These activities, because they are oriented to the present and not future development or economic contribution, may be judged as futile, trivial, or wasteful and are less likely to be supported or funded (Aitchison 2003; Gibson et al. 2009; Hammell 2004). In addition, we suggest that a move away from models that privilege normal bodies and development might open space to tap into children's own sense of their bodies, how their bodies work, and how they form connections with people, environments, and technologies to do the things they want to do (Gibson et al. 2012a).

Rethinking development creates spaces for alternate ways of living that contribute to the acceptance of differences. The centrality of work, productivity, and independence as markers of transitions and adult social status need to be balanced by openness to other ways of being in the world (Priestley 2003, pp. 132–42). As part of an expanded ethics of inclusion, rehabilitation could attend more to the nonproductive and nonmaterial contributions of individuals in society (Goodley and Runswick-Cole 2010). Moving away from thinking of development and transition as a delimited linear pathway opens up multiple other pathways that could also be explored with young people and their families. Similarly, rethinking mobility goals opens up a range of options. Early powered mobility, which gives access to the world at a younger age, has been advocated for a number of years (Butler 1986), but its uptake has been rather slow (Wiart and Darrah 2002). Promoting acceptance of difference also means helping families to feel comfortable with alternate modes of mobility like crawling, using assistive devices, and letting go of the pursuit of independent walking without guilt. These kinds of explorations help to create spaces for other ways of living that contribute to the acceptance of differences.

Creating space for parents and children to talk about these issues, and share their experiences, fears, and assumptions, is also important. Children's values and beliefs are forming at the same time that their parents' values and goals for their child may be shifting. Although negative views of disability and difference may be imposed on parents and disabled children, they may also resist these ideas depending on their own particular circumstances and their exposure to counter-narratives (Fisher and Goodley 2007; Landsman 2003). A challenge for rehabilitation is to determine how best to assist children in maintaining a positive disability identity while pursuing achievable therapeutic goals. Shifting away from viewing childhood as a critical period for interventions "at all costs" creates space for approaches that support disabled children to more fully enjoy and engage in the here and now of their daily lives.

Conclusion

In this chapter, we have suggested that a fundamental concept in children's rehabilitation, normal development, needs to be rethought on several levels. We are not suggesting that normal development be abandoned or replaced by something else. Rather, we are advocating for an opening up of possibilities for multiple ways of thinking about childhood that can collectively inform rehabilitation. Normal development became the dominant story of childhood because of a particular set of historical and cultural circumstances and beliefs that we have briefly reviewed. The point is not whether development is *true*, but does it *work*? Development is not the only way to understand the lives of children, nor is it the only way of conceiving of what is best, good, or right for disabled children and youth. Once we begin to question development, we open up new possibilities for working with children and families, for shared exploration of needs and hopes that will assist children and families to live well. This will sometimes involve ameliorating impairments, working on improvements of strength, speech, or walking; and it will sometimes involve advocating for acceptance of abilities, alternate ("abnormal") ways of moving or communicating, and letting go of the pursuit of normalcy. Our hope is that these ideas will contribute to the small but growing efforts of a few rehabilitation scholars and practitioners who are reimagining the field through critical thinking. In this vein, we look forward to ongoing debate and dialogue.

References

Aitchison, C. 2003. From leisure and disability to disability leisure: Developing data, definitions and discourses. *Disabil Soc* 18:955–69.

Armstrong, D. 1995. The rise of surveillance medicine. *Sociol Health Illn* 17(3):393–404.

Arnett, J. J., and Taber, S. 1994. Adolescence terminable and interminable: When does adolescence end? *J Youth Adolesc* 23:517–37.

Bottos, M. 2003. Ambulatory capacity in cerebral palsy: Prognostic criteria and consequences for intervention. *Dev Med Child Neurol* 45:786–90.

Bottos, M., Bolcati, C., Sciuto, L., Ruggeri, C., and Feliciangeli, A. 2001a. Powered wheelchairs and independence in young children with tetraplegia. *Dev Med Child Neurol* 43:769–77.

Bottos, M., Feliciangeli, A., Sciuto, L., Gericke, C., and Vianello, A. 2001b. Functional status of adults with cerebral palsy and implications for treatment of children. *Dev Med Child Neurol* 43:516–28.

Bourdieu, P. 2001. *Masculine Domination*. Stanford, CA: Stanford University Press.

Burman, E. 2001. Beyond the baby and the bathwater: Postdualistic developmental psychologies for diverse childhoods. *Eur Early Child Educ Res J* 9:5–22.

Burman, E. 2008. *Deconstructing Developmental Psychology*. New York: Routledge.

Burman, E. 2012. Deconstructing neoliberal childhood: Towards a feminist antipsycho-
logical approach. *Childhood* 19:423–8.

Burman, E. 2013. Desiring development? Psychoanalytic contributions to antidevelop-
mental psychology. *Int J Qual Stud Educ* 26:56–74.

Butler, C. 1986. Effects of powered mobility on self-initiated behaviours of very young
children with locomotor disability. *Dev Med Child Neurol* 28:325–32.

Davies, M. L. 1997. Shattered assumptions: Time and the experience of long term HIV
positivity. *Soc Sci Med* 44:561–71.

Edwards, C., and Imrie, R. 2003. Disability and bodies as bearers of value. *Sociology*
37:239–56.

Effgen, S. K. 2005. *Meeting the Physical Therapy Needs of Children*. Philadelphia, PA: F. A.
Davis.

Fisher, P., and Goodley, D. 2007. The linear medical model of disability: Mothers of disabled babies resist with counter-narratives. *Sociol Health Illn* 29:66–81.

Gall, C., Kingsnorth, S., and Healy, H. 2006. Growing up ready: A shared management
approach to transition. *Phys Occup Ther Ped* 38:47–62.

Gibson, B. E., Carnevale, F. A., and King, G. 2012a. "This is my way": Reimagining disability, in/dependence and interconnectedness of persons and assistive technologies.
Disabil Rehabil 34:1894–9.

Gibson, B. E., Mistry, B., Smith, B., Yoshida, K. K., Abbott, D., Lindsay, S., and Hamdani,
Y. 2014. Becoming men: Gender, disability, and transitioning to adulthood. *Health*
18:93–112.

Gibson, B. E., and Teachman, G. 2012. Critical approaches in physical therapy research:
Investigating the symbolic value of walking. *Physiother Theory Pract* 28:474–84.

Gibson, B. E., Teachman, G., Wright, V., Fehlings, D., and McKeever, P. 2012b. Children's
and parents' beliefs regarding the value of walking: Rehabilitation implications for
children with cerebral palsy. *Child Care Health Dev* 38:61–9.

Gibson, B. E., Young, N. L., Upshur, R. E. G., and McKeever, P. 2007. Men on the margin:
A Bourdieusian examination of living into adulthood with muscular dystrophy.
Soc Sci Med 65:505–17.

Gibson, B. E., Zitzelsberger, H., and McKeever, P. 2009. 'Futureless persons': Shifting life
expectancies and the vicissitudes of progressive illness. *Sociol Health Illn* 31:554–68.

Goodley, D., and Runswick-Cole, K. 2010. Emancipating play: Dis/abled children, development and deconstruction. *Disabil Soc* 25:499–512.

Goodley, D., and Runswick-Cole, K. 2011. Problematising policy: Conceptions of "child,"
"disabled" and "parents" in social policy in England. *Int J Inclus Educ* 15:71–85.

Goodley, D., and Runswick-Cole, K. 2013. The body as disability and possibility: Theorizing
the "leaking, lacking and excessive" bodies of disabled children. *Scand J Disabil Res*
15:1–19.

Hammell, K. W. 2004. Dimensions of meaning in the occupations of daily life. *Can J
Occup Ther* 71:296–305.

James, A., and James, A. 2004. *Constructing Childhood: Theory, Policy and Social Practice*.
New York: Palgrave Macmillan.

James, A., Jenks, C., and Prout, A. 1998. *Theorising Childhood*. Cambridge: Polity Press.

Landsman, G. 2003. Emplotting children's lives: Developmental delay vs. disability. *Soc Sci Med* 56:1947–60.

Matthews, S. 2007. A window on the "new" sociology of childhood. *Sociol Compass* 1:322–34.

Mayall, B. 1996. *Children, Health and the Social Order.* Buckingham: Open University Press.

McKeever, P., and Miller, K. L. 2004. Mothering children who have disabilities: A Bourdieusian interpretation of maternal practices. *Soc Sci Med* 59:1177–91.

Morss, J. R. 1995. *Growing Critical: Alternatives to Developmental Psychology.* New York: Routledge.

Motzkau, J. J. F. 2009. The semiotic of accusation: Thinking about deconstruction, development, the critique of practice, and the practice of critique. *Qual Res Psychol* 6:129–52.

Mulderij, K. 2000. Dualistic notions about children with motor disabilities: Hands to lean on or to reach out? *Qual Health Res* 10:39–50.

Piggot, J., Paterson, J., and Hocking, C. 2002. Participation in home therapy programs for children with cerebral palsy: A compelling challenge. *Qual Health Res* 12:1112–29.

Porr, S. M., and Rainville, E. B. 1999. The special vulnerabilities of children and families. In *Pediatric Therapy: A Systems Approach*, ed. S. J. Lane. Philadelphia, PA: F. A. Davis.

Priestley, M. 1998. Childhood disability and disabled childhoods: Agendas for research. *Childhood* 5:207–23.

Priestley, M. 2003. *Disability: A Life Course Approach.* Malden, MA: Polity Press.

Rose, N. 1999. *Governing the Soul: The Shaping of the Private Self*, 2nd ed. London: Free Association Books.

Thomas, C. 2007. *Sociologies of disability and illness: Contested ideas in disability studies and medical sociology.* New York: Palgrave MacMillan.

Transition Developmental Checklist. *Kentucky Commission for Children with Special Healthcare Needs.* Available from http://www.waisman.wisc.edu/cshcn/Health _IEP/packet%201Embedding%20Health%20Outcomes%20in%20the%20IEP /Transition_checklistKY.pdf (accessed September 11, 2014).

Walkerdine, V. 1993. Beyond developmentalism. *Theory Psychol* 3:451–69.

Wiart, L., and Darrah, J. 2002. Changing philosophical perspectives on the management of children with physical disabilities—Their effect on the use of powered mobility. *Disabil Rehabil* 24:492–8.

Wyn, J., and Woodman, D. 2006. Generation, youth and social change in Australia. *J Youth Stud* 9:495–514.

Young, N. L., Sheridan, K., Burke, T. A., Mukherjee, S., and McCormick, A. 2013. Health outcomes among youths and adults with spina bifida. *J Pediatr* 162:993–8.

II

Philosophy
in Action

5

Do Frogs Have Lips?—An Exploration of the Place of "Mind" in Rehabilitation

Richard J. Siegert

Matthew Maddocks

Even if we were aware of what was going on in people's central nervous systems, it is unlikely that we should cease to find a use for explaining their behaviour in terms of their conscious thoughts and feelings. Nor is it likely, in a case in which an inference drawn from one's physical condition conflicted with one's own awareness of one's own experience, that one should not continue to treat this awareness as the better authority.

Sir Alfred Ayer (Gregory 1998)
former Wyndham Professor of Logic, University of Oxford,
The Oxford Companion to the Mind

The aim of the present chapter is to explore one important theoretical issue for rehabilitation—namely, the place of the "mind" in rehabilitation. The relationship between the mind and the body, or the mind and the brain, is a longstanding one in philosophy and even has its own name, "the mind–body problem." Almost every undergraduate in philosophy, psychology, or neurosciences has heard of René Descartes and mind–body dualism. However, textbooks and journals in rehabilitation typically have little, if anything, to say about the mind itself. In contrast, with advances in neuroimaging techniques and improved understanding of neural plasticity, the brain is generally well represented. In this chapter, we will argue that the mind–brain–body problem is an important theoretical issue for rehabilitation to consider. First, some of the important stances that ancient and modern philosophers have taken to the mind–body problem will be briefly outlined, as these might be unfamiliar to some rehabilitation readers. Second, we provide examples of recent advances from psychology and rehabilitation that rely upon an understanding of human functioning and abilities at a cognitive (or mental) level of explanation rather than a purely neural one. These examples include: (i) current ideas about dual process or "two track" models of human cognition (Kahneman 2011); (ii) use of mindfulness meditation techniques for treating depression and other emotional and physical disorders (Chiesa and Serretti 2011); (iii) visual imagery in stroke rehabilitation (Zimmermann-Schlatter et al. 2008); and (iv) the use of cognitive behavioral therapy (CBT) to manage breathlessness in patients with chronic respiratory disease (Howard et al. 2010; Livermore et al. 2010). Third, the recent work of Denny Borsboom and colleagues will be introduced to illustrate one potential way forward in bringing the mind into rehabilitation (Kievit et al. 2011a). In this work, they have argued that recent advances in psychometrics now permit mathematical modeling of complex relationships between behavior and biology.

The epistemological stance taken in this chapter is best described as *scientific realism* (Fletcher 1996). Such a position assumes that a scientific approach is our best strategy for understanding and knowing the external world or universe. Within the social sciences, it lies somewhere between the two extremes of positivism and relativism. A realist position accepts that all human knowledge is inevitably socially constructed, but also argues that it is worth striving toward an objective knowledge of the world, while accepting the inevitable fallibility of scientific methods. In other words, the realist asserts the special nature of scientific method and considers it has profound advantages over other ways of knowing, such as art, literature, jurisprudence, or religion, while acknowledging that at best science offers us a rough approximation to reality.

The Mind–Body Problem and Some Positions on It

Phenomena such as the mind, consciousness, thinking, dreams, hallucinations, and déjà vu have historically been the concern of both psychology and that branch of philosophy known as *philosophy of mind*. While psychology has focused on operationalizing these phenomena to study them empirically, philosophy has been more concerned with identifying and unpacking the broader conceptual issues and assumptions that underpin our understandings of mind, body, and brain. An enduring concern of philosophers of mind has been the so-called mind–body problem. At its simplest, this debate centers on the nature of the relationship between a person's mind and body, although recently the focus has been mostly on the relationship between the mind and the brain. The debate on how these two phenomena are related has been a lively one since the French philosopher René Descartes first proposed *dualism*—the idea that mind and body were separate and qualitatively different, obeying different laws. Descartes argued that the mind (or spirit) was not physical in nature like the body, although the two were somehow connected or communicated at the pineal gland in the brain. While dualism is no longer defended by any serious philosophers of mind today, there exists a spectrum of viewpoints nonetheless. Some of the leading perspectives on the mind–body question are briefly outlined in the following section. These perspectives include physicalism, behaviorism, functionalism identity theory, and supervenience. For a more detailed exposition on these and other perspectives, the reader is referred elsewhere (Crumley II 2006).

Physicalism

Physicalism is the belief that all mental events are reducible to a physical or biological substrate. In other words, all the mental events we experience and "know" subjectively as sentient beings (including sensations, perception, emotions, thoughts, memories, ideas, beliefs, and imagery) are the direct result of biochemical and electrical events occurring in our body and the nervous system in particular. This approach to the mind–brain issue is characterized by a *reductionist* strategy to scientific understanding, in which complex events are broken down into constituent elements that can be studied in isolation. For example, we might try to understand a complex event like an olfactory hallucination by analyzing the molecular constituents of neurotransmitters in the brains of people who experience such hallucinations. In its strong version, physicalism holds that ultimately it should be possible to explain all mental phenomena at a physiological level.

Behaviorism

Behaviorism is an approach to understanding human and animal behavior that was strongly influential upon psychology for much of the twentieth century. Perhaps its most famous and sometimes controversial spokesperson was the American psychologist B. F. Skinner who pioneered and promoted operant psychology. Skinner described himself as a "radical behaviorist" and argued that the proper subject matter for a science of psychology was the study of observable behavior rather than unobservable mental phenomena. To this end, he advocated the study of other species' behavior, including the rat and pigeon, since it was possible to control the environment precisely and impossible to be influenced by introspective self-report. Skinner argued that the environment was paramount in determining our behavior and that schedules of reward and punishment were the key features of the environment in molding or *shaping* behavior. In fact, there are a number of different philosophical stances within behaviorism, but what they all share is a mistrust of explaining behavior in terms of unobservable mental phenomena. So some behaviorists dispute the very existence of mental phenomena (ontological behaviorism), whereas others simply argue that such phenomena are not amenable to observation and hence do not belong in any scientific explanation of behavior (Crumley II 2006).

Functionalism

Functionalism is an approach to understanding human behavior that has been influential in psychology since William James and has had a pervasive influence on modern cognitive psychology. A functionalist approach emphasizes the importance of understanding the biological or evolutionary purpose of mental abilities and modeling how different components of mental life operate as individual units and as part of a coordinated system. Contemporary accounts of functionalism have been greatly influenced by the information revolution, and an analogy is often drawn between computer hardware (neuroscience) and computer software (cognitive psychology). In other words, if one accepts the brain–computer analogy, a functionalist attempts to understand the software programs that operate the computer and a neuroscientist is concerned with the electronic circuits and microprocessors. Debates typically center on how well one can characterize human mental activity at one of these levels without reference to the other. A prime example of a functionalist tradition with close links to rehabilitation is cognitive neuropsychology, which attempts to understand brain function through careful analysis of the effects of localized brain lesions on individual patients.

Identity Theory

Subscribers to identity theory argue that we can correlate mental events with their physiological underpinnings. This approach does not dispute the existence of mental events but rather emphasizes their biological or neural foundations. Proponents of identity theory often distinguish between scientific explanations, which are framed in terms of the nervous system, and *folk psychology* explanations, that are framed in terms of beliefs, wishes, emotions, and desires. Consider, for example, the following imaginary scenario. My clinical colleague Dennis passes me on the ward at work and fails to respond to my usual cheery greeting. I might interpret this as follows: Dennis has a strong *desire* to be promoted to consultant. He *believes* his practice is far superior to my own. I was successful in the recent promotions whereas he was not. Dennis *feels* angry and resentful. He might even *believe* my cheery salutation to result from a *wish* on my part to humiliate him. It is this folk psychology that forms our everyday or common sense way of understanding and explaining our own and other people's behavior. Most identity theorists happily accept the existence of the folk psychological framework for explaining behavior but regard it as unscientific compared to an explanation at the level of the nervous system. The scientific explanation of behavior is concerned with identifying the neural correlates of mental phenomena.

Supervenience

Supervenience is a concept that is popular in philosophy to understand natural phenomena where multiple, hierarchical levels of explanation are required. For example, we might attempt to explain memory at several different levels. There is the level of the individual neuron where electrical and biochemical events occur. There is also the level of neural circuitry at which vast numbers of neurons coordinate their firing to transmit information from one brain region to another. At another level of explanation, we might speak in terms of large areas of the brain such as the prefrontal cortex, the basal ganglia, and the hippocampus. At still another level, we might explain a person's memory in terms of their observable behavior on a word recall task. If the person is the victim of a dementing illness such as Alzheimer's disease, we might enter a whole different level of explanation to describe the impact of events at all these "lower" levels on their daily comportment and family life. While philosophical stances on supervenience vary, the essential idea is that events at a lower level of explanation can have an effect or *supervene* at the next level up. Indeed any changes at a higher level must reflect changes at the next level down. Also important is the notion that salient phenomena can emerge at one level of explanation that are most meaningfully understood only at that level. For example, *personality* is often used to encapsulate an array of behavioral traits that characterize a person. Undoubtedly the behavior or traits reflected in personality dimensions such as "extraverted," "anxious," or "conscientious" are the direct manifestation of both genetic and neural factors at various levels. However, there remain certain aspects of personality, as a phenomenon worthy of scientific investigation, that cannot be explained at a genetic or neural level, for example, the developmental factors that predispose a person to high anxiety, or cultural differences in valuing and rewarding conscientiousness.

Working with People, Minds, and Damaged Brains

The central concern of the present chapter is that rehabilitation has embraced modern advances in understanding the structure and function of the brain but paid almost no attention to important philosophical questions about the mind and its relationship to the body. (See Chapter 9, for a reflection on the place of identity and self in rehabilitation.) For example, can mental events such as beliefs, wishes, and desires provide a scientific explanation of behavior? Some would argue that such phenomena are part of our folk psychology but have no place in a scientific explanation. Yet these issues are important because professionals in rehabilitation frequently work with people who have cognitive problems resulting from brain injury or neurological disease. While a sound understanding of brain anatomy and function is essential for professionals working with people with neurological disorders, so is an understanding of how the mind works. This begs the question of how the mind and the brain are connected or related to each other. In the following, we provide a brief introduction to four areas of research on uniquely mental phenomenon, each of which has relevance for rehabilitation.

Fast and Slow Thinking

The eminent psychologist and Nobel prize winner Daniel Kahnemann recently published a best-selling book titled *Thinking: Fast and Slow*, which summarizes much of what is currently known about human cognition or thinking (Kahneman 2011). As its title suggests, human cognition seems to be characterized by two qualitatively different types or styles of thinking. There appears to be one system for processing information that operates quickly, automatically, and largely outside of conscious awareness. For example, at a family gathering, we recognize faces and can interpret emotional expressions on faces almost instantaneously. This system is also primarily responsible for actions such as judging distance, alerting us to a hostile comment, making simple calculations, reading street signs, riding a skateboard, and detecting physical danger in our environment. There is also a second system or mode of thinking that is slower, more effortful and operates within our conscious awareness. For example, in the supermarket, we stop and deliberate over what to cook for tonight's meal and which aisles to negotiate to locate the various ingredients. Kahnemann's "system 2" is also largely in control of things like walking faster than normal, comparing the prices of two mobile phones, completing a tax return, and trying to recall the name of the person you sat next to on your first day in school.

Despite an overwhelming amount of evidence for this dual or parallel process model of human cognition to date, it has had little or no impact on rehabilitation theory, practice, or research. In other words, the best contemporary knowledge about fundamental processes that underpin human cognition or thinking are not yet part of the rehabilitation of people with cognitive problems. Transfer of this knowledge into rehabilitation practice could have direct and practical implications for care. For example, the parallel process model of human cognition could provide fertile ground for a more comprehensive assessment of patients and improved task prescription aimed at utilizing, developing, or compensating the most impaired type of thinking. It may also allow us to explore if a particular type of thinking is more acutely affected or resistant to disability, which could have implications for which approach to treatment is most relevant.

Mindfulness

Mindfulness-based approaches to self-regulation have been applied to a diverse range of conditions with mounting evidence for their value in reducing emotional distress and enhancing well-being. These conditions include chronic medical disease, major depression, psychiatric disorders, lower back pain, breast cancer, anxiety disorders, and eating disorders (Bohlmeijer et al. 2010; Chiesa and Serretti 2010, 2011; Cramer et al. 2012a,b; Vollestad et al. 2012; Wanden-Berghe et al. 2011). In essence, mindfulness training involves teaching people a form of meditation derived from traditional Buddhist teaching and practice but in a secular or nonreligious context. The person is instructed to focus their attention in a relaxed, nonjudgmental way on the sensations of breathing or the sounds in their environment or the moment-by-moment flow of their own conscious awareness. Indeed the precise object of focus is not that important, but rather the ability to stay in the present moment, observing but not judging or evaluating. When the individual finds that their mind has wandered from the focal object, they are instructed to note this fact and to redirect their attention but without self-criticism or judgment. While mindfulness practitioners frequently report achieving a state of relaxed awareness, the aim of mindfulness is not relaxation per se. For example, if a mindfulness practitioner becomes aware of anxious thoughts or feelings arising during meditation, then they are advised to observe these internal events in the same calm and nonjudgmental way. The actual mechanism by which mindfulness works in promoting well-being remains uncertain and demands further exploration. However, it is a simple and noninvasive procedure that has already been tested in trials on patients with brain injury, multiple sclerosis, stroke, and breast cancer with promising results (Azulay et al. 2012; Grossman et al. 2010). A unique aspect of mindfulness is that it directly aims to enhance a person's awareness of how their own mind functions. The mindful practitioner observes how thoughts constantly arise to distract them from an awareness of the "here and now" and is able to see thoughts as merely thoughts rather than absolute truths.

Visual Imagery in Motor Rehabilitation after Stroke

We will begin this section on mental imagery with what philosophers fondly call a "thought experiment." Before reading any further, take a few minutes to answer the following question and do so without referring to any texts, other people, Internet search engines, or other sources of information. Introspection is permitted. Here is the question—Do frogs have lips? Now what we are interested in here is not the correct answer to this question but rather the means by which you found the answer.* One answer to quite *how* you answered this question goes something like this—you created a mental picture of a frog, an image in your mind, and then inspected this mental image of a frog for the presence or absence of lips.† Interestingly in philosophy of mind, some schools of thought dispute the mere existence of mental imagery, and even where the existence of mental imagery is accepted, intense debate occurs around whether or not mental images can have causal properties.

This is not the case in the rehabilitation literature, where there has been increasing interest in recent years in the possibility that the human capacity for visual imagery might be employed to improve motor recovery after stroke (Ietswaart et al. 2011; Simmons et al. 2008). A systematic review of this topic concluded that, notwithstanding the inevitable need for larger studies and tighter research designs, "current evidence suggests that motor imagery provides additional benefits to conventional physiotherapy or occupational therapy" (Zimmermann-Sclatter et al. 2008, p. 1). The study of mental imagery to enhance physical performance in sport psychology is already well established and offers a promising heuristic link for rehabilitation with both sport psychology and cognitive neuroscience (Moran et al. 2012). The field of clinical psychology has also used mental imagery in the treatment of conditions such as post-traumatic stress disorder and phobia and has developed quite sophisticated approaches to the assessment of visual imagery abilities. Seen in this broader context, the interest within rehabilitation in the potential utilization of a person's imaging ability makes perfect sense. It seems important, then, that rehabilitation clinicians and researchers might also incorporate and participate in the discussion around the related philosophical issues.

Cognitive Behavioral Therapy for Breathless Patients with Chronic Respiratory Disease

In line with identity theory, the drive for an increased neuro-physiological understanding of breathlessness in chronic obstructive pulmonary disease (COPD) has led to the uptake of brain imaging techniques such as positron emission tomography (PET) and

---

* This example of a thought experiment is taken from Tye (1991).
† According to Wikipedia, frogs don't have lips, not in the same respect that we do, but they have a formation around the edge of their mouth (see http://wiki.answers.com/Q/Do_frogs_have_lips #ixzz1zoALXoF4).

functional magnetic resonance imaging (fMRI). Such techniques infer neural activation from changes in blood flow (Logothetis 2008). Experiments have provoked breathlessness in patients while capturing images of the brain to understand the underlying mechanisms and how the symptom is experienced (Parshall et al. 2012). Interestingly, these studies seeking a detailed biological explanation for breathlessness have primarily underscored the emotional elements of the symptom. Provocation of breathlessness has consistently led to activation of cortico-limbic structures that relate to feelings of threat, fear, pain, and anxiety (Banzett et al. 2000; Corfield et al. 1995; Evans et al. 2002; von Leupoldt et al. 2008, 2009). Using sophisticated methods in the search for a neurobiological explanation of breathlessness, studies have separated sensory from affective components and emphasized the role of the mind in how the symptom is experienced by patients (Evans 2010; Parshall et al. 2012).

To some extent, this understanding has translated into more widespread use of mind-based approaches that target the emotional, cognitive, and experiential processes brought about by breathlessness. This includes the use of CBT in the management of patients with chronic respiratory disease, particularly for patients who are optimally medically managed and still experiencing distress. A CBT approach emphasizes the role of emotional and cognitive processes in the shaping of the affective experience and subsequent behavior (Beck 1979). For example, a patient with COPD may experience high levels of anxiety due to a fear of being breathless. This perception of breathlessness and the inability to cope is linked with poor health outcomes, including lower levels of daily functioning (Arnold et al. 2005; Scharloo et al. 1998), engagement with rehabilitation programs (Fischer et al. 2009), and performance in tests of maximal exercise capacity (Fischer et al. 2012). As such, patients with negative perceptions around their illness are more likely to adopt a sedentary lifestyle that will exacerbate the deconditioning cycle associated with chronic disease (Jolley and Moxham 2009). By understanding and addressing negative emotion states such as high anxiety and their triggers, CBT aims to reduce the impact that emotions have on patient behavior by changing the way the patient thinks and responds to events. Early evidence for the effectiveness of CBT approaches in patients with COPD is promising (Howard et al. 2010; Livermore et al. 2010). However, studies remain small and underpowered, and further research is required.

Mind the Gap: A Psychometric Approach to the Reduction Problem

A recent paper by Kievit and colleagues has recommended a promising new research strategy for examining the relationship between neurological and psychological or behavioral variables. In their article, these authors are primarily concerned with the "reduction problem" or how to reconcile behavioral measures (e.g., scores on a multidimensional personality questionnaire) with physiological measures (such as fMRI data) (Kievit et al. 2011a,b). The approach they recommend and demonstrate uses statistical modeling developed in psychometrics and structural equation modeling (SEM) in particular. The authors argue that "mathematically tractable models with known statistical properties, developed largely in psychometrics, can map theoretical positions about the relationship between brain and behavioural measurements as developed in the philosophy of mind in impressive detail" (Kievit et al. 2011a, p. 69).

The use of SEM allows both latent variables (e.g., personality traits such as neuroticism) and manifest variables (e.g., heart rate, electroencephalography responses) to be employed in building explicit models that can be tested against actual data for goodness of fit to the model. Kievit et al. note that structural equation models can include both *summative* and *formative* models such as the two schematic models portrayed in Figure 5.1. Model 1, a reflective model, typifies the standard psychometric model or *latent variable* in which the measured responses on questionnaire items (variables A, B, and C) are summated to estimate the value of an underlying construct represented in the diagram by the large circle. For example, we cannot measure anxiety directly but an anxiety questionnaire works on the assumption that the responses on individual items are influenced by the underlying anxiety characteristic of a person. The smaller circles impinging on the three items represent error variance or variance not explained by the latent variable. Model 2 in Figure 5.1 represents the other situation, a summative model, in which we can measure variables more directly, and these variables might be combined to represent a theoretical construct. This usually takes the form of a regression model in which independent variables are used to predict or estimate a dependent variable. Kievit et al. give socioeconomic status (SES) as a classic example of such a variable. While it cannot be directly measured, SES is usually estimated by some combination of income and educational indices. In rehabilitation research, such summative models might employ

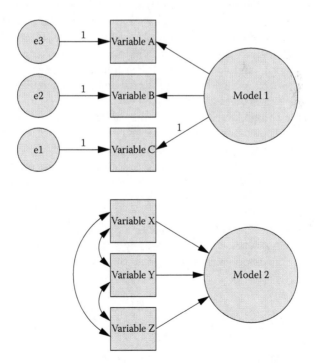

FIGURE 5.1 Schematic representation of a reflective (model 1) and a formative model (model 2).

physiological measures that could be used in conjunction with psychological variables in formative models to test increasingly complex theoretical models about the interaction of psychological and physiological variables. For example, in attempting to understand the mechanisms underpinning mindfulness, it might be possible to combine neurophysiological measures such as EEG, heart rate, and galvanic skin response with cognitive measures such as responses to a mindfulness questionnaire (Baer et al. 2008).

Kievit et al. advocate integrating both psychometric and neurological measurements in such models commenting that progress at the interface of psychology and neuroscience is dependent on increased development and testing of formal theories as mathematical models. They also suggest that this approach to integrating behavioral and neurological measures might permit specific tests of different approaches to the mind–body issue such as identity theory versus supervenience theory. They argue that reconciling psychological and neurophysiological phenomena is not just a conceptual or philosophical problem but equally a *measurement* problem.

Conclusion

The centuries-old mind–body problem remains an important concern for contemporary philosophy of mind, and vigorous debate continues over the best philosophical position to take on it. In the present chapter, we outlined some of the main philosophical approaches to this issue and argued that it is time for rehabilitation to seriously consider its relevance. In outlining some of the different perspectives on the mind–body issue, we have not favored one approach but simply argued that it is an important issue for rehabilitation. Recent research on fast and slow thinking, mindfulness, visual imagery in stroke rehabilitation, and CBT for breathlessness was presented as evidence that the mind already plays an important part in rehabilitation. One priority for research in this area may be to carefully explore the clinical and research implications of taking different philosophical stances to the mind–body problem. For example, von Leupoldt et al. (2011) recently highlighted the important role of positive and negative affective states in response to pulmonary rehabilitation, and Pbert et al. (2012) demonstrated that mindfulness meditation can be a useful intervention for people with asthma. It is our belief that as research on these topics advances, it will become increasingly important to consider some of the conceptual and philosophical issues we have outlined above. In addition, we have pointed to an important recent paper by Kievit and colleagues that suggests how a SEM approach might offer a robust method for quantitative testing of sophisticated theoretical models in rehabilitation that include both psychological and neurophysiological variables. Science proceeds by building a model, testing the model, and then refining the model—this is also a good pathway for rehabilitation to progress.

References

Arnold, R., A. V. Ranchor, M. J. DeJongste, G. H. Koeter, N. H. Ten Hacken, R. Aalbers, and R. Sanderman. 2005. The relationship between self-efficacy and self-reported physical functioning in chronic obstructive pulmonary disease and chronic heart failure. *Behavioral Medicine* 31 (3):107–15.

Azulay, J., T. Mott, K. D. Cicerone, and C. M. Smart. 2012. An open trial of mindfulness-based stress reduction (MBSR) with a mixed brain injury population. Paper read at International Brain Injury Association's Ninth World Congress on Brain Injury, in Edinburgh, Scotland.

Baer, R. A., G. T. Smith, E. Lykins, D. Button, J. Krietemeyer, S. Sauer, E. Walsh, D. Duggan, and J. M. Williams. 2008. Construct validity of the five facet mindfulness question-naire in meditating and nonmeditating samples. *Assessment* 15 (3):329–42.

Banzett, R. B., H. E. Mulnier, K. Murphy, S. D. Rosen, R. J. Wise, and L. Adams. 2000. Breathlessness in humans activates insular cortex. *NeuroReport* 11 (10):2117–20.

Beck, A. T. 1979. *Cognitive Therapy and the Emotional Disorders.* New York: International University Press.

Bohlmeijer, E., R. Prenger, E. Taal, and P. Cuijpers. 2010. The effects of mindfulness-based stress reduction therapy on mental health of adults with a chronic medical disease: A meta-analysis. *Journal of Psychosomatic Research* 68 (6):539–44.

Chiesa, A., and A. Serretti. 2010. Mindfulness based cognitive therapy for major depression: A systematic review and meta-analysis. *European Psychiatry* 25 (Suppl 1):1045.

Chiesa, A., and A. Serretti. 2011. Mindfulness based cognitive therapy for psychiatric disorders: A systematic review and meta-analysis. *Psychiatry Research* 187 (3):441–53.

Corfield, D. R., G. R. Fink, S. C. Ramsay, K. Murphy, H. R. Harty, J. D. Watson, L. Adams, R. S. Frackowiak, and A. Guz. 1995. Evidence for limbic system activation during CO2-stimulated breathing in man. *The Journal of Physiology* 488 (Pt 1):77–84.

Cramer, H., H. Haller, R. Lauche, and G. Dobos. 2012a. Mindfulness-based stress reduction for low back pain. A systematic review. *BMC Complementary and Alternative Medicine* 12:162.

Cramer, H., R. Lauche, A. Paul, and G. Dobos. 2012b. Mindfulness-based stress reduction for breast cancer—A systematic review and meta-analysis. *Current Oncology* 19 (5):e343–52.

Crumley II, J. S. 2006. *A Brief Introduction to the Philosophy of Mind.* Lanham, MD: Rowman and Littlefield.

Evans, K. C. 2010. Cortico-limbic circuitry and the airways: Insights from functional neuroimaging of respiratory afferents and efferents. *Biological Psychology* 84 (1):13–25.

Evans, K. C., R. B. Banzett, L. Adams, L. McKay, R. S. Frackowiak, and D. R. Corfield. 2002. BOLD fMRI identifies limbic, paralimbic, and cerebellar activation during air hunger. *Journal of Neurophysiology* 88 (3):1500–11.

Fischer, M. J., M. Scharloo, J. J. Abbink, A. J. van 't Hul, D. van Ranst, A. Rudolphus, J. Weinman, K. F. Rabe, and A. A. Kaptein. 2009. Drop-out and attendance in pulmonary rehabilitation: The role of clinical and psychosocial variables. *Respiratory Medicine* 103 (10):1564–71.

Fischer, M. J., M. Scharloo, J. Abbink, A. van 't Hul, D. van Ranst, A. Rudolphus, J. Weinman, K. F. Rabe, and A. A. Kaptein. 2012. Concerns about exercise are related to walk test results in pulmonary rehabilitation for patients with COPD. *International Journal of Behavioral Medicine* 19 (1):39–47.

Fletcher, G. J. O. 1996. Realism versus relativism in psychology. *American Journal of Psychology* 109 (3):409–29.

Gregory, R. L. (ed.). 1998. *The Oxford Companion to the Mind.* (2nd ed.). Oxford, UK: Oxford University Press.

Grossman, P., L. Kappos, H. Gensicke, M. D'Souza, D. C. Mohr, I. K. Penner, and C. Steiner. 2010. MS quality of life, depression, and fatigue improve after mindfulness training: A randomized trial. *Neurology* 75 (13):1141–9.

Howard, C., S. Dupont, B. Haselden, J. Lynch, and P. Wills. 2010. The effectiveness of a group cognitive-behavioural breathlessness intervention on health status, mood and hospital admissions in elderly patients with chronic obstructive pulmonary disease. *Psychology, Health & Medicine* 15 (4):371–85.

Ietswaart, M., M. Johnston, H. C. Dijkerman, S. Joice, C. L. Scott, R. S. MacWalter, and S. J. Hamilton. 2011. Mental practice with motor imagery in stroke recovery: Randomized controlled trial of efficacy. *Brain* 134:1373–86.

Jolley, C. J., and J. Moxham. 2009. A physiological model of patient-reported breathlessness during daily activities in COPD. *European Respiratory Review* 18 (112):66–79.

Kahneman, D. 2011. *Thinking: Fast and Slow.* London: Penguin.

Kievit, R. A., J. W. Romeijn, L. J. Waldorp, J. M. Wicherts, H. S. Scholte, and D. Borsboom. 2011a. Mind the gap: A psychometric approach to the reduction problem. *Psychological Inquiry* 22 (2):67–87.

Kievit, R. A., J. W. Romeijn, L. J. Waldorp, J. M. Wicherts, H. S. Scholte, and D. Borsboom. 2011b. Modeling mind and matter: Reductionism and psychological measurement in cognitive neuroscience. *Psychological Inquiry* 22 (2):139–57.

Livermore, N., L. Sharpe, and D. McKenzie. 2010. Prevention of panic attacks and panic disorder in COPD. *The European Respiratory Journal* 35 (3):557–63.

Logothetis, N. K. 2008. What we can do and what we cannot do with fMRI. *Nature* 453 (7197):869–78.

Moran, A., A. Guillot, T. MacIntyre, and C. Collet. 2012. Re-imagining motor imagery: Building bridges between cognitive neuroscience and sport psychology. *British Journal of Psychology* 103:224–47.

Parshall, M. B., R. M. Schwartzstein, L. Adams, R. B. Banzett, H. L. Manning, J. Bourbeau, P. M. Calverley, A. G. Gift, A. Harver, S. C. Lareau, D. A. Mahler, P. M. Meek, and D. E. O'Donnell. 2012. An official American Thoracic Society statement: Update on the mechanisms, assessment, and management of dyspnea. *American Journal of Respiratory and Critical Care Medicine* 185 (4):435–52.

Pbert, L., J. M. Madison, S. Druker, N. Olendzki, R. Magner, G. Reed, J. Allison, and J. Carmody. 2012. Effect of mindfulness training on asthma quality of life and lung function: A randomised controlled trial. *Thorax* 67 (9):769–76.

Scharloo, M., A. A. Kaptein, J. Weinman, J. M. Hazes, L. N. Willems, W. Bergman, and H. G. Rooijmans. 1998. Illness perceptions, coping and functioning in patients with rheumatoid arthritis, chronic obstructive pulmonary disease and psoriasis. *Journal of Psychosomatic Research* 44 (5):573–85.

Simmons, L., N. Sharma, J. C. Baron, and V. M. Pomeroy. 2008. Motor imagery to enhance recovery after subcortical stroke: Who might benefit, daily dose, and potential effects. *Neurorehabilitation and Neural Repair* 22 (5):458–67.

Tye, M. 1991. *The Imagery Debate.* Cambridge, MA: MIT Press.

Vollestad, J., M. B. Nielsen, and G. H. Nielsen. 2012. Mindfulness- and acceptance-based interventions for anxiety disorders: A systematic review and meta-analysis. *British Journal of Clinical Psychology* 51:239–60.

von Leupoldt, A., T. Sommer, S. Kegat, H. J. Baumann, H. Klose, B. Dahme, and C. Buchel. 2008. The unpleasantness of perceived dyspnea is processed in the anterior insula and amygdala. *American Journal of Respiratory and Critical Care Medicine* 177 (9):1026–32.

von Leupoldt, A., T. Sommer, S. Kegat, H. J. Baumann, H. Klose, B. Dahme, and C. Buchel. 2009. Dyspnea and pain share emotion-related brain network. *NeuroImage* 48 (1):200–6.

von Leupoldt, A., K. Taube, K. Lehmann, A. Fritzsche, and H. Magnussen. 2011. The impact of anxiety and depression on outcomes of pulmonary rehabilitation in patients with COPD anxiety/depression and pulmonary rehabilitation. *CHEST Journal* 140 (3):730–6.

Wanden-Berghe, R. G., J. Sanz-Valero, and C. Wanden-Berghe. 2011. The application of mindfulness to eating disorders treatment: A systematic review. *Eating Disorders* 19 (1):34–48.

Zimmermann-Schlatter, A., C. Schuster, M. A. Puhan, E. Siekierka, and J. Steurer. 2008. Efficacy of motor imagery in post-stroke rehabilitation: A systematic review. *Journal of Neuroengineering and Rehabilitation* 5:8.

6

Rethinking Movement: Postmodern Reflections on a Dominant Rehabilitation Discourse

David A. Nicholls

Barbara E. Gibson

Joanna K. Fadyl

Introduction

Movement is a word that expresses a great many different ideas. Economists talk of the free movement of goods and services; musicians talk of musical movements; people "move on" when they leave a long-held sadness behind; and people literally move in their hundreds through emigration and diaspora; there are labor movements and civil rights movements, clock movements, and biological movements like diffusion and osmosis. Given this diversity of meaning, it is reasonable to ask why it is that in the field of physical rehabilitation, we have taken such a narrow view of movement?* If we acknowledge all the possible ways that we might think about movement, why have we restricted our views of the phenomenon to the language of levers and forces, flexion and extension, and the physical displacement of a body from point A to point B? Where did all the other meanings of movement go?

To some extent we already know some of the answers to this question. From a historical point of view, modern-day physical rehabilitation emerged in parallel with medicine and surgery. Originally a part of the physician or surgeon's craft, physical rehabilitation became a distinct subdiscipline of medicine in World War I and has been a growing branch of orthodox health care ever since (Guthrie Smith 1952; Krusen 1964; Linker 2011). In recent years, medicine has been the subject of widespread sociological scrutiny with authors like Elliot Freidson, Deborah Lupton, Brian Turner, and Simon Williams offering powerful critiques of biomedical science (Freidson 1970; Lupton 1997; Turner 2008; Williams 2006). Since the 1980s, a large volume of literature has emerged that has identified some now familiar motifs about biomedicine: Medicine is hegemonic; medicine is concerned with the body part not the person; medicine marginalizes more holistic ways of thinking about health and illness; medicine commodifies the body; etc. (Annadale 1998; Gabe et al. 1994; Kennedy 1981; Larkin 1983; Osborne 1997; Petersen and Bunton 1997; Rose 1994; Samson 1999). Thus, some of the reasons for our particular view of movement in physical rehabilitation today stem from the influence of medicine and its concern for the body-as-machine (Lawler 1997; Nettleton 2005; Nicholls and Gibson 2010; Shilling 2003; Synnott 1992; Turner 2008).

There are several reasons why an alternative understanding of movement is now needed in physical rehabilitation: first, because health care is changing and so are the desires of the people who deliver, use, and consume its services; second, because some of the aspects of movement that are marginalized by a biomedical view seem to be becoming increasingly important to people's experience of health and well-being; and third, because the social and political context in which physical rehabilitation now operates is no longer the same as that which existed when it first came into being. Health professionals can no longer simply expect state protection of title or territory, access to publicly funded training, or the largely unquestioned ability to apply discipline-specific views of notions of the body, movement, and rehabilitation to patient populations in the name of expert knowledge. A prevailing sense among those who work in health care is that the stability that was once offered by "the health care system" is unraveling. Far from being just a problem, however, the current state of health care also provides a hitherto unforeseen opportunity to question some of the basic assumptions that underpin our practices, and challenge ourselves to think and practice differently.

---

* It should be noted here that our focus in this chapter is on what we have termed *physical rehabilitation*. This umbrella term refers to those forms of rehabilitation that have come to focus upon the form and function of the body and its various anatomical, biomechanical, kinesiological, pathological, and physiological forms. We recognize this is a somewhat arbitrary and undertheorized distinction, but were concerned to distinguish it from the forms of rehabilitation found particularly in addictions and mental-health work, or areas of occupational science where notions of recovery are favored, in which we have little professional experience.

One opportunity open to us relates to the way we have habitually been taught to think as health professionals and service providers. In Western science, and particularly biomedicine, we have learned to value logic and reason. Indeed, it would be literally *unreasonable* to propose thinking otherwise. But learning to think only in terms of logic and reason may stifle creativity and free expression. This, in part, explains why some of the newly emerging medical humanities programs being added to undergraduate medical degrees now focus strongly on nonscientific ways of thinking common in the visual arts, poetry, and music (see, for example, Bleakley et al. 2006).

What benefit could these "alternative" modes of thinking bring to physical rehabilitation? A simple answer to this would be to acknowledge that we do not actually know. We can no more predict the outcome of our investigations with new forms of language, new pedagogies, and new philosophies of movement than we could have anticipated the myriad uses for electricity when it was newly available. But that should not stop us from trying. Indeed, our efforts may need to be a little unreasoned if we are to respond to the complex and diverse questions now being posed by clients, consumers, and service users of physical rehabilitation services.

Fortunately, our endeavors do not operate in a vacuum, and in recent years, a growing number of writers and theorists have begun to explore the territory at the margins of orthodox practice. Our plan for this chapter then is to utilize some postmodern approaches and apply these to current ways of thinking and practicing physical rehabilitation. We acknowledge that the approach to movement that is questioned here is indeed dominant in physical rehabilitation, and that responses such as the ones we suggest here are partial. Our hope is that it gives others the encouragement to seek playful postmodern responses to their own concerns, offering opportunities for thinking and practicing differently.

Starting with a broadly linguistic reappraisal, we look at what might be possible if we restore some of the language of movement to its fullest meaning. What, for example, might change in our practice if we reappraise how we have come to use words like *articulate*, *eccentric*, and *mobilize*? From here we look at the possibilities for a broader philosophically and sociologically informed approach to physical rehabilitation, before exploring the works of five postmodern writers: Paul Virilio, Erin Manning, Gilles Deleuze, Felix Guattari, and Michel Foucault. We draw on these authors to examine how aspects of their work might be applied to rehabilitation thinking and practice.

Throughout the chapter, we have tried to offer a plurality of viewpoints in order that we might remain open to many different responses to the problems now being posed of physical rehabilitation. This is a deliberate approach in keeping with some of the fundamental principles of postmodernism: an approach that we believe offers some interesting and powerful ways in which we might rehabilitate movement. We begin the chapter with a précis of some of the reasons for taking an unreasoned approach to the meaning of movement.

Limits of Current Biomedical Approach to Rehabilitation

Critique of biomedical reasoning generally, and more specifically applied to the field of rehabilitation, is now extensive and can be explored in much greater detail elsewhere (see, for example, Armstrong 1983; Fusco 2006; Kendall and Clapton 2006; Lupton 1997;

Markula and Pringle 2006; Seymour 1998; Sullivan 2005; Turner 2008; and other chapters in this book including Chapters 3, 4, 7, 10, and 14). Broadly speaking, this criticism is both epistemological and ontological in that it addresses some of the limitations in the way people's experience of illness and injury is understood in biomedicine. Critics argue that the "reality" created by biomedicine and its application to rehabilitation is dominated by a scientific interest in the evaluation, determination, and objectification of disabled people and the value-neutral pursuit of objectivity (Corker and French 1999; Fisher and Goodley 2007; Goering 2008; Shildrick and Price 2002; Tremain 2001). It is argued that biomedicine:

- Sees the person/patient/other in primarily mechanistic terms—be it through allopathic, anatomical, biomechanical, physiological, or pathological principles
- Privileges a focus on the biological body rather than the unique individual in a particular context and places too much emphasis on abstract, detached understandings of health
- Pays insufficient heed to cultural, economic, environmental, existential, geographical, political, and social understandings of health and illness
- Has resulted in the formation of a set of privileged groups within society—the health professionals—who claim specialist knowledge of the functioning of the body, which in turn perpetuates notions of health and health care to fit that model (Abbott and Meerabeau 1998; Evetts 2006; Freidson 2001; Light 2000)

Notwithstanding the contributions made in the name of biomedicine over the last century, and despite changes in medicine and rehabilitation that have shown a greater concern for patient- and family-centered care, quality of life outcomes, and function-versus impairment-based interventions, the central tenets of biomedicine and the "patient as a constellation of symptoms" still dominate physical rehabilitation. In this chapter, we take a different line—to apply a postmodern approach to the problems of rehabilitation as presented to us by biomedicine—to see if, in doing so, we might inspire others to explore the possibility of engaging with rehabilitation differently.

Our approach centers on the ways in which we have come to understand movement—movement being one of the quintessentially reductive qualities most highly valued in physical rehabilitation. Importantly, we are not attempting to discard a biomedical view of movement and rehabilitation only to replace it with another equally problematic approach. Rather, we are attempting to create options for those who engage in physical rehabilitation, explore the possibility of thinking otherwise, enrich the experience of rehabilitation, and reveal some of the fullness of its potential as a vital component in the lives of thousands, possibly millions, of people. To this end, we begin the main body of our paper by playing with some of the ascribed meanings within the language of physical rehabilitation.

Moving Language

If thought corrupts language, language can also corrupt thought.

George Orwell

Our first response to the meaning of movement considers the language of rehabilitation and asks what might be possible if we subvert the commonly accepted meanings given to its vocabulary. If we acknowledge that language plays a vital role in defining the way we think and practice, it would not be unreasonable to argue that new practices might emerge from either new language or reinterpretation of existing terms.

If we look first at the way our present practice lexicon has been formed, we find that the language of movement and rehabilitation has its own specific etymology. David Armstrong's work, for example, explores the politics of posture and movement (Armstrong 2002). By examining how firstly posture, and then later exercise, became political concerns for health reformers in the early part of the twentieth century, Armstrong shows that the control over people's bodies became an integral part of industrial, military, and social reform. Importantly, the language of posture and movement played a vital role in linking bodily form and function with the emerging disciplines of psychology and sociology. Schoolyard drilling and synchronized group exercise were used extensively to both inculcate a specific set of moral and social values, but also to reveal those whose posture or movements fell outside the norm. A child's ability to stand to attention—a highly abstracted "normal" posture—inferred a child's attitude toward authority. Attention and attitude subsequently became highly significant terminology in the fields of psychology—inferring not only a particular physical form and function but also acting as a pointer to "problem" children who needed to be brought into line. Armstrong's work is particularly interesting because it exemplifies a growing body of scholarship that challenges our familiarity with ideas that have their own very specific historical and social conditions of possibility. Much, if not all, of the language now in use in physical rehabilitation deserves the same critical scrutiny, because we may find that there are cultural, moral, political, and social "truths" implicit in the historical emergence of notions such as exercise, fitness, manipulation, and treatment that are worthy of further examination.

As with Armstrong's work, we are interested in the historical and social significance of language—what words make it possible to think and do—rather than a linguistic analysis of the language of rehabilitation. We do, however, recognize that for words to become meaningful to practice, they must be both specific to rehabilitation *and* carry some general meaning. Words like *balance, manipulate, occupation, stability,* and *treat* are commonplace in rehabilitation, but they are words that also carry broader social meanings that sometimes blur and bleed into one another, and this blurring shapes our thoughts and practices. The word *balance,* for instance, means a person's ability to maintain their center of mass over their center of gravity, but it also infers personal or psychological stability and an ability to remain upright and steady. This more general definition may at times subtly influence the context in which the word is used in physical rehabilitation.

We are always searching for definitional clarity in rehabilitation, always looking for clearer definitions for terms to define anatomical and physiological signs, patient symptoms, disease classifications, and modes of assessment and treatment. Underpinning this search are the reductive tendencies of biomedicine, which draw on over four centuries of Enlightenment thinking that, in turn, values the search for the fundamental laws of nature. Postmodernism seeks to challenge the assumptions that underpin the ingrained principles of western science. One way it achieves this is to reverse the tendency to reduce, clarify, and finalize language, preferring instead to encourage diversity of meanings, inclusiveness, and complexity.

While developing this chapter, we engaged colleagues in a postmodern experiment. We gave them words that were common to rehabilitation and asked them to define how they used the particular word in their rehabilitation practice. We then asked them to put these definitions aside and to think instead about possible alternative meanings for the words. We asked them to be playful and unconstrained in their interpretations. Some of the results of this exercise are included in Table 6.1.

What becomes clear from such an exercise is that words used in rehabilitation are drawn from common language and have common meanings that are broader than their use in movement sciences. It is our belief that apart from asking how the words used in rehabilitation became delimited—as David Armstrong (2002) and others have done— we should ask what is gained and what is lost when we strip words of the fullness of their meaning and choose to use only part of the whole. In rediscovering the social history of this process of linguistic "refinement," we can discover how words and phrases have come to be dominant and what the effects of this domination might be. And most importantly, given the present economic and political tensions surrounding rehabilitation, we should ask what might be gained by broadening the meanings of the words we use in rehabilitation and introducing new vocabulary that we might have rejected before.

There are many parallels for this process, possibly no more apposite than in the world of design engineering, which is a discipline whose sole purpose is to create things that *will be* from the things that currently *are*. At the heart of this process is a creative playfulness that relies on certain openness toward instability and uncertainty. As H. G. Wells (1904) wrote:

> Think of arm chairs and reading chairs and dining-room chairs, and kitchen chairs, chairs that pass into benches, chairs that cross the boundary and become settees, dentists' chairs, thrones, opera stalls, seats of all sorts, those miraculous fungoid growths that cumber the floor of the Arts and Crafts Exhibition, and you will perceive what a lax bundle in fact is this simple straightforward term. In cooperation with an intelligent joiner I would undertake to defeat any definition of chair or chairishness that you gave me.

Following on from H. G. Wells, we could argue that we have many "chairs" in rehabilitation: in other words, terms that carry the weight of specific meanings, and we have sought, over the years, to limit the function of these chairs to specific purposes. In our first response to the meaning of movement, we have advocated a different approach: an approach that looks to "defeat the definition of chairishness" that we have been given and ask what might be possible if we created the space for some new chairs to emerge in rehabilitation. We now move on to explore how movement might be thought of as something more than physical displacement. How might our approach to movement vary if we expanded our interest in the specific language of rehabilitation to take in the sociopolitical possibilities of movement?

TABLE 6.1   Comparing the Use of Established Rehabilitation Words with Some Possible Expanded Meanings

Word	Established Usage in Rehabilitation	Possible Expanded Meanings
Abduct	To move a limb or body part away from the midline	Kidnap, steal away, run away with someone/something, to snatch; to decrease or remove something; to twist or curve; something auxiliary; moving away from the midline position or convention; modes of resistance and deviance; something taken or stolen
Articulate	To have joints or jointed segments; to form a joint or to be connected by joints	To state your position clearly, venting, letting off steam, explaining things, quick witted, parrot, disambiguate; articulations allowing for movement or a significant change of direction; intersections, junctions, places that allow us to flex, extend, rotate, or change direction, often in the middle of otherwise rigid structures (like rehabilitation itself perhaps?)
Eccentric	A muscle that elongates while under tension	Being outside of your own center, ex-centric, radical, different, abnormal quirky; when forces bringing about movement are greater than the forces of resistance; a controlled "letting go"; acceding to greater forces, partial control; a braking force and an exuberance that defies authority and jurisdiction; acceptance of the need for new positioning
Exercise	An activity requiring physical effort, carried out especially to sustain or improve health and fitness	A test of one's skill, a task designed to allow for practice, military drilling, or training; an activity designed specifically to increase one's fitness; to achieve an optimal state; to use a faculty, or a right or process—to exercise one's mind; an activity that applies equally to mental and physical work; purposeful movement, deliberate displacement, capable, but capable of being optimally healthy and also worried or perplexed
Fit	In good health, especially because of regular physical exercise	Something more than merely "well"; lean, tanned, and sexy; attractive, appealing, good-looking; better than normal, athletic, superior, but also, at times, simply appropriate for purpose; the right size or shape to occupy a position or situation, competent, correct, suitable, and of an acceptable standard; fit for task; to fix something into place, or to attach something new to a body or component; join two things together to make a whole, produce a match to bring about harmony, a correlation
Manipulate	To feel and passively adjust the alignment of segments of the body with the hands or with the use of tools, implements, or machines	To handle or control someone cleverly, unfairly, or unscrupulously; to get what you want through devious means; to control a passive object like a tool; to adjust something to achieve a more desirable outcome than was achieved spontaneously or naturally

(Continued)

TABLE 6.1 (CONTINUED)   Comparing the Use of Established Rehabilitation Words with Some Possible Expanded Meanings

Word	Established Usage in Rehabilitation	Possible Expanded Meanings
Mobilize	To move a body part, region, or structure	To organize a group for active service or to act in a concerted way in order to bring about a particular objective or goal; to move something, to displace it, to take it from its original position and disturb or perturb it; make something movable or capable of movement, free something from bondage, liberate it either partially or wholly use one's own energy to free something, or set something adrift; make something able to be transported by or as a liquid, carried away when once it was fixed or static
Perturb	To displace an object from its normal state or path	To unsettle, bother; influenced to change direction with the application of external force—physical or otherwise; force something to become irregular—to deviate, to be different
Posture	A particular static arrangement of the body that conforms to accepted norms for bodily alignment	A static position, fixed in time, a precursor to movement; a statue; an attitude or visible expression of one's inner thoughts; affect expressed through the alignment of body structures; nonverbal communication; a signal; a pose, as with a model, designed to be alluring, bold, passive, threatening, etc.; a discipline imposed on the body for social reasons (aesthetics, status); a stance one takes on particular issues, a point from which one communicates one's views to others; a deceit; a behavior designed to convey a false impression—being defiant in the face of a challenge, for example
Stretch	The act of elongating one's limbs or body	Something elastic and giving, something that can be elongated without tearing or breaking (a relationship, or piece of material, perhaps); yielding; achieve full length, enduring the extra tension involved; purposeful movement toward something, requiring will and effort; straightening; spreading out, increasing one's size and reach over an area over time; an expanse of territory (water or land); something taken to extreme (stretch limo)—an exaggeration on an original design, something interposed that exaggerates the original form; plasticity of form and time; lived time and space; marks left on the skin, signs of growth and expansion; distortion; never returning to one's original state; an idea that goes a little too far; to have more surface area but less depth
Treat	Provide skillful therapeutic intervention designed to bring about relief or cure	Gifting something to someone at one's own expense; an attitude toward someone or something—a way of behaving; a treaty—a document that sets out the way parties will be treated; out of the ordinary, a surprise or special event that serves to remind someone of their particular state or status; look after, work with, reward; doing something to someone for fair means or foul; to manipulate X to obtain Y

Multiple Movements

If we open up to multiple movements; to movements that are not only physical, but social, cultural, emotional, and even musical, we begin to see that our present rehabilitation approaches are largely restricted to biomechanical interpretations of the physical displacement of the body or its various parts. Engaging with the question common to critical and postmodern approaches to research, we are encouraged to ask "why is it so?" How has it become historically and socially possible to speak about movement in such limited terms? What "machinery" (i.e., structures, systems, conventions, etc.) needs to be in place to encourage us to continually privilege *these* biomechanical interpretations of movement over *other* possibilities? And more politically, whose interests are served in perpetuating these narrow linguistic conventions?

Postmodernism seeks to critique the narrowing of meaning and open spaces for more diversity. Thus, our second exercise extended out from the earlier word play, to consider other ways in which we might diversify our approach to movement. How might we reconceive the value of movement in physical rehabilitation through, for example, an exploration of other movements in the arts, cultural studies, history, philosophy, politics, or sociology? In cultural political studies, for example, a great deal of attention has focused on the notion of migration and diaspora—or the dispersion of people from their original homeland. In recent times, much greater interest has focused on the consequences of migration on people's health and the effects of migration on indigenous populations (Falge et al. 2011; Liamputtong 1999). But as yet, little work has been undertaken in the health sciences to examine how migration represents a movement of peoples: a collective, purposeful energy flow, driven by need or want, that carries with it notions of loss and displacement. Somehow we seem to have lost the ability to see the metaphorical significance of notions like detachment and dislocation in physical rehabilitation, preferring instead to reduce these ideas down to the level of bones and joints, muscles and tendons. And while we would argue these are entirely valid ways of thinking about structural integrity, we might also gain something from thinking about how these words carry cultural, existential, philosophical, political, social, and spiritual significance for individuals and communities that, if acknowledged, could open up entirely new dimensions to our rehabilitation practices.

Importantly, we would need to do very little work to reimagine how we have come to use these ideas, because a great deal of work of exploration has already been undertaken by people in other disciplines. For alternative notions of migration, for instance, we might turn to fictional writing, cinema, music, and the theater for which notions of travel, discovery, and the possibilities or anxieties over new encounters are constantly recurring motifs. Why is it that such works rarely touch on the everyday practices and thinking that takes place in physical rehabilitation? Why is it that biomechanical definitions of movement often displace and marginalize these more humanistic or artistic expressions of movement? What might be gained, therefore, from opening our practices up to artistic, historical, and social notions of movement?

Movement as a (Postmodern) Metaphor

Our third exercise expands our notion of movement even further to consider how we might think of movement as a metaphor. In physical rehabilitation, we frequently use metaphors to express the complexity of what could not be captured otherwise. In many ways, metaphors represent the art of physical rehabilitation and they approximate with the more experiential dimensions of movement. When we speak of momentum, goals or progress, or growth and development, we are often speaking of the material objective reality of a person's physical aptitude or performance, but we are also keeping one eye on the more ethereal aspects of these phenomena, and thus drawing on the power of metaphor to give the particular notion additional weight of meaning. Momentum, for instance, is meaningful to us because it infers a measurable and ongoing displacement of the body in space, but also forward progress, improvement, and recovery. Thus, the language is useful because it allows practitioners to connect with notions of scientific objectivity while also connoting the possibility of broader future achievement.

Not surprisingly, given the weight of significance that they carry, metaphors are used widely by postmodern theorists. But rather than preserving their original meanings, it is more common for them to be appended or subverted and given new, subaltern, and multiple meanings. These alternative uses of metaphors are deliberately designed to subvert and disrupt common meanings and understandings, and force us to confront the often unquestioned obviousness of our language. In postmodern research, words are deployed for their double meanings, which encourage us to consider the possibility of thinking otherwise. Words like *conduct* and *discipline* have been used extensively by authors like Michel Foucault for their double meaning. Foucault (1977) argued that many of his earlier writings were concerned with the "conduct of conduct," and that he was interested in discipline as a "technology" as it had been applied to the discipline of medicine. Elsewhere, Christina Papadimitriou and David Stone refer to their work with wheelchair users and their experiences of becoming "enwheeled" (Papadimitriou 2008), and we, ourselves, have written about the transgression of notions of touch and new possibilities for "connectivities" in the field of disability (Gibson 2006; Gibson et al. 2012a; Nicholls and Holmes 2012).

What these various metaphors point to is a concern with liberating movement from its present ossified position and, in doing so, decentering it, displacing it, and creating uncertainty that carries with it its own kinetic energy. To pursue this idea further, we briefly explore the work of five postmodern philosophers, who have all used metaphors of movement. Our belief is that these works, and others besides, have a place in new thinking about rehabilitation. We should be clear here that these are only brief samples from the extensive works of these philosophers and are not attempts to offer tidy summaries of their entire oeuvres.

Virilio and Perpetually Increasing Speed

Speed means a great deal in rehabilitation science. From the speed of a muscle fiber twitch to the speed at which a person can perform an assessment task like the Timed Up and Go or the Six Minute Walk Test, speed is one of the defining characteristics by which we assess people's functional capacity. However, our understanding of speed as a measurable, quantifiable experience has been challenged in recent years by phenomenologists who argue that science has given us an abstract and limited measure of time passing, which seems inadequate when one considers how elastic time appears to us in the "real" world. Sometimes the seconds race by and at other times they drag because our perception of our engagement with the world progresses at a different speed to that which can be measured by a clock or stopwatch. This notion of "lived time" has been an important focus for the study of people's lived experience of illness (Churchill and Wertz 2001; Van Manen 1990). Writers like Barbara Adam (1995) have written about "shadow times"—the times lived through activities such as caring about, for, and with others. Adam argues that these times are constituted outside of the commodified time of the clock and are oriented less to the progression of measurable future goals than to the demands and rhythms of the present. Davies (1997) and others have used the term "the philosophy of the present" to describe how some people come to develop an appreciation of the present that resists dominant notions of linear time and embraces uncertainty when coming to terms with illness (Ezzy 2000; Fisher and Goodley 2007; Gibson et al. 2009).

Paul Virilio, an existential phenomenologist, has built on the notion of lived time by reflecting on the effects of technology on our lives. Virilio's notion of *dromology*, or the critical question of perpetually increasing speed that has beset our society, raises questions about how important speed has become in our hypermodern times. Virilio asks what it is like to live as a human "caught up in the technocultural vectors of ever increasing speed" (Armitage 2000). Our desire to be ever more efficient in our work, ever more ambitious in our exploration of our world, and ever more aggressive in pursuit of our goals, has resulted in us dashing from one activity to another, traversing the globe at the click of a mouse, and condensing the journey between our objectives. The journey between starting and ending points now occupies the smallest possible time and is compressed until, ultimately, there will be no distinction between where we are now and where we want to be next.

Virilio's "relentless logic of speed" (Armitage 1999, p. 6) offers a number of challenges to rehabilitation theorists and researchers. It asks how our fantasy of a hypermodern world is being reflected in our expectations of people in rehabilitation. How, for example, do people engage meaningfully in the world when their ability to move at the required speed has been hampered by injury or impairment (and here movement is both the movement of the body in space and also the metaphorical movement toward a full and meaningful engagement in society)? At the same time, how are we using people's inability to move at the required speed as a means to differentiate between ability and disability, or good and bad outcomes?

Zygmunt Bauman argues: "Speed of movement has become today a major, perhaps the paramount factor in the social stratification and the hierarchy of domination" (Bauman 2001, p. 247). How does the rehabilitation rhetoric of *goals* and *progress*—even the notion of *re*-habilitation itself—allow for and even perpetuate social segregation, that sees disabled people constantly excluded from the nondisabled world? Hansen and Philo (2007), in researching disabled people's experiences of "timings and spacings," showed how individuals are disadvantaged when their bodies do not conform to the putatively "normal" ways of doing things embedded in everyday places and spaces. The authors underlined the difficulties encountered when deceptively straightforward tasks like dressing take up large amounts of time, or if rushed cause pain and stress. Participants found little tolerance for their altered rhythms and timings in the workplace and other environments. Similarly, Kevin Paterson, who has a communication impairment secondary to cerebral palsy, has eloquently described how culturally embedded norms of timing and movement privilege certain bodies while excluding others (Paterson and Hughes 1999). Using the example of being posed a question in an elevator and the impossibility of answering quickly enough before having to exit, Paterson suggests that the "duration norm of this particular communication was not commensurate with my carnal needs" (Paterson and Hughes 1999, p. 606).

These examples reflect how imposed notions of normal time and rhythms embedded in common everyday tasks serve to marginalize and exclude individuals as pervasively as accessibility barriers like curbs and stairs. Present rehabilitation practices do little if anything to challenge these norms and may even reinforce them through interventions aimed toward normalizing performance on the basis of time- and space-dependent outcomes. Imposing temporal and spatial restrictions that reflect the needs of the able-bodied, dromologically dominated world may be reinforcing normalized notions of bodily differences instead of embracing diversity and inclusiveness. (See Chapter 14 for a related discussion of human temporalities and the implications for rethinking rehabilitation.)

Manning and Pre-Acceleration

When two experienced dancers come together, it sometimes appears as if they move as one. Their movements flow, their gestures mirror each other, and they seem to be able to anticipate what the other person will do and respond accordingly. The same thing occurs, at times, in martial arts and in manufacturing, but it can also occur in interactions between a therapist and a patient. We are talking here specifically about what happens in the brief instance immediately before a movement or mobilization occurs. In this moment—which may pass in the blink of an eye—we experience astonishing complexity, as both parties silently process thousands, possibly millions, of physical, physiological, emotional, and communicative cues. In many ways, we can say that when movement actually occurs, we have resolved this complexity into a single act and the period of "undecidability" has passed. It is in the moment immediately *before* movement occurs, therefore, where the greatest creativity exists and where our options are virtually limitless. This moment is what Manning (2009) calls the moment of "pre-acceleration."

Pre-acceleration is the "virtual momentum of a movement's taking form before we actually move" (Manning 2009, p. 6), and it folds together our thoughts, feelings, memories, and concepts of being embodied, and moving into the biological/physical processes that convert incipient movement into actual displacement. It is a moment of infinite possibility when our creative potential is at its highest. When I move to take a step, anything is possible, "[B]ut as the step begins to actualize, there is no longer much potential for divergence: the foot will land where it lands" (Manning 2009, p. 7). Pre-acceleration, therefore, concerns the assemblages—the constantly changing connections—we make between our "selves" and the world of beings and things around us, and the opening to possibility that exists immediately before we act; "Pre-acceleration [is] a movement of the not-yet that composes the more-than-one that is my body" (Manning 2009, p. 14). Of course, nothing remains static at that point of creative potential. There comes a moment when the "not yet" actualizes. But what Manning makes us aware of is that too often we miss the opportunity at the point of pre-acceleration, because we are always already at the moment of actualization.

Often in rehabilitation, we concern ourselves with the therapeutic acts and interventions that, in Manning's terms, represent only the most prosaic, terminal, and least creative stage in the process of therapeutic engagement. Our focus in rehabilitation is often on becoming skilled in the execution of techniques or strategies, and our emphasis is on quantifying the effect of those actions by evaluating their outcomes, rather than the processes that made them possible. In our orientation toward achieving a range of defined outcomes (such as range of movement, return to work, discharge goals), we are closing off possibilities and creative solutions by presuming what will be, often even before it happens.

One way to challenge ourselves to reconsider the possibilities in the process, and take the focus away from having particular outcome expectations, is to consider different ways of evaluating what we do. To highlight one possible direction, some practitioners and researchers propose that we put more focus on what opportunities rehabilitation opens up, shifting the emphasis from aiming for specific, measurable outcomes to thinking more about enabling possibilities within the context of people's lives. For example, Gibson et al. (2012b) offer a different way of looking at what constitutes a suitable home environment for people experiencing disability, evaluating the person–environment match in terms of opportunities for access to things such as meaningful relationships, community life, and self-expression. This type of evaluation reorients us to the idea that opportunity and possibility are important objectives, rather than focusing on achieving certain defined "outcomes." (For a further discussion of other ways to rethink evaluation, see Chapter 3.)

Deleuze and Guattari and Assemblages

Erin Manning's notion of pre-acceleration makes a strong connection with the philosophical writings of Gilles Deleuze and Felix Guattari who, for their part, draw heavily on movement metaphors in their writings (Deleuze and Guattari 1983, 1987). The authors discuss the virtues of "nomadic" and "rhizomatic" approaches to thinking, "smooth and striated" space, and the lines of flight that become possible when one resists the constraints of motionless, or territorialized, thinking that has become prevalent in Western science.

Deleuze and Guattari play on the notion of smooth and striated muscle, but instead use smooth and striated space to refer to the way in which we as human beings experience and engage with the worlds we occupy. For them, striated spaces are hierarchical, rule-intensive, strictly bounded, and confining, whereas smooth spaces are open and dynamic and allow for transformations to occur (Deleuze and Guattari 1987, p. 486). Striated space dominates present-day health care with its systems and structures designed to label, examine, identify, and normalize difference. Smooth space, on the other hand, removes these restrictions and allows for people to engage in multiple assemblages, which make possible a thousand "becomings," or new and multiple ways of being that we perpetually move in and out of. For Deleuze and Guattari (1987), "all becoming occurs in smooth space" (p. 486).

By rethinking the assemblages we make in rehabilitation, it may be possible to allow for new becomings for our patients and ourselves. In assemblage theory, entities make connections with one another and, as with Manning's dancers, open up the possibility of new becomings (DeLanda 2006). Thinking of ourselves as therapeutic *explorers*— being open to previously unimaginable possibilities offered by moments of therapeutic assemblage—may encourage us to resist the urge, so powerful in rehabilitation dominated by Western science, to restrict these possibilities with the rule-bound structures of striated space. In doing so, we might allow for, indeed encourage, new cultural, personal, political, and social possibilities that resist our limited existing narratives surrounding illness and disability.

For example, discourses of dependence and independence are pivotal to the present operation of physical rehabilitation. These discourses rely on temporally stable definitions of functional performance that are used to classify people as "disabled" or "able-bodied." The idea of multiple assemblages that we move in and out of, however, provides an alternative to these discourses (Gibson et al. 2012a). Assemblages formed between a body, a wheelchair, and a place are not permanent states but temporary connections that enable certain actions and activities and constrain others. To refer to a person as "wheelchair dependent" carries with it stable notions of ability and disability that belie the myriad possibilities of a body–wheelchair–place assemblage. A wheelchair can provide significant advantages in one time/place, i.e., making movement easier or faster, and disadvantages in others, if movement is limited by stairs, for example. Thus, talk of dependence and disability becomes reconfigured as moments of possibility realized through multiple assemblages. Assemblages are fluid states that are neither inherently negative nor positive but rather *movements* with effects. Rehabilitation shifts to a process whereby people are helped to live well through the multiple assemblages they make with the world.

Foucault and Resistance

The notion of resistance is one of the defining features of postmodernism (Cheek 2000; Fox 1999), and Michel Foucault offers one of the most compelling explanations for the positive function of resistance in society. Contrary to the way many people have understood resistance in the past, Foucault saw it as the corollary of power, where power is not a negative oppression or control over people, but a creative force that makes things possible. According to Foucault, power and resistance are present where people have choices, and the ability to convince people to choose one way over another, without recourse to force, is the power that people contest. Much power derives from the ability to persuade someone that one viewpoint, one argument or act, is preferable to another. Thus, control of certain discourses, or particular forms of knowledge, lies at the heart of Foucault's notion of power. Power, therefore, exists where choices are made, and can only exist where people are free to choose. Power only exists where there is the possibility for resistance (Foucault 1977).

This is a very different notion of power than the more common idea that one group has power at the expense of others. Marxists and feminist scholars, for example, have long argued that there are people in society who have power and there are others who do not. Some speak of "empowering" others and giving voice to the "powerless." Some strands of the disability rights movement assert that the medical system creates binary states where some are able bodied and others disabled; some are healthy and others sick; and some who are mad and others sane. The medical system, it is argued, reinforces social order by giving power to those who create and maintain these social structures (doctors and other orthodox health professionals) and creates powerlessness in those who they administrate (patients, subjects in medical experiments, etc.) (Swain et al. 2003). Foucault challenged these ways of thinking about power and argued instead that power is everywhere and cannot be possessed by one at the expense of the other; that no one is ever completely powerless.

Foucauldian scholars look for the ways in which power, resistance, and knowledge operate in society to construct what we have come to see as the "truth"—or what Foucault (1984) called the "truth games" that we play. Foucault saw the truth as something that was historically and socially constructed and not as a set of empirical facts. The truth was the assimilation of a number of competing, overlapping, sometimes symbiotic discourses that made some things thinkable, while denying others.

As an example, some people have drawn on Foucault's work to critique the notion of "disability," opening space for resistance of this notion, and for stimulation of other ways of approaching the issues it broaches. In "On the Government of Disability," Shelley Tremain used Foucault's work to show how the methods by which equal participation in society has been pursued can be seen to reinforce the idea that disabled people are different from other people (Tremain 2001). She argued that the approach of defining what disability is and isn't, *identifying* certain people as "disabled," and making special allowances and provisions based on that identity actually makes "disability" *more* distinct and visible (Tremain 2001). If allowances and provisions are based on whether someone is labeled as disabled, this obliges people to constantly ask questions like "what is disability" and "is this person disabled"—keeping the notion of "disability" at the forefront of people's thinking and interactions. Rehabilitation is an illustration of this concept because eligibility for rehabilitation is often based on whether somebody is considered disabled. For example, speech therapy and communication supports may be offered to somebody who has been given a diagnosis of a speech disorder. However, a person who has an accent that people find difficult to understand may experience similar difficulties with verbal communication but is not eligible for the same services. What can be seen here is that because the emphasis is on ensuring that people identified as disabled are not disadvantaged, rather than identifying situations of disadvantage and mitigating them, it is common to find that solutions focus on disabled individuals rather than the situations that disable them. This can have the effect of highlighting and separating out disabled people, and overlooking situations of disadvantage that do not fit the definition of disability.

There have been movements in recent years to try and shift the focus away from disability and promote the idea that the emphasis should be on acceptance of and allowance for the diverse range of human functioning. An article by Patston (2007), for example, introduces a framework of "constructive functional diversity," offering an alternative to the idea that certain people are "impaired" in their functioning (and therefore disabled) and others are not. Constructive functional diversity promotes the idea that some ways of functioning are more common than others, and that as a society it is important to recognize and accept those that are less common and allow for people to function in diverse ways. When people encounter situations in which their way of functioning causes them difficulty or disadvantage, they experience "functional dissonance" (Patston 2007). In Foucauldian terms, this can be seen as a competing discourse and resistance to the notion of disability. As rehabilitation practitioners, we too have the opportunity to pause and consider how notions of disability structure our practices, and how concepts that resist disability, such as functional dissonance and accessibility, might move our practices in different directions. These ideas might call into question who is considered eligible for rehabilitation, or what rehabilitation could focus on addressing.

Discussion

The meaning of movement is the very movement of meaning.

Erin Manning

Envisaging a new future and new ways to think about the body, health, and physical rehabilitation is challenging. Our approach here has been to take a postmodern approach to the possibility of thinking otherwise about our practice. Change in rehabilitation can be facilitated through changes in language. Old words and phrases can be reimagined to realize the shift in our thinking. And so our first section played with the idea that we might see new value in our existing lexicon of movement, and through it imagine new practices and approaches. We also imagine that we will be required to see rehabilitation in a broader context, and so we have tried to imagine more philosophically- and sociologically-informed approaches to practice. This culminated in our exploration of the works of five notable postmodern philosophers and a brief encounter with some of their ideas. These authors offer two things: firstly, they provide some innovative ways of thinking about the world to stimulate our creativity; and secondly, and equally as valuably, they provide concrete clues as to how we might practice in the future.

As with all the research work that we have cited here, the opportunity to think "otherwise" runs through the heart of this chapter. In the past, professionals who have engaged in physical rehabilitation have been encouraged to follow some quite heavily defined pathways that have enhanced the consistency and reliability of our work, but have stifled some of its creativity. From standardized curricula and limiting scopes of practice to prescriptive patient care pathways and evidence-based practice, our work has developed a certain logic that has enhanced the predictability of our work and helped to reassure the public and our colleagues that we can be trusted, but has, at the same time, reduced our capacity for innovation and enterprise.

Two of the cornerstones of postmodernism are a criticism of the dogma that sometimes underpins people's unreflective adherence to order and predictability, and an opening toward diversity and inclusiveness. So it is inevitable that much of what has been written in this chapter will run counter to the conventional way of seeing things in present-day physical rehabilitation. But we have arrived at such a point of tension in health care now, where people are much less enamored by the orthodox health professionals and are actively seeking alternatives, where the welfare state that was the bedrock of twentieth-century rehabilitation is being slowly dismantled, and where people are increasingly engaged in diverse "body projects" in order to stay youthful and active in old age, that some people are beginning to talk about a future for rehabilitation that is quite different from its past.

Conclusion

In this chapter, it has been our contention that one way we might reimagine rehabilitation is through the use of postmodern ideas and approaches. We have looked at ways in which we might redefine the language of movement. We have also examined how we might re-examine the historical and sociopolitical possibilities of movement. And we have used the work of key postmodern philosophers to explore ways in which we might move movement forward.

Throughout the chapter, we have tried to open space to think differently about movement. Our intention is not to dismiss current ways of thinking only to impose new restrictions or replace one system of knowledge with another. Rather, our goal has been to critique the way we presently think about movement in a way that creates the possibility of a thousand new responses, a thousand plateaus from which we might now leap (Deleuze and Guattari 1987).

Acknowledgments

In completing this chapter, Barbara E. Gibson was supported by a Canadian Child Health Clinician Scientist Program Career Development Award and Joanna Fadyl was supported by a Health Research Council of New Zealand Disability Placement Programme PhD Award. The authors acknowledge the support of a range of colleagues and students who contributed to the production of Table 6.1.

References

Abbott, P., and L. Meerabeau. 1998. Professionals, Professionalization and the Caring Professions. In *The Sociology of the Caring Professions*, eds. P. Abbott, and L. Meerabeau, 1–19. London: UCL Press.
Adam, B. 1995. *Timewatch*. Cambridge: Polity Press.
Annadale, E. 1998. *The Sociology of Health and Medicine*. Cambridge: Polity Press.
Armitage, J. 1999. Paul Virilio: An Introduction. *Theory Culture Society* 16:1–23.
Armitage, J. 2000. The Uncertainty Principle: Paul Virilio's 'The Information Bomb'. *M/C: A Journal of Media and Culture* 3(3). Available at http://journal.media-culture.org .au/0006/virilio.php (accessed October 23, 2014).

Armstrong, D. 1983. *Political Anatomy of the Body: Medical Knowledge in Britain in the Twentieth Century*. Cambridge: Cambridge University Press.

Armstrong, D. 2002. *A New History of Identity*. London: Palgrave Macmillan.

Bauman, Z. 2001. *The Individualized Society*. Cambridge: Polity Press.

Bleakley, A., R. Marshall, and R. Brömer. 2006. Toward an Aesthetic Medicine: Developing a Core Medical Humanities Undergraduate Curriculum. *Journal of Medical Humanities* 27:197–213.

Cheek, J. 2000. *Postmodern and Poststructural Approaches to Nursing Research*. Thousand Oaks, CA: Sage.

Churchill, S. D., and F. J. Wertz. 2001. An Introduction to Phenomenological Research in Psychology: Historical, Conceptual and Methodological Foundations. In *The Handbook of Humanistic Psychology: Leading Edges in Theory, Research, and Practice*, eds. J. F. T. Bugental, K. J. Schneider, and J. F. Pierson, 247–262. London: Sage.

Corker, M., and S. French. 1999. *Disability Discourse*. Buckingham: Open University Press.

Davies, M. L. 1997. Shattered Assumptions: Time and the Experience of Long Term HIV Positivity. *Social Science and Medicine* 44:561–571.

DeLanda, M. 2006. *A New Philosophy of Society: Assemblage Theory and Social Complexity*. London: Continuum.

Deleuze, G., and F. Guattari. 1983. *Anti-Oedipus: Capitalism and Schizophrenia*. Minneapolis, MN: University of Minnesota Press.

Deleuze, G., and F. Guattari. 1987. *A Thousand Plateaus—Capitalism and Schizophrenia* (B. Massumi, Trans.). Minneapolis, MN: University of Minnesota Press.

Evetts, J. 2006. Short Note: The Sociology of Professional Groups: New Directions. *Current Sociology* 54:133–143.

Ezzy, D. 2000. Illness Narratives: Time, Hope and HIV. *Social Science and Medicine* 50:605–617.

Falge, C., C. Ruzza, and O. Schmidtke. 2011. *Migrants and Health: Political and Institutional Responses to Cultural Diversity in Health Systems*. London: Ashgate.

Fisher, P., and D. Goodley. 2007. The Linear Medical Model of Disability: Mothers of Disabled Babies Resist with Counter-Narratives. *Sociology of Health and Illness* 29:66–81.

Foucault, M. 1977. *Discipline and Punish: The Birth of the Prison*. London: Allen Lane.

Foucault, M. 1984. The Order of Discourse. In *Language and Politics*, ed. M. Shapiro, 108–138. Oxford: Basil Blackwell.

Fox, N. 1999. *Beyond Health: Postmodernism and Embodiment*. London: Free Association Books.

Freidson, E. 1970. *Profession of Medicine: A Study of the Sociology of Applied Knowledge*. New York: Dodd Mead.

Freidson, E. 2001. *Professionalism—The Third Logic*. Chicago: University of Chicago Press.

Fusco, C. 2006. Inscribing Healthification: Governance, Risk, Surveillance and the Subjects and Spaces of Fitness and Health. *Health and Place* 12:65–78.

Gabe, J., D. Kelleher, and G. Williams, eds. 1994. *Challenging Medicine*. London: Routledge.

Gibson, B. E. 2006. Disability, Connectivity and Transgressing the Autonomous Body. *Journal of Medical Humanities* 27:187–196.

Gibson, B. E., F. A. Carnevale, and G. King. 2012a. "This is My Way": Reimagining Disability, In/Dependence, and Interconnectedness of Persons and Assistive Technologies. *Special Issue of Disability and Rehabilitation: Promoting Participation and Engagement in Rehabilitation* 34:1894–1899.

Gibson, B. E., B. Secker, D. Rolfe, F. Wagner, B. Parke, and B. Mistry. 2012b. Disability and Dignity-Enabling Home Environments. *Social Science and Medicine* 74:211–219.

Gibson, B. E., H. Zitzelsberger, and P. McKeever. 2009. "Futureless" Persons: Shifting Life Expectancies and the Vicissitudes of Progressive Illness. *Sociology of Health and Illness* 31:554–568.

Goering, S. 2008. "You Say You're Happy, But...": Contested Quality of Life Judgements in Bioethics and Disability Studies. *Bioethical Inquiry* 5:125–135.

Guthrie Smith, O. F. 1952. *Rehabilitation, Re-education and Remedial Exercise*. London: Bailliere, Tindall, and Cox.

Hansen, N., and C. Philo. 2007. The Normality of Doing Things Differently: Bodies, Spaces and Disability Geography. *Tijdschrift Voor Economische en Sociale Geografie* 98:493–506.

Kendall, E., and J. Clapton. 2006. Time for a Shift in Australian Rehabilitation. *Disability and Rehabilitation* 28:1097–1101.

Kennedy, I. 1981. *The Unmasking of Medicine*. London: George Allen and Unwin.

Krusen, F. H. 1964. *Concepts in Rehabilitation of the Handicapped*. Philadelphia, PA: W. B. Saunders.

Larkin, G. 1983. *Occupational Monopoly and Modern Medicine*. London: Tavistock.

Lawler, J. 1997. *The Body in Nursing*. London: Churchill Livingstone.

Liamputtong, P. 1999. *Living in a New Country: Understanding Migrants' Health*. Sydney: Ausmed Publications.

Light, D. W. 2000. The Medical Profession and Organisational Change: From Professional Dominance to Countervailing Power. In *Handbook of Medical Sociology* (5th ed.), eds. C. E. Bird, P. Conrad, and A. Fremont, 201–216. Upper Saddle River, NJ: Prentice Hall.

Linker, B. 2011. *War's Waste: Rehabilitation in World War I America*. Chicago: University of Chicago Press.

Lupton, D. 1997. Doctors on the Medical Profession. *Sociology of Health and Illness* 19:480–497.

Manning, E. 2009. *Relationscapes: Movement, Art, Philosophy*. Cambridge, MA: MIT Press.

Markula, P., and R. Pringle. 2006. *Foucault, Sport and Exercise: Power, Knowledge and Transforming the Self*. London: Routledge.

Nettleton, S. 2005. The Sociology of the Body. In *The Blackwell Companion to Medical Sociology*, ed. W. C. Cockerham, 43–63. London: Blackwell.

Nicholls, D. A., and B. E. Gibson. 2010. The Body and Physiotherapy. *Physiotherapy Theory and Practice* 26:497–509.

Nicholls, D. A., and D. Holmes. 2012. Discipline, Desire, and Transgression in Physiotherapy Practice. *Physiotherapy Theory and Practice* 28:454–465.

Orwell, G. 1946. *Politics and the English Language*. London: Horizon. Available in eText from http://orwell.ru/library/essays/politics/english/e_polit (accessed October 24, 2014).

Osborne, T. 1997. Of Health and Statecraft. In *Foucault, Health and Medicine*, eds. A. Petersen, and R. Bunton, 173–188. London: Routledge.

Papadimitriou, C. 2008. Becoming En-Wheeled: The Situated Accomplishment of Re-Embodiment as a Wheelchair User after Spinal Cord Injury. *Disability and Society* 23:691–704.

Paterson, K., and B. Hughes. 1999. Disability Studies and Phenomenology: The Carnal Politics of Everyday Life. *Disability and Society* 14:597–610.

Patston, P. 2007. Constructive Functional Diversity: A New Paradigm Beyond Disability and Impairment. *Disability and Rehabilitation* 29:1625–1633.

Petersen, A., and Bunton, R. 1997. *Foucault, Health and Medicine*. London: Routledge.

Rose, N. 1994. Medicine, History and the Present. In *Reassessing Foucault: Power, Medicine and the Body*, eds. C. Jones, and R. Porter, 48–72. London: Routledge.

Samson, C. 1999. Biomedicine and the Body. In *Health Studies: A Critical and Cross Cultural Reader*, ed. C. Samson, 3–21. Oxford: Blackwell.

Seymour, W. 1998. *Remaking the Body: Rehabilitation and Change*. London: Routledge.

Shildrick, M., and J. Price. 2002. Bodies Together: Touch, Ethics and Disability. In *Disability/Postmodernity: Embodying Disability Theory*, eds. M. Corker, and T. Shakespeare, 63–75. London: Continuum.

Shilling, C. 2003. *The Body and Social Theory* (2nd ed.). London: Sage.

Sullivan, M. 2005. Subjected Bodies: Paraplegia, Rehabilitation, and the Politics of Movement. In *Foucault and the Government of Disability*, ed. S. Tremain, 27–44. Ann Arbor, MI: University of Michigan Press.

Swain, J., S. French, and C. Cameron. 2003. Practice: Are Professionals Parasites? In *Controversial Issues in a Disabling Society*, eds. J. Swain, S. French, and C. Cameron, 131–140. Buckingham: Open University Press.

Synnott, A. 1992. Tomb, Temple, Machine and Self: The Social Construction of the Body. *British Journal of Sociology* 43:79–110.

Tremain, S. 2001. On the Government of Disability. *Social Theory and Practice* 27:617–636.

Turner, B. S. 2008. *The Body and Society: Explorations in Social Theory*. London: Sage.

Van Manen, M. 1990. *Researching Lived Experience*. New York: SUNY Press.

Wells, H. G. 1904. Scepticism of the Instrument. *Mind* 13(51):379–393, 23.

Williams, S. J. 2006. Medical Sociology and the Biological Body: Where Are We Now and Where Do We Go From Here? *Health* 10:5–30.

7

Therapeutic Landscape: Rethinking "Place" in Client-Centered Brain Injury Rehabilitation

Pia Kontos

Karen-Lee Miller

Cheryl Cott

Angela Colantonio

Introduction

Rehabilitation facilities increasingly advocate a best practice care philosophy known as "client-centered care" (CCC) (Armstrong 2008; Cott 2004; Maitra and Erway 2006), which is predicated on client autonomy and choice in goal-setting, optimizing client–practitioner partnership, and breaking down status hierarchies between patients and staff (Law et al. 1995). Despite the proven benefits of CCC (Law et al. 1995; Pegg et al. 2005; Restall et al. 2003; Stewart et al. 2000), barriers to its systematic implementation into practice persist (Sumsion and Smyth 2000; van den Broek 2005; Wilkins et al. 2001). Efforts to understand what facilitates and/or impedes CCC within the rehabilitative context have for the most part focused on the processes and influences within practice contexts, such as a hierarchical model of care that reinforces sick role behavior of clients

(Cott 2004; Maitra and Erway 2006), and the reluctance of therapists to share the power of decision making with clients (Rebeiro 2000), favoring instead therapist-driven therapeutic activities (Mortenson and Dyck 2006). The interfaces between physical spaces, social roles, and rehabilitation treatment practices within inpatient rehabilitation settings, and their interactive roles in determining CCC, have been largely unaddressed. To this end, insights of critical health geography (Andrews 2004; Andrews and Moon 2005; Curtis et al. 2007; Evans et al. 2009; Gesler 1992; Kearns and Moon 2002; Watson et al. 2007)—specifically the notion of "therapeutic landscape" theory—have much to offer explorations of CCC in rehabilitation settings.

"Therapeutic landscape" theory was first introduced in health geography to identify how physical, social, and symbolic aspects of place contribute to well-being of individuals and communities (Gesler 1991). Landscapes of treatment and healing (Gesler 1992) include attributions of *cure* through restorative properties of thermal springs, spa towns, and shrines (Gesler 1998; Williams 2010a), as well as attributions of *care* in homes (Dyck 1998; Williams 1999), communities (Gesler 1992; Parr et al. 2005), and summer camps (Kearns and Collins 2000; Thurber and Malinowski 1999). Hybrid *cure–care* service sites (Gesler 2003) include clinics and hospitals (Curtis et al. 2007; Gesler 2003; Kearns and Barnett 2000), and psychiatric and addiction treatment facilities (Gesler 1992; Moon et al. 2006).

Therapeutic landscape theory facilitates understanding of "how the healing process works itself out in places" (Gesler 1992, p. 743) as well as when it does not, and to varying degrees (Wakefield and McMullan 2005; Williams 2002, 2010b; Wilton and DeVerteuil 2006). In determining whether a landscape may or may not be experienced as therapeutic, scholars have recently attended to the presence of ambiguity and contingency (Cutchin 2007; Davidson and Parr 2007) given the significant roles played by personal experience (Conradson 2005), interpretation (Milligan 2007), relations between service users and providers (Donovan and Williams 2007; Kennedy et al. 2004; McLean 2007), and the presence of subtle "unhealthy elements" in otherwise reputed environs (Cutchin 2007) that may diminish or contradict intrinsic or intended healing attributes of places.

The aim of this chapter is to understand rehabilitative care in traumatic brain injury (TBI) using therapeutic landscape as a novel framework. Traditionally, rehabilitation studies concerned with environmental influences on care focused either on built environment, such as the impact of hospital/unit aesthetics or design (Codinhoto et al. 2009), or social environment in terms of "rehabilitation enculturation" (Thompson 1990) and meanings of health and wellness assigned to transition from hospital to community settings (Turner et al. 2009; Warren and Manderson 2008). Yet interfaces between these physical and social spaces and their interactive roles in determining therapeutic care have been left woefully unaddressed. Thus, the therapeutic landscape framework offers a more critical approach to understanding context than has yet been considered in studies of neurorehabilitation. In turn, TBI offers a novel context for the exploration and application of therapeutic landscape, serving as an important empirical case to broaden and enrich theory. (Other approaches to rethinking rehabilitative care in TBI are described in Chapters 8, 9, 12, and 14.)

This chapter first provides an overview of therapeutic landscape theory. Next, we review literature that emplaces rehabilitation as both a physical space and as an ideology of care. A description follows of TBI and its presumed treatment challenges. We present qualitative findings from two inpatient rehabilitation hospitals with a focus on their sociospatial design and care delivery characteristics. We conclude with a discussion of the implications of the client–practitioner–place triad for therapeutic landscape theory.

Therapeutic Landscape

Therapeutic landscapes have been traditionally defined as those places, situations, locales, settings, and milieus that encompass physical, social, and psychological environments associated with treatment or wellness (Gesler 1992; Williams 1998). Therapeutic landscapes generally have an affinity or a strong sense of place to those experiencing them (Williams 1999), and an "enduring reputation for achieving physical, mental, and spiritual healing" (Gesler 1993, p. 171).

Since its development, therapeutic landscape theory has been applied to a range of health-enhancing and/or medicalized *sites*, from hospital respite and retreat, to home care (Conradson 2005; Curtis et al. 2007; Dyck 1998; Gesler 2003; Pascal 2010; Williams 1999); *services*, from conventional medicine to holism (Moon et al. 2006; Williams 1998); and *approaches*, from illness treatment to health promotion (Conradson 2005; Curtis et al. 2007; Milligan et al. 2004; Pascal 2010; Wilton and DeVerteuil 2006). Critics have called for increasing examination of health-affirming places that are not conducive to healing and that may have contradictory, or worse, effects (Curtis et al. 2007; Wakefield and McMullan 2005; Williams 2002, 2010b). The initial theory made clear distinctions between landscapes understood to be therapeutic (health promoting) and those perceived as non-therapeutic (health-detracting) (Donovan and Williams 2007). However, recent critiques of the linear effect of place on well-being or ill-health have accompanied nuanced understandings that the same setting may be differently experienced by individuals due to personal self-landscape interpretation (Conradson 2005; Curtis and Riva 2010). Thus, effects of place are neither universal nor reflective of attributes of the landscape but instead may be influenced by complex individual characteristics and social relations (Kennedy et al. 2004; McLean 2007). Additionally, microdesign aspects such as informal messages in hospital waiting rooms may introduce points of unhealthy disruption in otherwise health-affirming spaces (Crooks and Evans 2007).

Therapeutic landscape theory has extended beyond the natural and social boundaries of physical sites to examine emotion (English et al. 2008; Parr et al. 2005), bodies (English et al. 2008), and the mind (Andrews 2004; Williams 1998) as landscapes of healing. While this broadens the theory to include less topophiliac or tangible spaces, these efforts have, regrettably, often failed to re-emplace the bodies, feelings, and faculties of imagination they describe. The unintended consequence is a kind of disembodied placelessness, which destabilizes broader theoretical attempts to articulate the emergent and often messy interplay between self-body, place, experience, and well-being. In contrast, neurorehabilitation offers an intriguing opportunity to simultaneously examine the conventional preoccupation in therapeutic landscape with institutional health care space, as well as more recent foci on sociorelational geography, and the physical and emotional emplacement of ill and healing bodies.

The Built Environment of Rehabilitation

Therapeutic landscape theory identifies hospitals as "extraordinary" places since they exist outside most people's everyday geographies, are associated with catastrophic or uncommon events, and "are not intended to support longer-term connections to health and healing" (English et al. 2008, p. 70). While rehabilitation hospitals share with acute-care hospitals the first two characteristics, they distinctly break with the third. Reintegration into the community is the driving ethos of rehabilitation, and work in-hospital is often premised on continuing therapy and healing in the community (Warren and Manderson 2008).

Most rehabilitation settings mimic the "medicalized hospital tradition" of a regimented institutional routine (Moon et al. 2006, p. 142). Waking, physician visits, inroom tray meals, and visiting hours occur at strictly predetermined times (Papadimitriou 2008; Warren and Manderson 2008). Typically initiated following an acute care hospital stay once the patient is medically stable, rehabilitative length of stay for the brain-injured patient is usually six weeks during which time the goals are to regain competence in self-care activities, mobility, and independent living (Cullen et al. 2007). Discharge is not equated with "cure" and often involves some measure of ongoing outpatient therapy (Warren and Manderson 2008).

Rehabilitation hospitals are primarily designed around a "gymnasium model" of rehabilitation (McClusky 2008; Warren and Manderson 2008). Clients enter treatment rooms filled with exercise equipment to strengthen muscles with repetitive weights; activity tables to improve memory and coordination through game participation; and platforms, support bars, and model stairways "that lead nowhere" to develop balance (McClusky 2008, p. 80). Treatment rooms consist of simulated kitchens, bathrooms, and bedrooms in order to improve performance on activities of daily living and the multiple sequential activities associated with meal times, sleep/wake times, and toileting.

Brain-injured clients are generally accommodated in sparsely furnished private or semiprivate rooms since multiple bed ward-style rooms provide too great mental and social stimulation. Clients are encouraged to decorate their rooms with pictures and other personalized items, which function as memory aids for encouragement to return to the skills, hobbies, and social and family networks in place before the brain injury.

The Ideological Landscape of Rehabilitation

Rehabilitation refers to not only the therapeutic activities seeking to restore physical functioning but also the care ideologies underpinning the practices and techniques employed in doing so (Ory and Williams 1989). Care ideologies are shaped by the larger social and economic landscape of health-care provision and, in turn, come to organize the microbehaviors of staff and clients. Care ideologies are of particular interest to therapeutic landscape scholars since they emplace care practices in the institutional and regulatory structures from which they are formulated (see, for example, Crooks and Evans 2007; McLean 2007). In traditional rehabilitation care ideology, therapy is focused on immediate management of the new disability and/or adjustment to functional losses (Papadimitriou 2008; Warren and Manderson 2008). To this, however, has been added the ideology referred to as "client-centered care" (CCC) (Cott 2004). CCC heralds client autonomy and choice in goal-setting, and the benefits of the client–practitioner partnership (Law et al. 1995). By design, CCC is intrinsically aligned with the ideal therapeutic hospital "social environments" outlined by therapeutic landscape scholars, since it is inclusive of patient participation and communication, and the breaking down of status hierarchies (Gesler 2003). Nonetheless, it has been suggested that CCC is fraught with tensions with a lack of congruence between its reality and ideal (Mortenson and Dyck 2006; Townsend et al. 2003; see also Chapters 3 and 14). This has been attributed to limitations of the current health-care climate, as well as to the discomfort experienced by some therapists with CCC's ideological shift toward client decision-making power, which may circumvent profession-specific rehabilitative aims should clients disagree (Corring and Cook 1999; Mortenson and Dyck 2006; Townsend 1999).

The Personal Landscape of Traumatic Brain Injury

The body is "the smallest and most personal landscape" of those injured or unwell (English et al. 2008, p. 76). In TBI, the body is often experienced as a landscape of loss with infrequent recovery of full cognitive or physical function (Warren and Manderson 2008). TBI is defined as "an alteration in brain function, or other evidence of brain pathology, caused by an external force" (Menon et al. 2010, p. 1637). TBI is first characterized by a period of altered consciousness (amnesia/coma) of mild, moderate, or severe duration. Neuropsychological impairment (e.g., difficulties with memory), behavioral issues (e.g., aggression), and cognitive limitations (e.g., difficulties with "real-world tasks" such as activities of daily living) often accompany impairments in physical functioning, sensation, and coordination. Reasoning, relating, walking, and working may all have to be relearned to some degree (Menon et al. 2010).

Given that CCC is grounded in principles of *cognitively* based choice, autonomy, and negotiation, the neuropsychological and behavioral issues of brain-injured clients are believed to limit or exclude their full participation. Subsequent policy and practice mechanisms for improving brain-injured clients' capacity to be decisive copartners in care have, for the most part, singularly focused on increasing clients' self-awareness of their cognitive and physical deficits in order to obtain agreement and cooperation with therapist-proposed treatment goals. Common approaches to enhance self-awareness in clients with TBI include nonconfrontational dialogue videotaped feedback, counseling, and education on deficits (Barco et al. 1991; Bieman-Copland and Dywan 2000).

Methods

The study was a three-year (2008–2011) qualitative-based evaluation of an educational intervention to improve client-centered neurorehabilitation care (Gray et al. 2011; Kontos et al. 2012). Postevaluation, we conducted a secondary analysis (Gladstone et al. 2007) of baseline or preintervention data to examine sociospatial aspects that promote and hinder positive therapeutic outcomes in neurorehabilitation.

The Settings

Data were collected in similarly sized neurorehabilitation units (Facility A, 32-bed; Facility B, 27-bed) of two inpatient rehabilitation hospitals in Ontario, Canada.

Participants

Research ethics boards of participating hospitals approved the study; thirty-eight participants ($n = 38$) took part. They were purposively selected (Patton 1990) using criterion-based selection (Le Compte and Preissle 1993), which was directed by a literature search that identified nursing and allied health as the most common neurorehabilitation professions, and management and chaplaincy as providing unique insight into unit organization and the life worlds of the clients. Thus, licensed direct care practitioners with the most and least years' experience from nursing (RN, RPN; $n = 11$), occupational therapy (OT; $n = 5$), physical therapy (PT; $n = 5$), speech language pathology (SLP; $n = 6$), social work (SW; $n = 3$), recreational therapy (RT; $n = 1$), and psychology ($n = 1$) were included as well as nurse administrators (RN leader $n = 5$) and a nondenominational chaplain ($n = 1$).

Participants consented to the intervention and data collection; 72.73% of participants had worked in the facility for four years or more. Participants ranged in age from 27 to 63 years with a mean of 39.91 and a standard deviation of 9.85.

Interviews

Thirty-eight audiotaped, semistructured baseline interviews lasting approximately 60 min were conducted to explore the social, spatial, and organizational nature of institutional practice, specifically, (1) perceived impacts of TBI on clients' relationships, routines, and quality of life; (2) the hospital environment including social climate (staffing mix, interprofessional relations, patient–staff relations); clinical practice (culture of care, client–provider communication, therapeutic outcome measures); and rehabilitation features (descriptions of hospital environment, uniqueness of TBI to therapeutic practice, treatment and discharge decision-making mechanisms); and (3) the individual, structural, environmental, and cultural factors that constrain and enable CCC in neurorehabilitation.

Analysis

We conducted a secondary analysis (Gladstone et al. 2007) of baseline interviews with health-care practitioners. The purpose of the secondary analysis was to explore rehabilitative care in TBI using therapeutic landscape. Analysis of the interview transcripts began with thematic analysis techniques (Denzin and Lincoln 1998). Line-by-line analysis (Strauss 1987) of the verbatim transcripts identified important considerations of subjective notions of place (Davidson and Parr 2007; McLean 2007; Williams 1999) and emplacement (Laws 1995; Milligan et al. 2004) in relation to TBI rehabilitative care. Subjective notions of place include a sense of place or place identification, for example, "the felt value given to a place" (Williams 1999, p. 73) such as attributions of safety, comfort, or danger to particular places, including homes or communities independent of actual risk (Cristoforetti et al. 2011; Manzo et al. 2008). Emplacement refers to the spatiality of personal identity through landscape, which is then constituted in embodied form, for example, gated retirement communities that both mutually constitute and represent the suntanned golfer retiree identity (Laws 1995, p. 254), as the loft aesthetic comes to shape and represent young, hip urbanites (Podmore 1998).

Practitioners' accounts revealed a strong affinity with therapeutic landscape theory, which was drawn upon as a sensitizing concept in the secondary analysis (Blumer 1969; Gladstone et al. 2007). This approach involves combining inductive and deductive approaches whereby themes are identified in the data but are analyzed within the context of theory. Themes were investigated for interrelationships and conceptually refined moving from lower-order to higher-order themes as analysis progressed (Goldkuhl and Cronholm 2010). Finally, analytical categories were examined to determine points of intersection and conflict with therapeutic landscape theory and to assess themes vis-à-vis health-affirming and health-detracting considerations.

Findings

Findings highlight the significance of the interconnections among the client–practitioner–place triad in determining therapeutic aspects of TBI inpatient rehabilitation. We first describe health-detracting aspects of place including unsuccessful design attempts to simulate home environments, which impacted on clients' and practitioners' awareness and bodies, and how features such as practitioners' preference for quantitative outcome measures and the meagerness of psychological services have created an institutional emotional geography that produces and sustains the emotional suppression of clients. Next we describe the simultaneous presence of health-enhancing aspects such as the emplacement of identity and everyday geography-informed therapy.

An Artificial Place of Healing: Design, Impairment, and Awareness

Home elicits a strong sense of place and thus represents a focal center for healing (Williams 1999, 2002). Many health-care settings are now designed to emulate the home. Contemporary long-term care settings commonly resemble luxury high-rise apartment buildings with wood and upholstered furnishings that communicate "residence" rather than medical "institution" (Reed-Danahay 2001). Similarly, palliative-care units include kitchens where family members can prepare meals (Godkin 1980). Rehabilitation too has not escaped the mimicry of "homelike" design imperatives, which are intended to simulate the environments and activities of postdischarge life (Richardson et al. 2000).

Across both sites, treatment settings strove to imitate real-life situations by providing simulated kitchens where clients relearned how to select ingredients from a refrigerator and cupboards, follow a recipe, and prepare a meal. However, while each kitchen consisted of fully functional stovetop and oven, access was supervised and limited. The objective was to have clients engage in food preparation by reinforcing task sequencing rather than by evoking memory through the familiarity of habitual kitchen routines and the "smellscape" (Rodaway 1994)—the relationship between person and place through the olfactory senses—of the aroma of familiar home-cooked meals (for a similar critique, see McLean 2007). Consequently, participants described the homelike setup of the treatment rooms as artificial and unconvincing:

> It's more of a sterile environment here ... you're trying to make it like it would be at home but it's not. [Facility B, RT 1]

> I think the [rehabilitation] environment ... is difficult. It's not like being at home ... it's a little contrived. [Facility B, Nurse 1]

> This environment is very foreign to them ... they're not in their own [home] environment and ... it's difficult to set that situation that they're going to. [Facility A, SLP 1]

Lack of authenticity in the replication of home space was significantly implicated in clients' assessments, as well as personal understandings of the extent and etiology of poor performance:

> The relevance [of design] comes into play [during assessment] because they're not home. You know, it is a little artificial. [Clients rebut], "It's not my grocery store that I'm going to," "it's not my" whatever. [Facility B, SLP 2]

> We take them into our rehabilitation context which is foreign to them. We don't test them with materials they're even a tiny bit familiar with. [Facility B, PSYCH 1]

While brain-injured clients' lack of insight into their cognitive deficits is commonly attributed to neuropathology (Toglia and Kirk 2000), practitioners attentive to spatiality resisted the attribution of cognitive frailty. Instead, they recast clients' difficulties in recognizing cognitive impairment as a direct consequence of the poor environmental simulation of "home." This inaccuracy in self-assessment in turn led clients to ineffectively participate in goal setting and treatment decisions:

> We're in an artificial environment here with all kinds of artificial tools ... so it's a big problem trying to take people into an artificial lab world where they feel they're guinea pigs and try to get them to buy in and support you in your intentions and your initiatives. [Facility B, PSYCH 1]

Practitioners acknowledged that the artificiality of the simulated homelike setting made it difficult for clients to comprehend where unfamiliarity ended and the consequences of TBI impairment began. Consequently, practitioners argued for home assessments so that the clients' inability to access the tacit or embodied knowledge with which clients approached tasks in familiar settings would enable a more clear grasp on the nature and implications of their injuries:

> Occupational therapists will tell you [in] five out of six of their kitchen assessments that clients say, "Well, I would've done fine if it was my own kitchen." ... Have you ever tried cooking in someone else's kitchen?! ... So to go and do things in their [home] environment really hits home for them. It *really* helps with their awareness ... [Facility B, RT 1, original emphasis]

The artificiality of the rehabilitation setting was apparent not only in comparisons made between the hospital and clients' homes but also those made between units within the hospital. Nurses noted that simulated bathrooms and bedrooms of treatment rooms were dissimilar to the layout and equipment on the units where clients spent the majority of time, and the interaction of poor mechanical structures with nurses' and clients' bodies precipitated injury and fear:

> [Ward] bed[s are] not quite the same as the mat [in the treatment rooms] ... So [clients] could push off that mat, but when they came to push off the mattress [in their rooms], there's no way. So we said to the [physical therapists], "you come and show us from the client's bed how to do this and how to do that." They couldn't [when they arrived]. [Facility B, Nurse 2]

> Our bathroom is a big issue ... we don't have the [same] toilet [as that used by the occupational therapists] that could be accessed from any angle ... So we've already had some [nurse] injuries because of that ... a few weeks ago I had to call the [occupational therapist] up and say, "you need to help me. I've tried ... and we're over-stretching and they're panicking, the patients are panicking because I can't transfer them like they're doing downstairs because I don't have room for it." So ... now they're including that in their [activities of daily living] assessment, the bathroom up here. [Facility B, Nurse 3]

No Place to Grieve: The Repression of Emotion

Emotion is an emerging focus in place-based research including therapeutic landscape (Davidson and Parr 2007; Parr et al. 2005; Williams 2007). The articulation of emotion is "spatially mediated" (Davidson and Milligan 2004, p. 523), and thus sociospatial dynamics can serve to support or contain troubling emotions associated with illness and disability (Parr et al. 2005).

In our findings, as with the TBI literature in general, the emotional impact of TBI was profound. The nature of brain injury often results in significant losses for an individual (Blackmer 2003), causing anxiety (Al-Adawi et al. 2007), depression (Al-Adawi et al. 2007; Jorge et al. 2004), and post-traumatic stress (Bryant et al. 2001). Staff were highly cognizant of the difficulties with postinjury adjustment. The chaplain in Facility B explained: "They've lost their life, maybe not physically, but they've certainly lost their life that they knew and that's a big grief issue." Yet psychological and counseling services were meager in both study units. In large part due to fee arrangements associated with government funding, psychologists were restricted to cognition assessments only. A speech language pathologist noted:

> Our psychological services are very poor ... We all shake our heads at that because we see the emotional turmoil that [clients] go through, and really there's no outlet. [Facility, A, SLP 3]

Practitioners also noted macro-level pressures to demonstrate client improvement on standardized tests since "that's what receives the most weight" [Facility A, SLP 1]. For example, although she had earlier lamented inadequate psychological services, a speech language pathologist referred to her provision of emotional support as "going off track" and commented on the imperative of obtaining quantitative outcome measures to support treatment decision making:

> Evidence-based is the big thing now, right? Like, evidence-based practice is huge. So what you're doing has to be supported by research … [so] I feel like I couldn't adequately justify [going] off track, you know? … A lot of decisions [in interprofessional rounds] are made around [standardized scores]. [Facility A, SLP 3]

In addition to macrolevel barriers to addressing grief and loss issues, practitioners identified disciplinary factors such as rehabilitation foci and professional silos as reasons they were professionally and personally adverse to undertaking the "emotion work" (Miller et al. 2008) conducive to an emotionally supportive therapeutic landscape. For example, the exclusive focus of neurorehabilitation on physical functioning impaired emotional expression:

> We [are] too focused on function where sometimes the person just needs time to deal [emotionally] with what's happened … There's grieving that happens and that *needs* to happen. And sometimes we're so focused on moving forward with the therapy goals that the [client] can't grieve. [Facility B, OT 1, original emphasis]

Practitioners from multiple rehabilitation disciplines also commented on their lack of professional training and qualifications, which gave rise to palpable unease and inadequacy when clients broached emotional issues:

> I don't have the training to provide [emotional support] so I feel a little bit out of my element. [Facility A, PT 2]

> I'm not qualified to deal with if they're grieving about their life and, you know, their sense of self and they're depressed. [Facility A, SLP 3]

> I'm not *trained* to be a counsellor. [Facility B, SLP 4, original emphasis]

Finally, personal disinclination to be empathetic further suppressed emotional expression:

> I feel that sometimes I have to put up a bit of a wall with the patient because it's hard to be empathetic all the time, because you just get so overwhelmed with what they must be going through. [Facility A, SLP 1]

> I don't want to really know [the client's] life … I think it's just your instinct [to protect] your emotions. Like, you're gonna feel for the mom who has three kids and now can't even brush her own teeth. [Facility B, RN 1]

> Is that not our greatest fear? … We all think we're gonna turn into our patients, it's gonna happen to us … That's very frightening. [Facility B, RT 1]

This lack of support for emotional expression, evidenced by institutional and practitioner practices, significantly impacted clients' ability to fully participate in the range of therapies offered by neurorehabilitation practitioners:

> A lot of clients are not ready for [speech] therapy at all ... I think that especially with brain injury, there's so many emotions and grieving and, I don't know that they're adequately supported there to focus. So I find that a lot of my patients are really distracted by their own thoughts and by their feelings and attitudes toward, you know, their new self ... [Facility B, SLP 4]

Emplacing Identity: Therapy in the Context of the Life World of the Client

Therapeutic landscape theory suggests care environments should be evaluated against how well they "respect the personality, preferences, and cultural and religious mores of patients, especially when these may be seen to be partially modified due to the nature of their illness" (Curtis et al. 2007, p. 593). This is in keeping with a central tenet of CCC that successful rehabilitation involves tailoring interventions to clients' unique communicative, cognitive, physical, psychological, and behavioral profiles, stage of recovery, and goals (Wiseman-Hakes et al. 2010). (For further discussion of identity in relation to TBI, see Chapter 9.)

Many practitioners appeared keenly aware of the necessity of grounding therapeutic techniques in the self-identity of clients. Practitioners used ethnicity/culture, personal hobbies, and social placement to individually tailor therapy:

> One guy liked golf, like he was in his sixties and wanted to play golf, so [physical therapy involved using] a putting set and stuff. Another guy liked photography so I borrowed someone's camera, [and] he was worried about chopsticks and manipulating that with his hands, so ... [therapy utilized those items]. [Facility A, PT 3]

> I have a gentleman who was a web designer and ... doing anything paper and pencil he rejected. [H]e needed something more creative, more suited to his life, and that's what we did ... the goal was [to] develop a web page, a patient-centered web page for [our hospital] ... It was motivating! It was the one thing that actually engaged him. Nothing else, and we tried everything, nothing else worked. [Facility A, SLP 1]

At times practitioners were required to be creative in terms of breaking down the instrumental tasks of a client's occupation in order to identify discrete steps that could be relearned during therapy. For example, an occupational therapist related:

> [A client] shared that she was an actress and wanted to get back to acting. ... It did influence the speech language pathologist because she was able to say "okay, well let's work on voice projection, let's work on facial expression." [Facility A, OT 3]

The motivation clients derived from tasks associated with self-identity led to observable clinical improvements: "You just get better results" [Facility A, PT 2].

The degree to which practitioners tailored therapy to the everyday geographies of clients was illustrated by the commitment to balance outcome goals with clients' preinjury status and their acceptance of functional limitations. It began with practitioners recognizing their own practical embeddedness in therapeutic activities, and the recognition of differences in life worlds that clients will return to postrehabilitation. These practitioner efforts highlight their attention to the client–practitioner–place triad, which ensures correspondence with clients' life worlds rather than solely rehabilitative prerogatives.

> I talk to them about what they do and stuff, I try to take on *their* part of the world. So if they're a street person, you know, there are certain goals that I know aren't gonna be goals for them. They don't have a home. They probably have a routine. They typically do, you know, they go to certain places to eat, certain places to sleep ... If I get an Italian in here, I know a bit about the culture, so I know that food is important, big [families], lots of people around, so I take all those considerations in ... Like, I really try and come from their point of view. I try not to base everything on what's normal for me and what's common in my world. Because I know that everybody's world is different. [Facility B, RT 1, original emphasis]

> Many people say "oh, I was bad at that before," "I couldn't read before anyways." Well, that's huge. Literacy is huge. So we have to know this stuff so that when I'm preparing goals with them that it's done in a fair manner, and it makes sense, and I'm not teaching anyone English or how to read. [Facility A, SLP 3]

Discussion

TBI care, much like other forms of catastrophic injury rehabilitation, is a particularly spatial endeavor undertaken within and against a complex network of places, people, and clinical practices. Until now, however, inpatient rehabilitation has been overlooked as a real-world example of therapeutic landscape. The oversight is regrettable since our findings provide evidence that TBI rehabilitation can support the "continuing maturation" (Williams 2007) of the therapeutic landscape concept. In particular, it highlights the importance of the interconnectedness of the client–practitioner–place for structuring a therapeutic landscape. Theorizing this dynamic interconnection via the incorporation of other conceptual frameworks within health geography (Andrews and Moon 2005) and beyond (Curtis and Riva 2010) is recommended to further advance therapeutic landscape theory.

In our study, practitioners commented on embodiment and emplacement practices, which they believed would make the existing landscape in TBI a more positive one. They accorded weight to in-home assessments in order to better facilitate the re-embodying and re-enacting of clients' tacit knowledge to more accurately relay to professionals what is (still) known. They noted the importance and effect of environmental semblance among rehabilitation units since it is the repetition of bodily practice through which comfort, safety, and confidence in movement and routine arise. They further identified the importance of the emplacement of clients' unique identities within rehabilitation practice. Finally, staff practices highlighted the importance of attention to the emotional geography of rehabilitation institutions. Emotional geography examines how emotions are linked to spatialities across a variety of scales such as body, places, environments, and institutions. For example, emotions of fear, angst, and awe can be spatially mediated or induced through the presence of people or of natural wonders (Bondi et al. 2007, p. 2). In this study, staff practices identified—through omission—the importance of an emotional geography of place that enabled the expression of grief and loss in order to assist clients in establishing their new postinjury cognitive, physical, and emotional selves.

It appeared that neither hospital design, nor care ideology, nor social relations alone were sufficient to merit determination of whether and how the landscape of rehabilitation presented a therapeutic environment. This stands in marked contrast to existing therapeutic landscape studies that have tended to view therapeutic or nontherapeutic elements in isolation from one another, focusing, for example, on components of physical design (Crooks and Evans 2007), individual experience (Conradson 2005), or care provider (Donovan and Williams 2007; Williams 1999). While some authors have identified the coexistence of the components of the triad (see, for example, Donovan and Williams 2007; Gesler 2003; Kennedy et al. 2004), they have stopped short of exploring *how* these components are intertwined and in *what ways* the interrelationship between them serves to render places of care therapeutic or not. In contrast, this study demonstrated how various spatial and sociospatial aspects of client, practitioner, and the physical setting create an ambiguous therapeutic landscape that comprises TBI care. As such, it offers insight into the workings of complexity and contingency in therapeutic landscape.

In this study, practitioners directly linked the poor mimicry of home in hospital design to client performance and social behaviors. Practitioners' attentiveness to spatiality underscored an important interrelatedness between client and place. This is compatible with earlier research that supports the influence of environmental familiarity, with individuals performing better in domestic setting (Darragh et al. 1998), as well as clients' own complaints that the rehabilitation environment does not adequately prepare them for the built and social landscapes to which they will return (Cott 2004). More importantly, however, our research pointed to a change in practitioner consciousness regarding the etiology of client negation of deficits. Poor place mimicry was believed to lead to understandable confusion regarding cognitive capacity. Other studies have neglected the importance of design differences in the same institution as a contributing factor to dissonance in client's self-assessments, despite the significance of the cogency of the impairment–assessment relation in rehabilitation (see, for example, Barco et al. 1991; Bieman-Copland and Dywan 2000). Nonetheless, this should not be viewed as wholesale endorsement of home versus institutional assessments in the TBI population (Bottari et al. 2006), since contradictory evidence suggests the need for further investigation into the interaction of performance and environment to determine the most therapeutic environment for rehabilitation (Darragh et al. 1998).

Practitioners' attentiveness to client–place spatiality was also predicated on knowledge of the ways in which environmental deficits simultaneously acted upon their own bodies through nursing injuries. It is a novel research finding that nurses' physical bodies were implicated in the rehabilitation process in the moment of therapeutic disjuncture in client–place interaction, as well as social negotiations including difficulties with professional (nurse–client) and intraprofessional (nurse–occupational therapist; nurse–physical therapist) social relations. Design implications have elsewhere been identified as impeding social interactions among health practitioners (Oandasan et al. 2009;

Sinclair et al. 2009), but this is the first time that the client's body and the mutual impact of place on therapeutic engagement have been included as integral to the discussion of rehabilitation practices by, and among, professionals.

A biomechanical discourse has long defined physiotherapy's professional identity (Nicholls and Cheek 2006; Nicholls and Gibson 2010) with the related emphasis of rehabilitation being functional independence (Nicholls and Cheek 2006; see also Chapters 3 and 6). Our finding of the interference of such emphasis in resolving emotional issues is consistent with other rehabilitation studies (Cott 2004; Papadimitriou 2008). Nonetheless, until now, the lack of attentiveness to emotional issues by rehabilitation therapists has been viewed solely as a consequence of individual-level behavior or discipline specific training (Papadimitriou 2008). Certainly our work also found therapists across professions suggesting they lacked the skills and training to emotively engage clients, and attributing those skills instead to counselors despite evidence of emotion work by nonpsychological health professionals in other medical settings (see, for example, Miller et al. 2008). More importantly, however, our study found that the suppression of emotion was a spatial institutional artifact. That is, the emotional expression of TBI clients was organized through meager psychological services despite overwhelming recognition of need, the rehabilitation emphasis on physical functioning, and an emphasis on evidence-based research and quantitative outcome measures. Our finding that the sociocultural features of rehabilitation have created an institutional emotional geography that produces and sustains emotional suppression is a novel one. It starkly contrasts with the therapeutic landscape examined by English et al. (2008), in which practitioners consciously fostered the expression and release of emotion. That clients in our study were repeatedly described as requiring an outlet for grief, and often were disruptive in various therapies despite practitioners' suppression of clients' emotion, suggests that the institutional emotional geography of TBI rehabilitative care in the study sites is decidedly antitherapeutic. This is particularly egregious given the prevalence of emotional disorders such as anxiety, depression, and posttraumatic stress disorder in TBI survivors (Al-Adawi et al. 2007; Bryant et al. 2001; Jorge et al. 2004), and thus warrants further investigation.

The negotiation of decision-making power in the therapeutic relationship in terms of the insistence by clients for the emplacement of individuality, and practitioners' responsiveness in incorporating everyday geographies such as vocation and hobbies (e.g., web design, acting) into therapy, influenced the therapeutic landscape in a way that was consistent with the principles of CCC. Emplacement of client identity by practitioners effected changes in therapeutic space both physically, with the use of the tools and mechanisms of individual everyday geographies (e.g., chopsticks, golf clubs), as well as with the social relations of care in terms of client-directed therapy. Correspondingly, the use of therapeutic landscape in TBI research has proven useful in identifying how practitioners understand and ameliorate shortcomings in the ideological landscape of CCC in rehabilitation hospitals.

Our findings draw important links between place, practitioner, and client, with health-promoting and health-detracting properties emerging at the points of intersection of this triad. In offering a conceptual framework for understanding the interrelationship between objective and subjective aspects of care and caring in places, therapeutic landscape has the potential to generate knowledge that challenges presumptions in rehabilitation of the linearity of place and recovery. This illuminates important processes of interrelatedness, contingency, and emergence in relation to the spatial, ideological, practice, and reflexive factors pertinent to the therapeutic landscape of TBI rehabilitation. As such, future therapeutic landscape studies in rehabilitation settings may contribute significantly to evidence-based ameliorations of the unintentionally harmful aspects of neurorehabilitation in order to improve quality of care.

Simultaneously examining place, practice, and identity as suggested by critical health geographers offers important insights for understanding how the interactions between objective and subjective aspects of care and caring can influence health. If taken up in future research in neurorehabilitation and other clinical areas of rehabilitation, it is a perspective that can significantly contribute to the cumulative and systematic development of knowledge that informs policy and program interventions to improve CCC.

Acknowledgments

Support was received from the Canadian Institutes of Health Research Operating Grant (2008–2011, MOP-86624); Pia Kontos holds a Canadian Institutes of Health Research New Investigator Award (2009–2014, MSH-87726). Dr. Colantonio receives support through the Saunderson Family Chair at the Toronto Rehabilitation Institute and through a Canadian Institutes of Health Research Chair in Gender, Work and Health (CGW-126580).

References

Al-Adawi, S., Dorvlo, A.S.S., Al-Naamani, A., Glenn, M.B., Karamouz, N., Chae, H., Zaidan, Z.A. and Burke, D.T. (2007). The ineffectiveness of the Hospital Anxiety and Depression Scale for diagnosis in an Omani traumatic brain injured population. *Brain Inj* 21, 4, 385–393.

Andrews, G. (2004). (Re)thinking the dynamics between healthcare and place: Therapeutic geographies in treatment and care practices. *Area* 36, 3, 307–318.

Andrews, G. and Moon, G. (2005). Space, place, and the evidence base: Part II—Rereading nursing environment through geographical research. *Worldviews Evid Based Nurs* 2, 3, 142–156.

Armstrong, J. (2008). The benefits and challenges of interdisciplinary client-centred, goal setting in rehabilitation. *NZJOT* 55, 1, 20–25.

Barco, P.P., Crosson, B., Bolesta, M.M., Werts, D. and Stout, R. (1991). Training awareness and compensation in postacute head injury rehabilitation. In *Cognitive Rehabilitation for Persons with Traumatic Brain Injury: A Functional Approach* (J.S. Kreutzer and P.H. Wehman, eds.) pp. 129–146. Baltimore: Paul H. Brookes.

Bieman-Copland, S. and Dywan, J. (2000). Achieving rehabilitation gains in anosognosia after TBI. *Brain Cogn* 44, 1–18.

Blackmer, J. (2003). The unique ethical challenges of conducting research in the rehabilitation medicine population. *BMC Med Ethics* 4, 2.

Blumer, H. (1969). *Symbolic Interactionism: Perspective and Method*. Englewood Cliffs, NJ: Prentice Hall.

Bondi, L., Davidson, J. and Smith, M. (2007). Emotional turn. In *Emotional Geographies* (J. Davidson, L. Bondi and M. Smith, eds.) pp. 1–18. Burlington, VT: Ashgate.

Bottari, C., Dutil, E., Dassa, C. and Rainville, C. (2006). Choosing the most appropriate environment to evaluate independence in everyday activities: Home or clinic? *Aust Occup Ther J* 53, 98–106.

Bryant, R.A., Marosszeky, J.E., Crooks, J., Baguley, I.J. and Gurka, J. (2001). Posttraumatic stress disorder and psychosocial functioning after severe traumatic brain injury. *J Nerv Ment Dis* 189, 2, 109–113.

Codinhoto, R., Tzortzopoulos, P., Kagioglou, M., Aouad, G. and Cooper, R. (2009). The impacts of the built environment on health outcomes. *Facilities* 27, 3/4, 138–151.

Conradson, D. (2005). Landscape, care and the relational self: Therapeutic encounters in rural England. *Health Place* 11, 337–348.

Corring, D. and Cook, J. (1999). Client-centred care means that I am a valued human being. *Can J Occup Ther* 66, 2, 71–82.

Cott, C. (2004). Client-centred rehabilitation: Client perspectives. *Disabil Rehabil* 26, 24, 1411–1422.

Cristoforetti, A., Gennai, F. and Rodeschini, G. (2011). Home sweet home: The emotional construction of places. *J Aging Stud* 25, 3, 225–232.

Crooks, V.A. and Evans, J. (2007). The writing's on the wall: Decoding the interior space of the hospital waiting room. In *Therapeutic Landscapes* (A. Williams, ed.) pp. 165–180. Aldershot: Ashgate.

Cullen, N., Chundamala, J., Bayley, M. and Jutai, J. (2007). The efficacy of acquired brain injury rehabilitation. *Brain Inj* 21, 2, 113–132.

Curtis, S., Gesler, W.M., Fabian, K., Francis, S. and Priebe, S. (2007). Therapeutic landscapes in hospital design: A qualitative assessment by staff and service users of the design of a new mental health inpatient unit. *Environ Plann C* 25, 591–610.

Curtis, S. and Riva, M. (2010). Health geographies II: Complexity and health care systems and policy. *Prog Hum Geogr* 34, 4, 513–520.

Cutchin, M.P. (2007). Therapeutic landscapes for older people: Care with commodification, liminality, and ambiguity. In *Therapeutic Landscapes* (A. Williams, ed.) pp. 181–198. Aldershot: Ashgate.

Darragh, A.R., Sample, P.L. and Fisher, A.G. (1998). Environment effect on functional task performance in adults with acquired brain injuries: Use of the assessment of motor and process skills. *Arch Phys Med Rehabil* 79, 418–423.

Davidson, J. and Milligan, C. (2004). Embodying emotion sensing space: Introducing emotional geographies. *Soc Cult Geogr* 5, 4, 523–532.

Davidson, J. and Parr, H. (2007). Anxious subjectivities and spaces of care: Therapeutic geographies of the UK National Phobics Society. In *Therapeutic Landscapes* (A. Williams, ed.) pp. 95–110. Aldershot: Ashgate.

Denzin, N.K. and Lincoln, Y.S. (1998). *Collecting and Interpreting Qualitative Materials*. Thousand Oaks, CA: Sage.

Donovan, R. and Williams, A. (2007). Home as therapeutic landscape: Family caregivers providing palliative care at home. In *Therapeutic Landscapes* (A. Williams, ed.) pp. 199–218. Aldershot: Ashgate.

Dyck, I. (1998). Women with disabilities and everyday geographies: Home space and the contested body. In *Putting Health into Place: Landscape, Identity and Well-being* (A. Kearns and W.M. Gesler, eds.) pp. 102–119. Syracuse, NY: Syracuse University Press.

English, J., Wilson, K. and Keller-Olaman, S. (2008). Health, healing and recovery: Therapeutic landscapes and the everyday lives of breast cancer survivors. *Soc Sci Med* 67, 68–78.

Evans, J.D., Crooks, V.A. and Kingsbury, P.T. (2009). Theoretical injections: On the therapeutic aesthetics of medical spaces. *Soc Sci Med* 69, 716–721.

Gesler, W.M. (1991). *The Cultural Geography of Health Care*. Pittsburgh, PA: University of Pittsburgh Press.

Gesler, W.M. (1992). Therapeutic landscapes: Medical issues in light of the new cultural geography. *Soc Sci Med* 34, 7, 735–746.

Gesler, W.M. (1993). Therapeutic landscapes: Theory and a case study of Epidauros, Greece. *Environ Plann D* 11, 171–189.

Gesler, W.M. (1998). Bath's reputation as a healing place. In *Putting Health into Place: Landscape, Identity and Well-being* (A. Kearns and W.M. Gesler, eds.) pp. 17–35. Syracuse, NY: Syracuse University Press.

Gesler, W.M. (2003). *Healing Places*. Lanham, MD: Rowman and Littlefield.

Gladstone, B.M., Volpe, T. and Boydell, K.M. (2007). Issues encountered in a qualitative secondary analysis of help-seeking in the prodrome to psychosis. *J Behav Health Serv Res* 34, 4, 431–442.

Godkin, M.A. (1980). Identity and place: Clinical applications based on notions of rootedness and uprootedness. In *The Human Experience of Space and Place* (A. Buttimer and D. Seamon, eds.) pp. 73–85. London: Croom Helm.

Goldkuhl, G. and Cronholm, S. (2010). Adding theoretical grounding to grounded theory: Toward multi-grounded theory. *Int J Qual Methods* 9, 2, 187–205.

Gray, J., Rossiter, K., Colantonio, A., Kontos, P., Gilbert, J., Keightley, M., Gale, S., James, S., Nacos, M. and Prince, M. (2011). After the crash: A play about brain injury. *Can Theatre Rev* 146, 66–86.

Jorge, R.E., Robinson, R.G., Moser, D., Tateno, A., Crespo-Facorro, B. and Arndt, S. (2004). Major depression following traumatic brain injury. *Arch Gen Psychiatry* 61, 42–50.

Kearns, R.A. and Barnett, J.R. (2000). "Happy Meals" in the Starship Enterprise: Interpreting a moral geography of health care consumption. *Health Place* 6, 81–93.

Kearns, R.A. and Collins, D.C. (2000). New Zealand children's health camps: Therapeutic landscapes and the contract state. *Soc Sci Med* 51, 1047–1059.

Kearns, R.A. and Moon, G. (2002). From medical to health geography: Novelty, place and theory after a decade of change. *Prog Hum Geogr* 26, 5, 605–625.

Kennedy, H.P., Shannon, M.T., Chuahorm, U. and Kravetz, M.K. (2004). The landscape of caring for women: A narrative study of midwifery practice. *J Midwifery Womens Health* 49, 1, 14–23.

Kontos, P., Miller, K.L., Gilbert, J., Mitchell, G.J., Colantonio, A., Keightley, M. and Cott, C.A. (2012). Improving client-centered brain injury rehabilitation through research-based theater. *Qual Health Res* 22, 12, 1612–1632.

Law, M., Baptiste, S. and Mills, J. (1995). Client-centred practice: What does it mean and does it make a difference? *Can J Occup Ther* 62, 5, 250–257.

Laws, G. (1995). Embodiment and emplacement: Identities, representation and landscape in Sun City retirement communities. *Int J Aging Hum Dev* 40, 4, 253–280.

Le Compte, M. and Preissle, J. (1993). *Ethnographic and Qualitative Design in Educational Research*. San Diego, CA: San Diego Academic Press.

Maitra, K. and Erway, F. (2006). Perception of client-centered practice in occupational therapists and their clients. *Am J Occup Ther* 60, 3, 298–310.

Manzo, L.C., Kleit, R.G. and Couch, D. (2008). "Moving three times is like having your house on fire once": The experience of place and impending displacement among public housing residents. *Urban Stud* 45, 9, 1855–1878.

McClusky, J.F. (2008). Creating engaging experiences for rehabilitation. *Top Stroke Rehabil* 15, 2, 80–86.

McLean, A. (2007). The therapeutic landscape of dementia care: Contours of intersubjective spaces for sustaining the person. In *Therapeutic Landscapes* (A. Williams, ed.) pp. 315–332. Aldershot: Ashgate.

Menon, D.K., Schwab, K., Wright, D.W. and Maas, A.I. (2010). Position statement: Definition of traumatic brain injury. *Arch Phys Med Rehab* 91, 11, 1637–1640.

Miller, K.L., Reeves, S., Zwarenstein, M., Beales, J.D., Kenaszchuk, C. and Conn, L.G. (2008). Nursing emotion work and interprofessional collaboration in general internal medicine wards. *J Adv Nurs* 64, 4, 332–343.

Milligan, C. (2007). Restoration or risk? Exploring the place of the common place. In *Therapeutic Landscapes* (A. Williams, ed.) pp. 255–271. Aldershot: Ashgate.

Milligan, C., Gatrell, A. and Bingley, A. (2004). "Cultivating health": Therapeutic landscapes and older people in northern England. *Soc Sci Med* 58, 1781–1793.

Moon, G., Kearns, R.A. and Joseph, A. (2006). Selling the private asylum: Therapeutic landscapes and the (re)valorization of confinement in the era of community care. *Trans Inst Br Geogr* 31, 131–149.

Mortenson, B. and Dyck, I. (2006). Power and client-centred practice: An insider exploration of occupational therapists' experiences. *Can J Occup Ther* 73, 5, 261–271.

Nicholls, D.A. and Cheek, J. (2006). Physiotherapy and the shadow of prostitution: The Society of Trained Masseuses and the massage scandals of 1894. *Soc Sci Med* 62, 2336–2348.

Nicholls, D.A. and Gibson, B.E. (2010). The body and physiotherapy. *Physiother Theory Pract* 26, 8, 497–509.

Oandasan, I., Gotlib, C.L., Lingard, L., Karim, A., Jakubovicz, D., Whitehead, C., Miller, K.L., Kennie, N. and Reeves, S. (2009). The impact of time and space on interprofessional teamwork in Canadian primary care settings: Implications for health care reform. *Prim Health Care Res Dev* 10, 2, 151–162.

Ory, M.G. and Williams, T.F. (1989). Rehabilitation: Small goals, sustained interventions. *Ann Am Acad Pol Soc Sci* 503, 1, 60–71.

Papadimitriou, C. (2008). "It was hard but you did it": The co-production of "work" in a clinical setting among spinal cord injured adults and their physical therapists. *Disabil Rehabil* 30, 5, 365–374.

Parr, H., Philo, C. and Burns, N. (2005). "Not a display of emotions": Emotional geographies in the Scottish Highlands. In *Emotional Geographies* (J. Davidson, L. Bondi and M. Smith, eds.) pp. 87–116. Aldershot: Ashgate.

Pascal, J. (2010). Space, place and psychosocial well-being: Women's experience of breast cancer at an environmental retreat. *Illness Crisis Loss* 18, 3, 201–216.

Patton, M.Q. (1990). *Qualitative Evaluation and Research Methods*. Newbury Park, CA: Sage.

Pegg, P.O., Auerbach, S., Seel, R., Buenaver, L., Plybon, L. and Kiesler, D. (2005). The impact of patient-centered information on patients' treatment satisfaction and outcomes in traumatic brain injury rehabilitation. *Rehabil Psychol* 50, 4, 366–374.

Podmore, J. (1998). (Re) reading the "loft living" habitus in Montréal's inner city. *Int J Urban Reg Res* 22, 2, 283–302.

Rebeiro, K.L. (2000). Client perspectives on occupational therapy practice: Are we truly client-centred? *Can J Occup Ther* 67, 1, 7–14.

Reed-Danahay, D. (2001). "This is your home now!": Conceptualizing location and dislocation in a dementia unit. *Qual Res* 1, 1, 47–63.

Restall, G., Ripat, J. and Stern, M. (2003). Framework for strategies for client-centred practice. *Can J Occup Ther* 70, 2, 103–112.

Richardson, J., Law, M., Wishart, L. and Guyatt, G. (2000). The use of a simulated environment (easy street) to retrain independent living skills in elderly persons: A randomized controlled trial. *J Gerontol A-Biol* 55A, 10, M578–M584.

Rodaway, P. (1994). *Sensuous Geographies: Bodies, Sense and Place*. London: Routledge.

Sinclair, L.B., Lingard, L.A. and Mohabeer, R.N. (2009). What's so great about rehabilitation teams? An ethnographic study of interprofessional collaboration in a rehabilitation unit. *Arch Phys Med Rehab* 90, 7, 1196–1201.

Stewart, M., Brown, J.B., Donner, A., McWhinney, I.R., Oates, J., Wester, W.W. and Jordan, J. (2000). The impact of patient-centred care on outcomes. *J Fam Pract* 49, 9, 796–804.

Strauss, A.L. (1987). *Qualitative Analysis for Social Scientists*. New York: Cambridge University Press.

Sumsion, T. and Smyth, G. (2000). Barriers to client-centredness and their resolution. *Can J Occup Ther* 67, 1, 15–21.

Thompson, T.L.C. (1990). A qualitative investigation of rehabilitation nursing care in an inpatient rehabilitation unit using Leininger's theory. PhD dissertation. Detroit, MI: Wayne State University.

Thurber, C. and Malinowski, J. (1999). Summer camp as a therapeutic landscape. In *Therapeutic Landscapes: The Dynamics Between Place and Wellness* (A. Williams, ed.) pp. 29–51. Lanham, MD: University Press of America.

Toglia, J. and Kirk, U. (2000). Understanding awareness deficits following brain injury. *NeuroRehabilitation* 15, 57–70.

Townsend, E. (1999). *Good Intentions Overruled: A Critique of Empowerment in the Routine Organization of Mental Health*. Toronto: University of Toronto Press.

Townsend, E., Langille, L. and Ripley, D. (2003). Professional tensions in client-centered practice: Using institutional ethnography to generate understanding and transformation. *Am J Occup Ther* 57, 17–28.

Turner, B., Ownsworth, T., Cornwell, P. and Fleming, J. (2009). Reengagement in meaningful occupations during the transition from hospital to home for people with acquired brain injury and their family caregivers. *Am J Occup Ther* 63, 5, 609–620.

van den Broek, M.D. (2005). Why does neurorehabilitation fail? *J Head Trauma Rehabil* 20, 5, 464–473.

Wakefield, S. and McMullan, C. (2005). Healing in places of decline: (Re)imagining everyday landscapes in Hamilton, Ontario. *Health Place* 11, 299–312.

Warren, N. and Manderson, L. (2008). Constructing hope: Dis/continuity and the narrative construction of recovery in the rehabilitation unit. *J Contemp Ethnogr* 37, 2, 180–201.

Watson, D.B., Burtagh, M.J., Lally, J.E., Thomson, R.G. and McPhail, S. (2007). Flexible therapeutic landscapes of labour and the place of pain relief. *Health Place* 13, 865–876.

Wilkins, S., Pollock, N., Rochon, S. and Law, M. (2001). Implementing client-centred practice: Why is it so difficult to do? *Can J Occup Ther* 68, 2, 70–79.

Williams, A. (1998). Therapeutic landscapes in holistic medicine. *Soc Sci Med* 46, 1193–1203.

Williams, A. (1999). Place identity and therapeutic landscapes: The case of home care workers in a medically underserviced area. In *Therapeutic Landscapes: The Dynamics between Place and Wellness* (A. Williams, ed.) pp. 71–96. Oxford: University Press of America.

Williams, A. (2002). Changing geographies of care: Employing the concept of therapeutic landscapes as a framework in examining home space. *Soc Sci Med* 55, 141–154.

Williams, A. (2007). Introduction: The continuing maturation of the therapeutic landscape concept. In *Therapeutic Landscapes* (A. Williams, ed.) pp. 1–12. Aldershot: Ashgate.

Williams, A. (2010a). Spiritual therapeutic landscapes and healing: A case study of St. Anne de Beaupre, Quebec, Canada. *Soc Sci Med* 70, 10, 1633–1640.

Williams, A. (2010b). Therapeutic landscapes as health promoting places. In *A Companion to Health and Medical Geography* (T. Brown, S. McLafferty and G. Moon, eds.) pp. 207–223. Chichester: Wiley-Blackwell.

Wilton, R. and DeVerteuil, G. (2006). Spaces of sobriety/sites of power: Examining social model alcohol recovery programs as therapeutic landscapes. *Soc Sci Med* 63, 649–661.

Wiseman-Hakes, C., MacDonald, S. and Keightley, M. (2010). Perspectives on evidence based practice in ABI rehabilitation. "Relevant research": Who decides? *NeuroRehabilitation* 26, 1–14.

8

Rethinking Social-Relational Perspectives in Rehabilitation: Traumatic Brain Injury as a Case Study

Jacinta Douglas

Melanie Drummond

Lucy Knox

Margaret Mealings

Introduction

We would like to begin this chapter by asking you, the reader, to take some time to reflect upon what it is that brings a sense of meaning to your life. While several areas may have come to mind as sources of meaning in your life, it is likely that for most, if not all of us, relationships with significant others will feature high and large on our lists. In fact, research has shown that relationships with others are the most frequently reported source of life meaning across the lifespan (Baum and Stewart 1990; Delle Fave et al. 2013; De Vogler and Ebersole 1981, 1983; Emmons 2003; Hicks and King 2009; O'Connor and Chamberlain 1996, 2000). Unfortunately, we also know that it is this area of social relationships that can produce crucial and enduring challenges to people with

acquired disability. This challenge is clearly illustrated in the case of acquired neurological injury and particularly traumatic brain injury (TBI), which we use in this chapter to
demonstrate social-relational perspectives in rehabilitation.

Over the years, the social-relational domain after TBI has been generally characterized by poor outcomes, and, as a consequence, this area has provided relatively little
support for the effectiveness of rehabilitation (Hoofien et al. 2001; Temkin et al. 2009).
Indeed, no matter where in the world outcome is investigated, what tool is used to measure it, or when in the rehabilitation journey it is investigated, restrictions in the domain
of social living emerge as a common experience for the majority of survivors of TBI.
In the United Kingdom, Wade et al. (1998) used the Rivermead Head Injury Follow-up
Questionnaire and reported that 61% of 314 participants admitted with brain injury
of any severity continued to experience some social disability at six months follow-up.
Using the Functional Status Examination, Dikman et al. (2003), working in the United
States, found that 49% of 210 individuals with moderate-severe brain injury had some
level of difficulty in the domain of social integration, even three to five years after the
event. In Australia, Tate et al. (2005) used the Sydney Pyschosocial Reintegration Scale
to track the outcome of a consecutive series of 100 patients with severe TBI. Six years
after injury, restricted relationships were characteristic of outcome for 63.2% of participants (Tate et al. 1989), and at 23 years, this outcome continued to be the case for
the majority (54.4%) of participants (Tate et al. 2005). Quantitatively this is a daunting
prospect and one that comes achingly to life when the words of those living with the
consequences of TBI are considered.

Table 8.1 presents the insider's perspective on social outcome through a compila-
tion of comments made by participants with TBI who have generously worked with us
on a number of qualitative or mixed methods research projects (Douglas 2010b, 2012;
Douglas and Spellacy 2000; Shorland and Douglas 2010). These participants' insights
spanning years of living after injury provide a powerful reminder of the significance of
social relationships. Rachel (all participant names are pseudonyms) seems to capture the
essence of this personal challenge when she says, "sometimes a loneliness sort of washes
over me and I just want somebody to talk to." She creates the metaphor of drowning in
loneliness and longing for interaction, as well as referring to the critical lifeline assumed
by family, in this case her mother: "I only see mum in the afternoons when she comes
home from work or on the weekend."

With longstanding evidence of social isolation as background, we focus on the challenge of social relationships in this chapter. We argue that social relationships and interaction are important constructs in rehabilitation that can reveal novel approaches to
assessment and intervention when used to guide practice. The overall aim of the chapter is to illustrate the potential importance of attending to the social-relational aspects
of individuals' lives in the rehabilitation process. We do this through examination of
selected social theories in relation to the experiences of people whose lives have been
affected by TBI. We explore qualitative evidence that reflects a symbolic interactionist stance (Blumer 1969; Mead 1934) and has emerged from grounded theory inquiries
completed with people who know well the consequences of TBI. Our aim is to provide

TABLE 8.1 The Insider's Perspective: Reflections on Social Outcome after Severe TBI

Ben (27 years of age, 8 years after injury)
…to be accepted by other people, to have friends. I go out by myself on Saturdays just to tell people at work I've been out.

Katherine (31 years of age, 6 years after injury)
Find it very hard to make friends—not game enough. People will look at me and think I'm different. I feel that, Yes, I'm different even this far on. I need to feel that I'm loved.

Rachel (22 years of age, 2.9 years after injury)
I only see Mum in the afternoons when she comes home from work or on the weekend. So I just get a bit lonely sometimes in the house or sometimes when I'm at [university], sometimes a loneliness sort of washes over me and I just want somebody to talk to.

Dave (30 years of age, 15 years after injury)
I'm very much a one off sort of, or, whereas there's a big group here and a big group there and I'm sort of in the middle, sort of one on my own.

Michael (26 years of age, 6 years after injury)
I don't like being alone, when I'm alone I get sad thinking about life. Better when I am around other people.

Chris (24 years of age, 4.2 years after injury)
So that's the way it is for me—always being left out.

Ian (33 years of age, 3.5 years post injury)
After head injury you know everything just changes and your whole, your whole like social life just goes downhill you know, 'cause nobody wants to know you.

Note: The quotes included in this table are taken from interviews with participants in the following studies Douglas (2010b, 2012), Douglas and Spellacy (2000), and Shorland and Douglas (2010).

sufficient evidence to support a greater focus on considering the "relational self" in how we work with people in rehabilitation.

The Power of Social Relationships

The power of social relationships and the support afforded by them has long been acknowledged as a moderator of stress and well-being (Cohen and Wills 1985; Lin et al. 1986). More socially isolated people have been shown to be less healthy, both psychologically and physically, and have a higher risk of mortality (Cohen and Janicki-Deverts 2009; House et al. 1988; Sani 2012; Umberson and Montez 2010; Umberson et al. 2006). It is now more than a century since Durkheim (1951 [1897]) used census data from several European countries to explore the relationship between collective life (family, religious, political) and suicide. His finding that social life constitutes a shield against suicide has been not only replicated but also extended across a broad range of health and well-being outcomes. For example, more socially integrated individuals have been shown to live longer (Seeman 1996), have more resistance to disease ranging from the common cold (Cohen et al. 1997) to cardiovascular disease (Kop et al. 2005) including stroke (Rutledge et al. 2008), and show less age-related cognitive decline including dementia (Fratiglioni et al. 2004). House et al. (1988) and more recently Umberson et al. (2006) effectively summarized the case for social relationships as a health risk factor when they noted that the evidence linking social relationships to health, mortality, and morbidity was as strong as evidence linking smoking, obesity, blood pressure, and physical activity to health. Putnam (2000, p. 331) succinctly captured the essence of this link as well as the intervention potential of social relationships when he stated, "As a rough rule of thumb, if you belong to no groups but decide to join one, you cut your risk of dying over the next year in half."

Social relationships can be indexed by several variables, most notably the size of our social networks, the social support we perceive and receive, the quality and quantity of our social interactions, and our feelings of loneliness or isolation. These variables have all drawn research attention in the context of TBI with social isolation and loneliness emerging as a convergent finding (Douglas and Spellacy 2000; Eames et al. 1995; Hoofien et al. 2001; Lefebvre et al. 2008). In our own research (Douglas and Spellacy 2000), we investigated the role of social support as a factor influencing the long-term experience of depression in 35 survivors of severe TBI and their family caregivers. Findings clearly reinforced the importance of social support as a significant predictor of emotional adjustment for both the injured participants and their family members. Specific exploration of the component functions of social support revealed that strong-tie expressive support was particularly vital for maintaining the well-being of injured individuals and family caregivers. The expressive dimension emphasizes the emotional aspects of relationships (Lin et al. 1986). Sharing emotional problems, exchanging life experiences, and spending time with a person one likes or feels close to are examples of expressive social support actions. We all need close relationships in order to receive expressive social support. For the participants with TBI in this study, this need was often unmet. Further, as a result of lack of community involvement, they typically had a few if any opportunities to build new reciprocal friendships that could provide them with emotional or practical support. Consequently, there was little likelihood that this negative situation would diminish over time.

Relationships with significant others are not only an important vehicle for social support but also crucial in the evolution of self. Indeed, as we discuss in the section A Symbolic Interactionist Perspective, the notion that significant others influence the self is a long held tenet of psychological thought dating back to the concept of the "social me" first coined by William James in his seminal text *Principles of Psychology* (1890).

A Symbolic Interactionist Perspective

A man's [sic] *Social Self* is the recognition which he gets from his mates. We are not only gregarious animals, liking to be in sight of our fellows, but we have an innate propensity to get ourselves noticed, and noticed favorably, by our kind. No more fiendish punishment could be devised, were such a thing physically possible, than that one should be turned loose in society and remain absolutely unnoticed by all the members thereof. If no one turned round when we entered, answered when we spoke, or minded what we did, but if every person we met 'cut us dead,' and acted as if we were non-existing things, a kind of rage and impotent despair would ere long well up in us, from which the cruelest bodily tortures would be a relief; for these would make us feel that, however bad might be our plight, we had not sunk to such a depth as to be unworthy of attention at all. (James 1890, vol. 1, *Principles of Psychology*, p. 294)

In this excerpt, James evokes a powerful image of the emotional consequences of social isolation. Research findings have shown us that James's "impotent despair" is unfortunately not a rare experience for many survivors of severe TBI. This response is exemplified by two participants in our recent study (Douglas 2010b, see Table 8.1): Chris said, "So that's the way it is for me—always being left out"; and Ian explained, "After head injury you know everything just changes and your whole, your whole like social life just goes downhill you know, 'cause nobody wants to know you."

In a nutshell, James considered the "social self" as a key component of self essentially derived from an individual's internal conceptualization of the reactions of others to him/her. Following James's early consideration of the social self, many scholars have emphasized the personal significance of the interplay between the individual and society. Notable among these scholars are those working in the tradition of symbolic interactionism, considered to be one of the most enduring social theories of the twentieth century (Benzies and Allen 2001; Oliver 2012; Plummer 2000). Symbolic interactionism views the individual and the context in which the individual exists as inseparable and mutually constructed in the course of social interactions. Cooley (1902) captured this inseparability between the person and society by describing them as two sides of the same coin. He viewed the social self as the whole self and coined the term *looking glass self* to convey the idea that an individual perceives self through the eyes of others. Mead (1934), generally considered to be the founder of symbolic interactionism, views self as emerging from the responses of others. Mead describes the individual as "taking the attitudes of other individuals toward himself [*sic*] within a social environment or context of experience and behavior in which both he and they are involved" (p. 203). It is the individual's perception or interpretation of his/her own social world that influences the self.

Blumer, Mead's student, progressed the tradition of symbolic interactionism, particularly as an alternative to rigid behaviorism and extreme positivism. He said:

> The term "symbolic interactionism" refers of course to the peculiar and distinctive character of interaction as it takes place between human beings. The peculiarity consists in the fact that human beings interpret or "define" each other's actions instead of merely reacting to each other's actions. Their "response" is not made directly to the actions of one another but instead is based on the meaning which they attach to such actions. Thus, human interaction is mediated by the use of symbols, by interpretation, or by ascertaining the meaning of one another's actions. This mediation is equivalent to inserting a process of interpretation between stimulus and response in the case of human behavior. (Blumer 1962, p. 180)

Blumer (1962, 1969) went on to outline important tenets that underpin this theoretical stance. Individuals are seen as acting toward people and things based upon the meanings they have given to those people or things. Meaning arises in the process of interaction between people; it is a social process. People are assumed to have the capacity to negotiate meaning through symbols, giving rise to an interpretive process that is ever changing. In other words, human behavior is emergent and continually constructed.

People actively shape their own futures by forming new meanings and ways to respond (Benzies and Allen 2001; Blumer 1962, 1969). While symbolic interactionism is just one approach to understanding how people's actions are shaped, it provides a useful lens for considering important concepts relevant to rehabilitation.

Symbolic Interactionism in the Context of Rehabilitation

Symbolic interactionism has substantial potential to inform our thinking and rethinking about rehabilitation. From the symbolic interactionist perspective, truth is not fixed because meaning changes depending on the context for that individual. With this perspective in mind, during rehabilitation, people need to be viewed in the context of their own changing environments. Further, if the processes of interaction between people create meaning as we argue, then interaction itself represents the medium for understanding and new learning. Indeed, clear evidence of the importance of therapeutic alliance in psychotherapy (Ackerman and Hilsenroth 2003; Baldwin et al. 2007; Bordin 1979; Crits-Christoph et al. 2011; Hatcher 2011; Horvath 2006; Horvath et al. 2011; Hougaard 1994) speaks to the power of client–practitioner relationships to facilitate change. Recently,

TABLE 8.2 Symbolic Interactionism, Family Systems Theory, and Rehabilitation

Theoretical Principle	Rehabilitation Practice
Individuals act toward people and things based upon the meanings that they have given to those people or things.	Know how the person defines his or her world—objects, events, individuals, groups, structures.
The individual depends on and is influenced by the environment/s of which he or she is part.	Work within environments that influence the individual and in turn are influenced by the individual (e.g., family, neighborhood, school, work).
Meaning arises in the process of interaction between people. It takes place in the context of relationships.	Seek to understand the impact of the injury and its consequences on valued relationships.
	Support the maintenance and development of valued relationships.
	Deliver intervention within relational contexts.
Language gives humans a means by which to negotiate meaning through symbols.	Develop a common language reflective of personal values and meanings.
Self is developed through the process of interaction with others.	Facilitate social interaction, reflect on the process, and maximize positive encounters.
Families operate as social systems with interrelated elements (family members) creating a family structure with unique behaviors specific to the structure as a whole not simply reflective of individual members.	Work within the family system; identify patterns of interaction, messages, and rules within the family system.
Change in the family situation means readjustment of the total system and can pose problems and challenges for every single member of the family.	Pay particular attention to how the injury may influence the family's typical ways of being; for example, family roles (parent, spouse, clown, emotional one, tough guy) and subsystems may be disrupted.

research has begun to explore the potential influence of the therapeutic alliance in the context of rehabilitation (Bright et al. 2012; Schönberger et al. 2006, 2007; and see Chapter 13, for further discussion).

Coming to know the best way of dealing with a situation requires understanding the meaning of that situation from the perspective of individuals and their relationships. Knowledge cannot be fully developed through what is typically captured in reports or assessments; it is a complex process based on shared interactions with the injured individual and those with whom the individual relates. From the perspective of symbolic interactionism, the therapist benefits from direct consideration of the role/s of the client and viewing the world as much as possible from the client's perspective. To illustrate the potential application of symbolic interactionism to rehabilitation, basic theoretical principles with examples of their application in rehabilitation practice are outlined in Table 8.2. These principles are also broadly consistent with family systems theory (Bowen 2007; Gurman and Kniskern 1991; Vetere and Dallos 2003), some examples of which are also included in Table 8.2.

The Concept of the Relational Self

While theorizing about social influences on the self has fluctuated over the last century, interest in the impact of significant others on the self has been rekindled more recently. This upsurge of attention is exemplified in the work of Chen and colleagues (Anderson and Chen 2002; Chen et al. 2006), who have presented their conceptualization of the relational self to capture the "specific link between the self and significant others" (Chen et al. 2006, p. 153) or "the self that is experienced in relation to significant others in one's life" (p. 173). Chen et al. (2006) maintain:

> [M]ost people possess multiple relational selves and these selves exist at varying levels of specificity. ... a *relationship-specific relational self* designates the self in relation to a specific significant other, whereas a *generalized relational self* is akin to a summary representation of the self in the context of multiple relationships. These relationships may involve either a single normatively defined relationship domain (e.g., "me when I'm with family members") or idiosyncratic groupings of relationships (e.g., "me when I'm with close others of the same age" or "me when I'm with my poker buddies"). Finally, in addition to relationship specific and generalized relational selves, people may possess a *global relational self* which denotes conceptions and aspects of the self in relation to significant others as a general class of individuals. (p. 153)

According to this approach, our relational selves grow out of our interactions with significant others and specifically define the self in relation to others. Relational selves can be seen as being composed at least in part by attribute-based conceptions of the self linked to significant others. This view is consistent with the concept of self-with-other representations defined by Ogilivie and colleagues as "the set of personal qualities ... that an individual believes characterizes his or her self when with a particular other person" (Ogilivie and Ashmore 1991, p. 290). Like relational selves, self-with-other representations can be specific or generalized. Specific representations reflect a set of self-attributes linked to a single significant other, while generalized representations are linked to a group or groups of significant others.

Generalized self-other representations evolve through repeated experiences. For example, an individual may develop a generalized relational self as inept or lacking ability following repeated interactions with significant others that have triggered feelings of incompetence. Indeed, research on relational schemas or interpersonal scripts involving if–then contingencies has shown that repeated activation of these scripts can lead to negative and positive self-inferences and self-evaluations (Baldwin and Dandenaeau 2005; Baldwin and Main 2001). In the context of the above incompetence example, repeated experience of the contingency "If I don't do this properly, they will think I'm useless" can lead to a generalized negative self-inference, such as "I am incompetent."

Chen at al. (2006) propose that our relational selves are activated by contextual cues and when activated they influence our affective responses and behaviors including goal-directed, self-regulatory, and interpersonal behaviors (e.g., Berk and Anderson 2000; Higgins and May 2001; McGregor and Marigold 2003; Mikulincer et al. 2001). In addition, evidence suggests that aspects of the self associated with close relationships provide existential meaning and inform the individual of his or her place in society (Chen et al. 2006; Mikulincer et al. 2003). Finally while not yet empirically tested, it could also be expected that relational selves play a motivational function through their role in psychological well-being whereby significant others have been shown to influence need satisfaction (La Guardia et al. 2000).

Although we have long been aware of the enduring negative social consequences of TBI, there has been relatively little indepth research of the relationship domain, particularly when viewed in the context of the magnitude of research focused at the impairment level of disability. This relative shortage of research is broadly demonstrated by the results of database searches using "social relationships" and "traumatic brain injury" as search terms across article titles, abstracts, and keywords. For example, an illustrative quick search conducted in *Scopus* (Elsevier) at the time of writing this chapter yielded 19 articles published since 1995, while combining the search terms "mobility" and "traumatic brain injury" yielded 261 published articles. In other words, for each article published on social relationships, there are 14 articles published on mobility. Further, this limited research attention also appears to be mirrored in the context of rehabilitation, with intervention efforts directed toward relationships and social interaction continuing to lag considerably behind efforts focusing on impairment and activity.

Applying a Social-Relational Lens to TBI Rehabilitation

It is our contention that we have much to gain by incorporating a social-relational perspective more directly within our approach to brain injury research and rehabilitation. In particular, we have found the principles of social interactionism and the construct of the relational self to not only provide powerful insights into the experience of TBI but also reveal novel pathways for the development of specific intervention strategies and rehabilitation approaches more generally. Our work in this endeavor shares parallels with that of Bowen et al. (2010) and Wilson et al. (2009) and also builds on the seminal contribution of Ylvisaker and Feeney (1998).

In the sections First Scenario: TBI and the Relational Self, Second Scenario: Impairment Impacting the Relational Self When You Least Expect, and Third Scenario: Social Mediation of Cognitive Processes, we apply a social-relational lens to understanding the experience of severe TBI and illustrate implications for rehabilitation. Three contexts are explored. The first sheds light on the relational self and how it can be harnessed to facilitate change during rehabilitation. In the second case, we highlight the importance of including a social-relational perspective in assessment by illustrating how a sensory deficit can unexpectedly diminish an intimate relationship. The third scenario involves the construct of decision making and reveals its complex social nature. Each of these situational scenarios has been drawn from our own research conducted within a social interactionist framework using constructivist grounded theory methodology (Charmaz 2006, 2009).

First Scenario: TBI and the Relational Self

There is a growing body of work investigating self-related phenomena and their impact on psychosocial outcomes in the context of TBI (Cantor et al. 2005; Douglas 2010a, 2013; Gracey and Ownsworth 2012; Gracey et al. 2008; Levack et al. 2010; Muenchberger et al. 2008; Nochi 1998, 2000; Ownsworth 2014; Ylvisaker and Feeney 2000; Ylvisaker et al. 2008). (For further discussion, also see Chapter 9.) This body of literature not only supports the central role played by the construct of self in shaping post-injury adjustment but also serves to validate its importance as a rehabilitation relevant construct. A recent study we undertook to gain understanding of how 20 adults with severe TBI viewed "self" in the long term provided insight into the dynamic construction of self and also into the importance of maintaining self through social relationships (Douglas 2013). Grounded theory analysis of the data yielded three major themes labeled: *who I am*; *how I feel about myself*; and *staying connected*. For the purposes of the chapter, we focus on this latter theme.

Staying Connected

In our 2013 study, insights into the relational self (see Figure 8.1) emerged across all three themes but particularly from the theme labeled *staying connected*. The categories that emerged within this theme captured most of the sources of support and relationships defined within the "environmental factors" domain of the World Health Organization's (WHO) International Classification of Functioning, Disability and Health (ICF) (WHO 2001), and included family (immediate and extended), friends, caregivers (paid support workers), community members (acquaintances and neighbors), and pets (companion animals). Family was a particularly important social connector across all participants with some describing relationships with family as their only ("I have no friends anymore, only my family," Douglas 2013, p. 68) and most valued ("Family, the way they are …

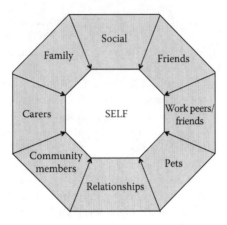

FIGURE 8.1   Self through social relationships.

they're just here and I know they're for me," p. 67) social relationships. While forming new friendships was commonly mentioned as a goal, relationships with existing friends provided valued social connections. These friendships included relationships with "old" (preinjury) friends ("A close group of friends that I see often … friends I had since before the accident—they stayed with me," Douglas 2013, p. 67) and "new" (postinjury) friends, some made through rehabilitation ("old group of guys from rehab who have had ABI's, we're all friends," Douglas 2013, p. 67) and others typically made through community activities ("Catching up with friends who don't have a disability … friends are from fly-ball," Douglas 2013, p. 67). For others, family (particularly, siblings and cousins) created additional links to social relationships within the domain of friendship ("Visits with my brother and some of my brother's friends—they're fun, they all know me," Douglas 2013, p. 67). Relationships with friends were maintained through various means including direct person-to-person interactions, phone contact, email correspondence, and *Facebook*.

Paid support workers (caregivers) also provided important social relationships for these severely injured adults. Participants enjoyed sharing their thoughts and feelings and doing things with their caregivers and some, who had enjoyed stability in caregiver arrangements, specifically described their paid support workers as friends ("carers are my friends," Douglas 2013, p. 68). The interview data also revealed the importance of companion animals in the context of both emotional and social well-being. For those with pets, caring for and enjoying the company of their pet helped maintain a sense of connection with society. Additionally, having a pet opened up new avenues to social relationships with community members ("I'm never lonely because I have Bess [dog]—she is who I confide in. Large group of friends are from dog obedience too … lady down the street, she has a terrier—really quite a few dog people around here," Douglas 2013, p. 68). Shopkeepers, restaurant staff, and lollypop people (crossing supervisors) were also identified as community members with whom participants regularly enjoyed social interaction showing them specific aspects of their relational selves (exemplified by another quote, "girls in the bakery love my sense of humor").

Our work with students returning to high school after injury has also supported the importance of staying connected (Mealings and Douglas 2010). A range of other studies investigating students' viewpoints of educational experience post–TBI across all levels of schooling, primary, secondary, and tertiary (college, university, technical, and further education) consistently reveal that students' concept of self and identity and their social relationships within the educational environment are crucial factors in shaping return-to-study outcomes (Backhouse and Rodger 1999; Hux et al. 2010; Kennedy et al. 2008; Mealings and Douglas 2010; Mealings et al. 2012; Sharp et al. 2006; Stewart-Scott and Douglas 1998; Todis and Glang 2008; Vaidya 2002).

The case study of Brett is presented here to illustrate the importance of strengthening and building relationships in order to facilitate participation in the school environment.

Brett's Return to School

Brett was 13 years of age and was settling into Year 7 at a local regional secondary school when he was injured in a motor vehicle accident. Brett lived at home with his parents and two older siblings and was active in a wide range of social activities including riding his bike, playing golf, and meeting up with friends. Brett had a general goal of becoming a builder in the future. In the accident Brett sustained a severe TBI with no other injuries, and his period of post-traumatic amnesia (PTA) was 45 days. Brett commenced a graded return to school about five months postinjury, joining in selected activities with his former class toward the end of the year with the aim of preparing for a fuller program in Year 8.

Brett's school journey was a difficult one after his injury. He was unable to settle into most of his mainstream classes, and alternative programs both within and external to the school were explored. Brett eventually participated in a community-based educational program with integration aide support, which was facilitated and monitored through his high school. Brett's changes such as his difficulty managing his anger, "I've got a lot more aggressive side now, (I) was just being, being a little shit," and his difficulties with thinking, "like I forgot so much that, oh, it was just too challenging," caused significant disruption to his school activities: "you see they've had to move me out of so many school environments, just one thing after another." The most significant implication of Brett's changed behavior was the effect on his social supports within the school context. While his relationship with friends remained one of the things Brett felt helped him to stay at school, his interactions with most teachers were difficult, "the way they treat you ... they don't understand what I've been through and they never will," resulting in Brett feeling that being at school was "frustrating, painful, mentally tiring, physically tiring, just too much." However, it was also the strength of two particular relationships, with an aide and a teacher, that assisted Brett to maintain his connection with the educational system. It took some time for Brett and the school to find the right aide, "one was old, one didn't like my attitude, one was too busy to have me" but "the aide I've got at the moment, he puts me right." This strong supportive relationship along with "a teacher that I know actually cares" enabled Brett to continue trying new programs and to persist in his community program, which is helping him to "get skills to doing stuff for work."

Brett's experiences serve to emphasize the potential benefits of applying a social-relational framework to return to study after brain injury. Intervention strategies need to be developed with direct consideration of the self in social context and the specific relationships the injured individual has with significant others in that context. Teacher education concerning the nature and consequences of TBI could reduce the likelihood of injured students sharing Brett's feeling that teachers did not understand what he had been through. His experience of supportive relationships with his teacher who "actually cared" and his aide who "put me right" illustrates how the sense of being understood could help injured individuals navigate their way through the challenge of change to personal goals. We contend that the characteristics of Brett's relational self, triggered with these significant others at school, facilitated interactions that were positive and contributed to his sustained engagement in the education environment.

Practice Implications

In the context of Chen et al.'s (2006) conceptualization of relational selves, relationships provide more than simple connections between self and other knowledge. They represent typical patterns of self–other relating, likely to be stored in procedural memory and capable of being activated by cues associated with the other or encounters in which the dynamics resemble those experienced with the other. They can evoke affective responses and support emotional adjustment. Positive aspects of these relational selves can also provide avenues for intervention, particularly in the domain of social interaction. For example, relational behaviors identified in the context of self-with-others can be targeted in therapy and naturally cued in new community situations to facilitate new social relationships. This type of approach enables clinicians to enhance positive/productive behaviors evident within the client's interpersonal repertoire and use these response patterns to replace negative/nonproductive behaviors. Positive or productive behaviors and coping strategies are defined as those that facilitate problem resolution (Frydenberg and Lewis 1997) and positive interpersonal interaction (Douglas et al. 2013). Further, according to relational self theory (Chen et al. 2006), the accessibility or baseline activation readiness of a relational knowledge construct (like any other construct) increases with its repeated activation (e.g., Higgins and Brendl 1995). Thus, frequent contextual activation of positive aspects of the relational self makes them more likely to be activated in any given context. In rehabilitation, this means that frequent activation of positive relational self structures (such as being a person who respects others) can increase a person's accessibility to productive interpersonal behaviors and thus provide a means of reducing nonproductive interpersonal behaviors, including those associated with impaired executive function. We have recently tested such an intervention (increasing the use of productive interpersonal responses to replace nonproductive responses) in the context of communication-specific coping and found it to be effective for adults with severe TBI (Douglas et al. 2013, 2014).

Within the *staying connected* theme evident in our previous research (Douglas 2013), it was not only the presence of significant other relationships (family, friends, caregivers, and pets) but also concrete reminders of these relationships that played a role in maintaining a sense of belonging to a larger social whole. Concrete reminders included tangible symbols like photos, diaries, certificates, greeting cards, letters, and emails that cue recall of shared experiences, which affirms a sense of connection with significant others and also evokes related affective responses (Gardner et al. 2005; Jonason et al. 2008; Twenge et al. 2007). Gardner et al. (2005) introduced the term *social snacking* in their research exploring unmet belonging needs. They described these indirect social strategies as providing temporary *social sustenance* through periods of social isolation or rejection. Given that social isolation is a persistent consequence of severe TBI, it is not surprising that the use of social snacks can be an important means of affirming connections for people with TBI. This being the case, provision of concrete reminders or social snacks is recommended as a beneficial strategy to employ during intervention and to maintain in home environments. Moreover, the presence of enduring memory deficits following TBI further supports the potential benefits of tangible social symbols because they can cue the recall of life experiences pivotal to the maintenance of the relational self and associated knowledge structures.

Second Scenario: Impairment Impacting the Relational Self When You Least Expect

Olfactory impairment as a result of TBI typically is referred to as post-traumatic anosmia (Costanzo and Miwa 2006). The impairment and its associated consequences are relatively underexamined in the context of brain injury rehabilitation. Incidence rates of post-traumatic olfactory impairment reported in the literature have varied markedly from 10% to 66.8% as a function of methodological variables (Drummond et al. 2007, 2013a). Despite this variation, incidence in samples of participants drawn specifically from TBI rehabilitation units has been shown to be particularly high. In the United States, Callahan and Hinkebein (2002) found that 56% of 122 consecutive newly admitted patients of mixed severity to a brain injury rehabilitation program were impaired, while 68% of 134 consecutively admitted patients to a brain injury unit in Australia were found to have diminished or absent sense of smell (Drummond et al. 2013b). In addition to olfactory dysfunction occurring with high frequency after moderate to severe TBI, it is associated with poor recovery with little to no regeneration of the olfactory system noted over time (Bromley 2000; Deems et al. 1991).

Research in the context of mixed etiology groups has shown that olfactory disturbance can have far-reaching impacts on daily living. There is general agreement that the priority for individuals living with anosmia is safety. However, other areas of living such as work, dietary behavior, personal hygiene, leisure, and general quality of life have been shown to be influenced negatively by the presence of an olfactory disorder (Haxel et al. 2008; Miwa et al. 2001; Santos et al. 2004; Temmel et al. 2001). In our recent study exploring the impact of post-traumatic olfactory impairment on daily activities and social participation from the perspective of adults living with the impairment, participants experienced a broad range of limitations (Drummond et al. 2013a). The limitations identified by participants fell into seven categories including eating and enjoyment of food, food preparation, personal safety, personal hygiene, work, leisure, and relationships. While we had expected a range of limitations to emerge across the interviews, we were somewhat taken aback by the magnitude of the hidden impact of the disorder, particularly on the domain of relationships. This hidden consequence was clearly and poignantly demonstrated by Jack's experiences.

Jack's Story

At the time of injury, Jack was 34 years old. He was a respected chef employed in a prestigious central business district restaurant. His world revolved around his love of food, a passion he shared with his wife. Their social circle was made up predominately of "foodies," people who enjoyed exploring, critiquing, and then attempting to recreate culinary delights. As a result of a motorcycle crash, Jack sustained moderate-severe TBI (duration of PTA 12 days) and significant orthopedic injuries. At the time of interview, approximately six months had elapsed since Jack's injury and he was attending rehabilitation on an outpatient basis.

Jack's rehabilitation program was intensive and for the most part focused on his physical recovery. His goal? To regain his physical strength and endurance so he could return to his much loved job. Following comments about the poor quality of food at the hospital alongside his noticeably increased use of salt, Jack's wife took him to one of his favorite restaurants. It was during this seemingly innocuous event that Jack reports his life changed forever: "go see friend ... and we have a steak with a pepper sauce ... I don't taste nothing ... (my wife) said the steak is perfect ... I don't taste nothing ... I use a lot of chili pepper and salt ... from there we start to understand that I have a problem" (Drummond et al. 2013a, p. 338). Jack had anosmia. He could not smell—neither pleasant odors nor noxious substances. As is usually the case, Jack also reported an absence of taste. He used condiments such as salt, pepper, and chili sauce—anything to elicit a response, something different, something identifiable—but nothing. He was unable to smell or taste.

Initial discussions with Jack centered around the implications of anosmia on his vocation. Being a chef he relied heavily on his senses to ensure his meals were of the highest standard. Jack reported that if he wasn't smelling the food, he was tasting the food—his two "tools of the trade," but this was no longer an option. While the consequence of Jack's anosmia continued to impact his ability to return to work, his focus soon turned to the social consequences of his impairment. Jack's lack of smell and taste meant that he no longer wanted to engage in the social dinners that he and his wife both loved. To him there was no point. He wasn't able to smell or taste the food and so had lost his ability to engage in the activity and social conversation associated with the food critiquing. Jack could no longer justify the costs of eating in gourmet establishments and he had lost the pleasure of eating: "I said to her, 'why we spend one hundred and twenty dollar for both and when I don't taste nothing and I complain when they bring the food' ... dinner before we go to a restaurant ... we both chef so we want something good ... now I go to McDonalds ... for me to go to McDonalds or go to a five start hotel for me it's the same thing. What's the difference?" (Drummond et al. 2013a, p. 338).

Jack's loss of enjoyment associated with shared dining led to the couple no longer meeting up with friends. Over time, they lost contact with many of their old friends and stayed home more and more. The newly imposed social isolation weighed heavily on Jack and his wife's relationship. This stress was increased by the additional and unexpected impact of Jack's olfactory impairment on the couple's sexual relationship. Jack spoke openly about his inability to "smell my wife" and pointed out that this was most distressing for him: "That's sad, even with my wife, I don't smell nothing to get me excited … the smell and taste they touch my life … my private life" (Drummond et al. 2013a, p. 341). Subtle cues such as his wife's perfume or her freshly washed hair could no longer be detected and to Jack: "it cannot get any worse than this."

While Jack felt he would always mourn the loss of his vocation, he was definite that the impact of the olfactory impairment upon his relationship weighed more heavily on him than any other injury-related problem. The sense of smell, so often overlooked or underrated, devastated Jack and his wife's life together. As Jack simply put it, "I've lost the pleasure of life."

Practice Implications

Jack's story is a powerful reminder that our rehabilitation efforts can miss crucial aspects of the social-relational domain of functioning. Indeed, the strain and distress within Jack's marital relationship and the role played by olfactory impairment became apparent as a result of Jack being involved in this research project. It was clear that Jack's social relationships, particularly his intimate relationship with his wife, were urgently in need of direct attention within his rehabilitation. Jack's relational self in the context of his marriage was being eroded, and the interactions within his relationship were becoming increasingly restricted. Social-relational goals and intervention were needed. Nevertheless, Jack's rehabilitation had been focused almost exclusively on vocational outcome. What is more, it may well have remained in that domain, missing the opportunity to support the relational domain at this crucial time, if Jack had not been given the opportunity to reflect on his experience in the research situation. The implication of Jack's story for practice is straightforward. We must always assess social-relational aspects of living routinely and directly and intervene as indicated across the rehabilitation continuum.

Third Scenario: Social Mediation of Cognitive Processes

In this scenario, we use the process of decision making to focus on the social nature of cognitive processes. Through our decisions, we define who we are and how others see us. Our decisions reflect our perceptions of our place in the world, what is important to us, and our goals for the future. This is apparent in the decisions we make throughout the course of each day, as well as decisions that may have significant consequences over the course of our lives (Markus and Nurius 1986).

Decision-making impairments are a common consequence of severe TBI. Assessments of decision-making capacity are therefore an important task for the multidisciplinary rehabilitation team. Traditionally, the assessment process has involved the evaluation of an individual's abilities on a range of clinical tasks, sometimes with limited attention paid to the context in which decision making may be occurring in everyday life. However, the importance of social contexts in the consideration of decision-making impairments is increasingly and necessarily being recognized (Fellows 2012; Knox et al. 2013a).

The social nature of the decision-making process is explored here through the case of Peter and his wife Karen. These data are drawn from our larger qualitative exploration of the experience of decision making after severe TBI from the perspective of individuals with severe TBI and those who participate in the decision-making process with them (Knox et al. 2013b). Peter and Karen separately participated in a series of interviews exploring their experience of decision making after Peter's injury.

Introducing Peter and Karen

Peter is a man in his forties who sustained a severe TBI as a result of a motor vehicle crash nine years prior to his first interview. Prior to his injury, he was employed full-time, was an active participant in several sports, and a member of a large social network. Peter described a range of changes in his life since his injury, including changes to his cognition (such as reduced idea generation and initiation), increased difficulties with communication, and a loss of physical independence.

Karen met Peter five years after his injury. They later married and have one child. Karen reported that when she met Peter, "there was more to him than meets the eye," and that learning about Peter's brain injury had been "a massive learning curve," which continued to present unique daily challenges in the context of their relationship.

A key theme emerging from interviews with Peter and Karen highlighted the socially mediated nature of the process of making decisions in their everyday lives. The impact of social interaction was evident in two ways: (1) the role that it plays in shaping perceived decision-making capability and (2) the role that it plays in supporting (and constraining) decision-making participation.

Shaping Decision-Making Capability

Peter and Karen's experience of decision making exemplified the way that social interactions can shape and reinforce perceptions about the individual's decision-making abilities. These interactions occurred at several levels: within their own relationship, in interactions with those in their social networks, with health and support professionals, and also with those in the broader community.

In our own work (Douglas 2013), we highlighted the importance of social connection and engagement in the development of self-concept after severe TBI. Individuals with TBI frequently report a sense of being viewed differently by others after their injury. Peter reported that he felt he was seen by others as a "big, sick, secret guy" following his brain injury. For Peter, a sense of perceived incompetence was experienced through these social interactions in the words and actions of those he encountered. For example, Peter recounted a situation where he was walking with his partner down the street and encountered a window washer, who he described: "didn't speak to me; he spoke to my wife, and said, 'What's wrong with him?'" Similarly, Karen noted that, when she met Peter, she observed that he was perceived by those around him as being incapable: "He was this, 6 foot 2 man, who, from what I could see, was quite capable of doing a lot more than he was being given credit for ... yet he was being treated by everyone around him, like he couldn't."

For rehabilitation professionals, a focus on identifying impairment and ensuring that the individual develops a greater sense of awareness of their limitations after injury may result in the individual being excluded from the decision-making process. Peter reported situations in rehabilitation where, rather than being encouraged to be an active participant in the decision-making process, he was expected to be a passive recipient of expert advice. For example, he reported, "all the doctors and nurses kept trying to convince my partner and myself to cancel the wedding." Rather than being engaged in a decision-making process whereby a range of options were considered and the risks and benefits associated with each weighed up, Peter felt that he was plainly being told what to do.

As already indicated in this chapter, empirical work on relational schemas and the relational self has shown that that repeated activation of experiences of being assumed to be incompetent or incapable can then be reflected in the individual's self-evaluation of being someone who is not capable of making reasonable decisions. This sense of incompetence can further lead to self-exclusion from making decisions across a range of life areas. Peter described that there had been times where "I wanted to ... but I didn't feel I was able to." This feeling of incompetence led Peter to the self-inference, "I've got nothing to offer that person, so I take a step back."

A sense of self-value and decision-making capability can be restored through relationships whereby the person with brain injury, in this case Peter, experiences himself as an individual whose views are important and who is able to contribute. Karen is able to play this role for Peter: "When I met him, I was like, 'Well there's a guy here who's got a life ... got something to say and he's got a voice.'" Karen's belief in Peter provides him with opportunities to develop an alternative positive self-inference, "I've got something to say" rather than his negative, "I've got nothing to offer."

Supporting Decision-Making Participation

Social connections are especially important for individuals with TBI who are likely to experience increased difficulties in managing social interactions, secondary to cognitive and communication impairments. Decision-making supporters come to understand the individuals with TBI (and their limitations and their need for support and accommodation) through their interactions with them. For example, Peter recognized that he required support with many aspects of decision making, "I'm reliant on, at this stage, my wife, telling me, things that to most other people would be common sense."

However, the nature of the interactions with significant others can either support or constrain decision-making participation. In Peter's case, his wife described that she took on a new role, as an advocate, to support decision-making participation by challenging the perceptions of others and creating a physical space in which Peter could participate in the decision-making process: "Peter's capable of speaking and people just have to take the time to listen to him a little bit more and I'm not his speaker." At times, this necessitated provision of direct feedback to others in order to modify their behavior: "I just said, 'You need to give Peter time ... you need to let Peter speak.'"

This advocacy role may not be easy for decision-making supporters to take on, and in many contexts, there may be little or no encouragement to do so. Decision-making supporters need to be able to reflect on their own actions and consider new ways of supporting the individual in order to maximize their participation in the decision-making process. For Karen, this has meant constant reflection on her own behavior and actions: "I've actually had to learn to step back and say to Peter, 'What do you want to do this weekend?'"

Practice Implications

Peter and Karen's insights into their experience of the decision-making process highlight the socially mediated and relational nature of decision making. In particular, their experience demonstrates that decisions occur within a social context, and the individual's ability to participate in decision making within that context is strongly influenced by the presence of support, the attitudes and beliefs held by decision-making supporters, and their approach in offering support and accommodation to the person with TBI. These issues are particularly relevant in the context of rehabilitation where interactions can introduce or reinforce relational schemas and if–then dynamics that lead to negative self-inferences influencing affective responses and goal-directed, self-regulatory, and interpersonal behaviors. Indeed, relational knowledge structures pertaining to decision making may be activated only rarely or not at all during rehabilitation, thus reducing their baseline activation readiness and consequently accessibility in future situations. Our knowledge of relational schemas indicates that interactional processes including decision-making activity need to be engaged in, monitored, and actively shaped through rehabilitation.

Conclusion

Rehabilitation in all its guises is a social-relational affair. It takes place (for better or worse) in a medium of social interaction that in turn shapes the self-inferences, which characterize the relational selves of the individual with TBI. (As noted earlier in this chapter, further discussion of these ideas is captured in Chapters 9 and 13.) Relational selves are activated by the contextual cues of the environments in which we participate and they influence how we feel and how we behave. The rehabilitation environment and the people in it are potent ingredients in the outcome recipe. Indeed, it is likely to be the quality of relationships with significant others in the injured individual's social contexts that proves to be one of the most powerful predictors of outcome following TBI. People with TBI have told us over and over again that relationships and interactions with those around them in health and rehabilitation settings, at home, in school, at work, and in their neighborhood are crucial to their long-term well-being and quality of life. In the words of symbolic interactionism, meaning is constructed through the process of interaction between people. It takes place in the context of relationships. Yet, there remains much work to be done to incorporate a social-relational perspective fully within the models that drive current rehabilitation efforts and underpin funding provision. We propose that this work can begin by directly applying a social-relational perspective to assessment and treatment activities across the rehabilitation continuum.

References

Ackerman, S. J. and M. J. Hilsenroth. 2003. A review of therapist characteristics and techniques positively impacting the therapeutic alliance. *Clinical Psychology Review*, 23(1), 1–33.

Anderson, S. and S. Chen. 2002. The relational self: An interpersonal social-cognitive theory. *Psychological Review*, 109, 619–645.

Backhouse, M. and S. Rodger. 1999. The transition from school to employment for young people with acquired brain injury: Parent and student perceptions. *Australian Occupational Therapy Journal, 46,* 99–109.

Baldwin, M. and S. Dandenaeau. 2005. Understanding and modifying the relational schemas underlying insecurity. In M. Baldwin (Ed.), *Interpersonal Cognition.* New York: Guilford Press, pp. 33–61.

Baldwin, M. and K. Main. 2001. Social anxiety and the cued activation of relational knowledge. *Personality and Social Psychology Bulletin, 27,* 1637–1647.

Baldwin, S. A., B. E. Wampold and Z. E. Imel. 2007. Untangling the alliance-outcome correlation: Exploring the relative importance of therapist and patient variability in the alliance. *Journal of Consulting and Clinical Psychology, 75*(6), 842–852.

Baum, S. and R. Stewart. 1990. Sources of meaning through the life span. *Psychological Reports, 67,* 3–14.

Benzies, K. and M. Allen. 2001. Symbolic interactionism as a theoretical perspective for multiple method research. *Journal of Advanced Nursing, 33,* 541–547.

Berk, M. and S. Anderson. 2000. The impact of past relationships on interpersonal behavior: Behavioral confirmation in the social-cognitive process of transference. *Journal of Personality and Social Psychology, 79,* 546–562.

Blumer, H. 1962. Society as symbolic interaction. In A. Rose (Ed.), *Human Behavior and Social Processes: An Interactionist Approach.* Boston: Mifflin, pp. 179–192.

Blumer, H. 1969. *Symbolic Interactionism: Perspective and Method.* Berkeley, CA: University of California Press.

Bordin, E. 1979. The generalizability of the psychoanalytic concept of the working alliance. *Psychotherapy: Theory, Research and Practice, 16,* 252–260.

Bowen, C. 2007. Family therapy and neurorehablitation: Forging a link. *International Journal of Therapy and Rehabilitation, 14,* 344–349.

Bowen, C., G. Yeates and S. Palmer. 2010. *A Relational Approach to Rehabilitation: Thinking about Relationships after Brain Injury.* London: Karnac.

Bright, F., P. Boland, S. Rutherford, N. Kayes and K. McPherson. 2012. Implementing a client-centred approach in rehabilitation: An autoethnography. *Disability and Rehabilitation, 34*(12), 997–1004.

Bromley, S. 2000. Smell and taste disorders: A primary care approach. *American Family Physician, 61,* 427–436.

Callahan, C. and J. Hinkebein. 2002. Assessment of anosmia after traumatic brain injury: Performance characteristics of the University of Pennsylvania Smell Identification Test. *Journal of Head Trauma Rehabilitation, 17,* 251–256.

Cantor, J., T. Ashman, M. Schwartz et al. 2005. The role of self-discrepancy theory in understanding post-traumatic brain injury affective disorders: A pilot study. *Journal of Head Trauma Rehabilitation, 20,* 527–543.

Charmaz, K. 2006. *Constructing Grounded Theory: A Practical Guide Through Qualitative Analysis.* London: Sage.

Charmaz, K. 2009. Shifting the grounds: Constructivist grounded theory methods. In J. M. Morse, P. N. Stern, J. Corbin, B. Bowers, K. Charmaz and A. Clarke. (Eds.), *Developing Grounded Theory: The Second Generation.* Walnut Creek, CA: Left Coast Press, pp. 127–193.

Chen, S., H. Boucher and M. Tapias. 2006. The relational self revealed: Integrative conceptualization and implications for interpersonal life. *Psychological Bulletin, 132,* 151–179.

Cohen, S., W. Doyle, D. Skoner, B. Rabi and J. Gwaltney Jr. 1997. Social ties and susceptibility to the common cold. *Journal of the American Medical Association, 277,* 1940–1944.

Cohen, S. and I. Janicki-Deverts. 2009. Can we improve our physical health by altering our social networks? *Perspectives on Psychological Science, 4,* 375–378.

Cohen, S. and T. A. Wills. 1985. Stress, social support, and the buffering hypothesis. *Psychological Bulletin, 98,* 310–357.

Cooley, C. 1902. *Human Nature and Social Order.* New York: Scribner.

Costanzo, R. and T. Miwa. 2006. Posttraumatic olfactory loss. *Advances in Oto-Rhino-Laryngology, 63,* 99–107.

Crits-Christoph, P., M. Gibbons, J. Hamilton, S. Ring-Kurtz and R. Gallop. 2011. The dependability of alliance assessments: The alliance-outcome correlation is larger than you might think. *Journal of Consulting and Clinical Psychology, 79*(3), 267–278.

De Vogler, K. and P. Ebersole. 1981. Adults' meaning in life. *Psychological Reports, 49,* 87–90.

De Vogler, K. and P. Ebersole. 1983. Young adolescents' meaning in life. *Psychological Reports, 52,* 427–431.

Deems, D., R. Doty, G. Settle et al. 1991. Smell and taste disorders, a study of 750 patients from the University of Pennsylvania Smell and Taste Center. *Archives of Otolaryngology Head and Neck Surgery, 117,* 519–528.

Delle Fave, A., A. Brdar, M. Wissing and D. Vella-Brodrick. 2013. Sources and motives for personal meaning in adulthood. *The Journal of Positive Psychology, 8*(6), 517–529.

Dikman, S., J. Machamer, J. Powell and N. Temkin. 2003. Outcome 3 to 5 years after moderate to severe traumatic brain injury. *Archives of Physical Medicine and Rehabilitation, 10,* 1449–1457.

Douglas, J. 2010a. Placing brain injury rehabilitation in the context of the self and meaningful engagement. *Seminars in Speech and Language, 31,* 197–204.

Douglas, J. 2010b. *"Always being left out":* The Impact of Communication Impairment on Social Living after Severe TBI. Presentation at the *Eighth World Congress on Brain Injury, Washington, DC. Brain Injury, 24*(3), 153.

Douglas, J. 2012. Social linkage, self-concept and well-being after severe traumatic brain injury. In J. Jetten, C. Haslam and S. A. Haslam (Eds.), *The Social Cure: Identity, Health and Wellbeing.* Hove: Psychology Press, pp. 237–254.

Douglas, J. 2013. Conceptualizing self and maintaining social connection following severe traumatic brain injury. *Brain Injury, 27,* 60–74.

Douglas, J., L. Knox and C. Mitchell. 2014. Evaluating the efficacy of communication-specific coping intervention for adults with traumatic brain injury (TBI). *Journal of the International Neuropsychological Society, 20*(s1), 239.

Douglas, J., L. Knox, C. Mitchell and H. Bridge. 2013. Improving communication-specific coping after traumatic brain injury: Evaluation of a new treatment using single case experimental design. *Brain Impairment, 14*(2), 344.

Douglas, J. M. and F. J. Spellacy. 2000. Correlates of depression in adults with severe traumatic brain injury and their carers. *Brain Injury, 14,* 71–88.

Drummond, M., J. Douglas and J. Olver. 2007. Anosmia after traumatic brain injury: A clinical update. *Brain Impairment, 8,* 61–70.

Drummond, M., J. Douglas and J. Olver. 2013a. "If I haven't got any smell...I'm out of work." Consequences of olfactory impairment following traumatic brain injury. *Brain Injury, 27*(3), 332–345.

Drummond, M., J. Douglas and J. Olver. 2013b. The incidence of olfactory impairment following traumatic brain injury. *Brain Impairment, 14*(1), 153.

Durkheim, E. 1951 [1897]. *Suicide.* New York: Free Press.

Eames, P., G. Cotterill, T. Kneale, A. Storrar and P. Yeomans. 1995. Outcome of intensive rehabilitation after severe brain injury: A long term follow up study. *Brain Injury, 10,* 631–650.

Emmons, R. A. 2003. Personal goals, life meaning, and virtue: Wellsprings of a positive life. In C. L. M. Keyes and J. Haidt (Eds.), *Flourishing: Positive Psychology and the Life Well-Lived.* Washington, DC: American Psychological Association, pp. 105–128.

Fellows, L. K. 2012. Current concepts in decision-making research from bench to bedside. *Journal of the International Neuropsychological Society, 18,* 937–941.

Fratiglioni, L., S. Pallard-Borg and B. Winblad. 2004. An active and socially integrated lifestyle in late life might protect against dementia. *Lancet Neurology, 3,* 343–353.

Frydenberg, E. and R. Lewis. 1997. *Coping Scale for Adults.* Melbourne: ACER.

Gardner, W., C. Pickett and M. Knowles. 2005. "Social snacking" and "social shielding": The use of symbolic social bonds to maintain belonging needs. In K. Williams, J. Forgas and W. von Hippel (Eds.), *The Social Outcast: Ostracism, Social Exclusion, Rejection and Bullying.* New York: Psychology Press, pp. 227–242.

Gracey, F. and T. Ownsworth. 2012. The experience of self in the world: The personal and social contexts of identity change after brain injury. In J. Jetten, C. Haslam and S. A. Haslam (Eds.), *The Social Cure: Identity, Health and Wellbeing.* Hove: Psychology Press, 273–295.

Gracey, F., S. Palmer, B. Rous et al. 2008. "Feeling part of things": Personal construction of self after brain injury. *Neuropsychological Rehabilitation, 18,* 627–650.

Gurman, A. and D. Kniskern. 1991. *Handbook of Family Therapy.* New York: Brunner/Mazel.

Hatcher, R. L. 2011. Alliance theory and measurement. In C. Muran and J. P. Barber (Eds.), *The Therapeutic Alliance: An Evidence-based Guide to Practice.* New York: Guilford Press, pp. 7–28.

Haxel, B., L. Grant and A. Mackay-Sim. 2008. Olfactory dysfunction after head injury. *Journal of Head Trauma Rehabilitation, 23,* 407–413.

Hicks, J. and L. King. 2009. Positive mood and social relatedness as information about meaning in life. *The Journal of Positive Psychology, 4*(6), 471–482.

Higgins, E. and C. Brendl. 1995. Accessibility and applicability: Some "activation rules" influencing judgment. *Journal of Experimental Social Psychology, 31,* 218–243.

Higgins, E. and D. May. 2001. Individual self-regulatory functions: It's not "we" regulation, but it's still social. In C. Sedikides and M. Brewer (Eds.), *Individual Self, Relational Self, Collective Self.* Philadelphia, PA: Psychology Press, 47–67.

Hoofien, D., A. Gilboa, E. Vakil and P. J. Donovick. 2001. Traumatic brain injury (TBI) 10–20 years later: A comprehensive outcome study of psychiatric symptomatology, cognitive abilities and psychosocial functioning. *Brain Injury, 15,* 189–209.

Horvath, A. O. 2006. The alliance in context: Accomplishments, challenges, and future directions. *Psychotherapy, 43*(3), 258–263.

Horvath, A. O., A. C. Del Re, C. Flückiger and D. Symonds. 2011. Alliance in individual psychotherapy. *Psychotherapy, 48*(1), 9–16.

Hougaard, E. 1994. The therapeutic alliance—A conceptual analysis. *Scandinavian Journal of Psychology, 35*(1), 67–85.

House, J., K. Landis and D. Umberson. 1988. Social relationships and health. *Science, 241,* 540–545.

Hux, K., E. Bush, S. Zickefoose, M. Holmberg, A. Henderson and G. Simanek. 2010. Exploring the study skills and accommodations used by college student survivors of traumatic brain injury. *Brain Injury, 24*(1), 13–26.

James, W. 1890. *The Principles of Psychology* (Vol. 1). Cambridge, MA: Harvard University Press.

Jonason, P., G. Webster and E. Lindsey. 2008. Solutions to the problem of diminished social interaction. *Evolutionary Psychology, 6,* 637–651.

Kennedy, M., M. Krause and L. Turkstra. 2008. An electronic survey about college experiences after traumatic brain injury. *NeuroRehabilitation, 23,* 219–231.

Knox, L., J. Douglas and C. Bigby. 2013a. Whose decision is it anyway? How clinicians support decision making participation after acquired brain injury. *Disability and Rehabilitation, 35*(22), 1926–1932.

Knox, L., J. Douglas and C. Bigby. 2013b. "There's a lot of things I just know I can't influence": The experience of adults with severe TBI and their parents in making decisions about life after injury. *Brain Impairment, 14*(1), 169–170.

Kop, W., D. Berman, H. Gransar et al. 2005. Social network and coronary artery calcification in asymptomatic individuals. *Psychosomatic Medicine, 67,* 343–352.

La Guardia, J., R. Ryan, C. Couchman and E. Deci. 2000. Within-person variation in security of attachment: A self-determination theory perspective on attachment, need fulfillment and well-being. *Journal of Personality and Social Psychology, 79,* 367–384.

Lefebvre, H., G. Cloutier and J. Levert. 2008. Perspectives of survivors of traumatic brain injury and their caregivers on long-term social integration. *Brain Injury, 22,* 535–543.

Levack, W., N. Kayes and J. Fadyl. 2010. Experience of recovery and outcome following traumatic brain injury: A metasynthesis of qualitative research. *Disability and Rehabilitation, 32*(12), 986–999.

Lin, N., A. Dean and W. Ensel. 1986. *Social Support, Life Events, and Depression.* Orlando, FL: Academic Press.

Markus, H. and P. Nurius. 1986. Possible selves. *American Psychologist, 41*(9), 954–969.

McGregor, I. and D. Marigold. 2003. Defensive zeal and the uncertain self: What makes you so sure? *Journal of Personality and Social Psychology, 85,* 838–852.

Mead, G. H. 1934. *Mind, Self and Society: From the Standpoint of a Social Behaviorist.* Chicago: University of Chicago Press.

Mealings, M. and J. Douglas. 2010. "School's a big part of your life…": Adolescent perspectives of their school participation following traumatic brain injury. *Brain Impairment, 11*(1), 1–16.

Mealings, M., J. M. Douglas and J. Olver. 2012. Considering the student perspective in returning to school after TBI: A literature review. *Brain Injury, 26*(10), 1165–1176.

Mikulincer, M., V. Florian and G. Hirschberger. 2003. The existential function of close relationships: Introducing death into the science of love. *Personality and Social Psychology Review, 7*, 20–40.

Mikulincer, M., G. Hirschberger, O. Nachmias and O. Gilleth. 2001. The affective component of the secure base schema. *Journal of Personality and Social Psychology, 81*, 305–321.

Miwa, T., M. Furukawa, T. Tsukatani, R. Costanzo, L. DiNardo and E. Reiter. 2001. Impact of olfactory impairment on quality of life and disability. *Archives of Otolaryngology—Head and Neck Surgery, 127*, 497–503.

Muenchberger, H., E. Kendall and R. Neal. 2008. Identity transition following traumatic brain injury: A dynamic process of contraction, expansion and tentative balance. *Brain Injury, 22*, 979–992.

Nochi, M. 1998. "Loss of self" in the narratives of people with traumatic brain injuries: A qualitative analysis. *Social Science and Medicine, 46*, 869–878.

Nochi, M. 2000. Reconstructing self-narratives in coping with traumatic brain injury. *Social Science and Medicine, 51*, 1795–1804.

O'Connor, K. and K. Chamberlain. 1996. Dimensions of life meaning: A qualitative investigation at mid-life. *British Journal of Psychology, 87*, 461–477.

O'Connor, K. and K. Chamberlain. 2000. Dimensions and discourses of meaning in life: Approaching meaning from qualitative perspectives. In G. T. Reker and K. Chamberlain (Eds.), *Exploring Existential Meaning: Optimizing Human Development across the Life Span*. Thousand Oaks, CA: Sage, pp. 75–91.

Ogilivie, D. and R. Ashmore. 1991. Self-with-other representation as a unit of analysis in self-concept research. In R. C. Curtis (Ed.), *The Relational Self: Theoretical Convergencies in Psychoanalysis and Social Psychology*. New York: Guilford Press, pp. 282–314.

Oliver, C. 2012. The relationship between symbolic interactionism and interpretive description. *Qualitative Health Research, 22*, 409–415.

Ownsworth, T. 2014. *Self Identity after Brain Injury*. Hove: Psychology Press.

Plummer, K. 2000. Symbolic interactionism in the twentieth century. In B. Turner (Ed.), *The Blackwell Companion to Social Theory*. Malden, MA: Blackwell, pp. 193–222.

Putnam, R. 2000. *Bowling Alone: The Collapse and Revival of American Community*. New York: Simon and Schuster.

Rutledge, T., S. Linke, M. Olsen et al. 2008. Social networks and incident stroke among women with suspected myocardial ischemia. *Psychosomatic Medicine, 70*, 282–287.

Sani, F. 2012. Group identification, social relationships and health. In J. Jetten, C. Haslam and S. A. Haslam (Eds.), *The Social Cure: Identity, Health and Wellbeing*. Hove: Psychology Press, pp. 21–38.

Santos, D., E. Reiter, L. DiNardo and R. Costanzo. 2004. Hazardous events associated with impaired olfactory function. *Archives of Otolaryngology—Head and Neck Surgery, 130*, 317–319.

Schönberger, M., F. Humle and T. Teasdale. 2006. The development of the therapeutic working alliance, patients' awareness and their compliance during the process of brain injury rehabilitation. *Brain Injury*, 20(4), 445–454.

Schönberger, M., F. Humle and T. Teasdale. 2007. The relationship between clients' cognitive functioning and the therapeutic working alliance in post-acute brain injury rehabilitation. *Brain Injury*, 21(8), 825–836.

Seeman, T. 1996. Social ties and health: The benefits of social integration. *Annals of Epidemiology*, 6, 442–451.

Sharp, N., R. Bye, G. Llewellyn and A. Cusick. 2006. Fitting back in: Adolescents returning to school after severe acquired brain injury. *Disability and Rehabilitation*, 28(12), 767–778.

Shorland, J. and J. Douglas. 2010. Understanding the role of communication in maintaining and forming friendships following traumatic brain injury. *Brain Injury*, 24(4), 569–580.

Stewart-Scott, A. and J. M. Douglas. 1998. Educational outcome for secondary and post-secondary students following traumatic brain injury. *Brain Injury*, 12(4), 317–331.

Tate, R., G. Broe, I. Cameron, A. Hodgkinson and C. Soo. 2005. Pre-injury, injury and early post-injury predictors of long-term functional and psychosocial recovery after severe traumatic brain injury. *Brain Impairment*, 6, 81–96.

Tate, R., J. Lulham, G. Broe, B. Strettles and A. Pfaff. 1989. Psychosocial outcome for the survivors of severe blunt head injury: The results from a consecutive series of 100 patients. *Journal of Neurology, Neurosurgery, and Psychiatry*, 52, 1128–1134.

Temkin, N., J. Corrigan, S. Dikmen and J. Machamer. 2009. Social functioning after traumatic brain injury. *Journal of Head Trauma Rehabilitation*, 24, 460–467.

Temmel, A., C. Quint, B. Schickinger-Fischer, L. Klimek, E. Stoller and T. Hummel. 2001. Characteristics of olfactory disorders in relation to major causes of olfactory loss. *Archives of Otolaryngology—Head and Neck Surgery*, 128, 635–641.

Todis, B. and A. Glang. 2008. Redefining success: Results of a qualitative study of postsecondary transition outcomes for youth with traumatic brain injury. *Journal of Head Trauma Rehabilitation*, 23(4), 252–263.

Twenge, J. M., L. Zhang, K. R. Catanese, B. Dolan-Pascoe, L. F. Lyche and R. F. Baumeister. 2007. Replenishing connectedness: Reminders of social activity reduce aggression after social exclusion. *British Journal of Social Psychology*, 46, 205–224.

Umberson, D. and J. Montez. 2010. Social relationships and health: A flashpoint for health policy. *Journal of Health and Social Behavior*, 51, 554–566.

Umberson, D., K. Williams, D. Powers, H. Liu and B. Needham. 2006. You make me sick: Marital quality and health over the life course. *Journal of Health and Social Behavior*, 47(1), 1–16.

Vaidya, A. 2002. Issues related to school re-entry following traumatic brain injury. *International Journal of Cognitive Technology*, 7(1), 38–45.

Vetere, A. and R. Dallos. 2003. *Working Systemically with Families: Formulation, Intervention and Evaluation*. London: Karnac.

Wade, D., N. King, F. Wenden, S. Crawford and F. Caldwell. 1998. Routine follow-up after head injury: A second randomized controlled trial. *Journal of Neurology, Neurosurgery, and Psychiatry*, 65, 177–183.

Wilson, B., F. Gracey, J. Evans and A. Bateman. 2009. *Neuropsychological Rehabilitation: Theory, Models, Therapy and Outcome.* Cambridge: Cambridge University Press.

World Health Organization (WHO). 2001. *International Classification of Functioning Disability and Health.* Geneva: WHO.

Ylvisaker, M. and T. Feeney. 1998. *Collaborative Brain Injury Interventions: Positive Everyday Routines.* San Diego, CA: Singular Publishing.

Ylvisaker, M. and T. Feeney. 2000. Reconstruction of identity after brain injury. *Brain Impairment, 1,* 12–28.

Ylvisaker, M., K. McPherson, N. Kayes and E. Pellett. 2008. Metaphoric identity mapping: Facilitating goal-setting and engagement in rehabilitation after traumatic brain injury. *Neuropsychological Rehabilitation, 18,* 713–741.

9

Rehabilitation and Recovery of Self-Identity

Emily J. Thomas

William M. M.
Levack

William J. Taylor

Piglet sidled up to Pooh from behind. "Pooh" he whispered. "Yes, Piglet?" "Nothing" said Piglet, taking Pooh's paw, "I just wanted to be sure of you." (Milne 1992, p. 265)

Purpose and Scope of Chapter

Reports of "loss of self" or "change in self-identity" are becoming increasingly common in research looking at the lived experience of illness or disability. The purpose of this chapter is to explore this concept in more depth, to look at what people might mean when they report a change in self-identity, to examine what the mechanisms underlying this might be, and to present the argument that perhaps greater attention should be paid by rehabilitation services to problems with self-identity resulting from disability or illness.

Clearly, a detailed or definitive discussion on the nature of self is beyond the scope of this chapter (it being a topic that has perplexed philosophers for millennia), so we instead refer primarily to some pertinent theories of self-identity as they have been applied to processes of personal change. We shall not be considering the effects of congenital disability or pediatric disability in this chapter, but will focus on the effects of disability acquired as an adult (i.e., after 18 years of age). Most of the discussion will apply equally to sudden onset traumatic events as to progressive disability. Self-identity problems can occur following any major injury or illness, but it is worthwhile to consider the specific effects of cognitive impairment and the unique challenges that this brings.

Introduction

In the context of surviving severe illness or trauma, people have been known to say "I don't know who I am anymore." In people with a sudden onset of severe impairment, this may not seem surprising, intuitive almost. However, what leads us to examine this statement more closely are the exceptions: the people who may have relatively little in the way of objective impairment but who still question the continuity of their identity, or those with severe impairment who profess to have "remained the same" despite extensive changes to their physical body, observed behavior, or cognitive processes.

This loss or change of self-identity is a complex phenomenon, highly individual and multifactorial. In order to try to understand what individuals might mean when they report that they no longer know themselves, we need to consider at the very least the nature of their impairments, the nature of their sense of self, and how these two interact.

After a physical injury, when people say that they do not feel like the same person anymore, or that they no longer have a strong sense of who they are, they may be referring to any number of things including but not limited to

- Physical changes (e.g., loss of physical abilities or characteristics, including aspects of their physical appearance that they previously considered "core" to their sense of self)
- Interoceptive changes (i.e., a change in the sense of connection they have with their body, which may include strange bodily feelings that are difficult to interpret, such as spasms or pain that are not associated with observable physical pathology)
- Changes in sense of agency (i.e., not feeling completely in control of one's own actions or emotions)
- Changes in life trajectory (e.g., loss of a previously assumed future self, perhaps encompassing for instance work and family aspirations)
- Changes in connection with their past or ongoing life history (e.g., resulting from memory impairment, particularly loss of autobiographical memories)
- Changes in how they feel they are treated by others such as friends, family, health professionals, work colleagues, and members of the general public
- Development of discrepancies in their life (e.g., "what I am now" versus "what I was" versus "what I consider 'normal'")
- Loss of purpose or an inability to make sense of what has happened in their life, resulting in a sense of emptiness or worthlessness

Thus, a report of a loss of self-identity can cover a multitude of different experiences and is a very individual phenomenon. It is probably helpful then to start with a few definitions of key concepts before considering some of the prior theories that have been used to attempt to explain reports of loss or change in self-identity after disabling conditions.

Definitions of Self, Self-Identity, and Personhood

Research into change in self-identity is hampered by a confusing array of terms that are not always clearly defined. Definitions of self-identity vary (and can even conflict with one another) across and within academic disciplines. In many ways, this is not unexpected: self-identity has been a notoriously difficult concept to describe. After all, William James, a highly influential nineteenth-century psychologist, referred to it as "the most puzzling puzzle" (James 1890, p. 330). So we find, for instance, that a recent study looking at synthesizing the available evidence on identity change in dementia reported a great deal of difficulty in integrating and interpreting data from different studies due to the wide range of concepts, models, and methodologies used (Caddell and Clare 2013). Indeed, many studies in this review failed to outline a clear theoretical framework underpinning their research at all (Caddell and Clare 2013).

Furthermore, *self* and *identity* are terms that "are sometimes used interchangeably and other times to refer to different things" (Leary and Tangney 2012, p. 71). One potentially useful distinction that has been drawn between self and identity is that self is "a process and organization born of self-reflection" (Owens 2003, p. 206), whereas identity is a "tool (or in some cases perhaps a stratagem) by which individuals or groups categorize themselves and present themselves to the world" (Owens 2003, p. 206). However, following the recommendation of Leary and Tangney (2012), we shall in this chapter avoid use of the term *self* (in isolation, without disambiguation) as much as possible because of its breadth of meaning, and shall instead refer preferentially to the terms *sense of self* and *self-identity*. We shall use both of these terms to refer to the *subjective* sense of one's own being, with sense of self referring to the *process* of producing this subjective sense and self-identity referring to the *product* of this process. To further clarify our intent, when referring to a third-party opinion of someone else's essential characteristics (i.e., what might be considered an *objective* view of their character), we shall use the term *personality*, with *personality change* referring to perceived alterations of this objectively determined character.

Another historical distinction related to the topic of self-identity is worth mentioning. Ricoeur (1995, p. 374) drew a distinction between identity as *selfhood*, which he labelled *ipse*, and identity as *sameness*, to which he applied to term *idem*. This highlights two distinct ways in which the term *self-identity* can be used. This perspective is reflected in two of the definitions for identity found in the Oxford English Dictionary (http://www.oed.com):

1. Who or what a person or thing is; a distinct impression of a single person or thing presented to or perceived by others; a set of characteristics or a description that distinguishes a person or thing from others
2. The sameness of a person or thing at all times or in all circumstances; the condition of being a single individual

Another distinction has been drawn between the psychological concept of self and the anthropological concept of person (Harris 1989). The Western concept of person gives legal, moral, and social significance to an individual human being, with society conceptualized as being composed of these equal, autonomous units (LaFontaine 1985). LaFontaine (1985) draws on Mauss's definitions to describe *self* as an individual's awareness of having a unique identity and *person* as society's acceptance of that identity.

Some Key Ideas Related to Change in Self-Identity

There are many theories and models relevant to self-identity. Four broad concepts from these various models and theories have been selected for discussion below because of their particular relevance to change and reconstruction of self-identity after crises. These are as follows:

1. Self-identity as being composed of egocentric and sociocentric components
2. Self-identity as development
3. Self-identity as narrative
4. Self-identity as a composite of multiple selves

See Chapter 8 for a reflection on the "relational" self.

Self-Identity as Being Composed of Egocentric and Sociocentric Components

While, as suggested to above, the construct of self has eluded a widely accepted definition, it is generally viewed as being multifaceted. Classically, James (1890) divided the self into the "I-self" and the "me-self." Here, the I-self is viewed as the self-as-subject, while the me-self is viewed as self-as-object. The I-self is the experiencing entity—the position from which an individual views the world. This facet of the self-identity encompasses features like continuity, volition, and uniqueness. The me-self in contrast incorporates statements that a person can make about his or her self-identity, such as traits, beliefs, and attributes. This facet of self-identity can be considered analogous to self-concept. Another way of considering this is to think about the question, "what kind of person am I?" In this context, the "answer" to that question (maybe referring to one's gender, physical characteristics, ethnicity, personal values, and so forth) is the self-concept or me-self, while the entity that is actually doing that thinking about those characteristics and attributes is the I-self. The former, being easier to study, has received much more attention and is sometimes (erroneously) used synonymously with self-identity as a whole, neglecting the more elusive "I" aspects.

Another useful distinction between the various facets of self-identity is that they can, broadly speaking, be divided into *egocentric* (or private components) and *sociocentric* (or public components). Historically, psychological research has focused on the egocentric self (i.e., the features of a person considered to make them unique as an individual), yet there is growing recognition of the contribution of social factors into one's self-identity. Social identity theory (Tajfel and Turner 1979) and self-categorization theory (Turner 1985; Turner et al. 1987) consider social identity as being the sense of self that people derive from their membership in social groups (e.g., family, work, community groups). These are combined in the social identity approach, which incorporates the view that "we can all define ourselves not just as an I and a Me but also as a We and an Us" (Jetten et al. 2012, p. 4).

From this perspective, cultural differences in views on the concept of self-identity might be explained in part by the varying emphasis placed on its egocentric and sociocentric elements. We might recognize the Western view as being more individualistic, thus emphasizing the contribution of egocentric aspects to the construction of self-identity (e.g., uniqueness from others). In contrast, we might view Chinese, Japanese, Korean, Pacific Island, and Māori cultures (to name a few) as being more collectivist—placing more emphasis on the sociocentric element of self-identity. Within these cultures, in comparison to Western cultures, individuals might view themselves more in relation to their position in a group rather than on an individual level. Current literature on social identity in Western societies, however, shows us that this distinction is not fixed, and our perspective (me versus we) can shift according to the situation we are in (Oyserman and Lee 2008). In fact, there is increasing recognition in Western literature on the role that social markers of identity play in health-related outcomes (Jetten et al. 2012). Social identities may be particularly pertinent in times of stress, as this may involve a change in our relationships with others and an increased need for social support.

Social constructionist views of self-identity, such as that proposed by Mead (1934), take this one step further and view self-identity as being formed at the *interface* of the person and society. From this perspective, the self is entirely social in origin, and therefore individual self-identity cannot exist without a social context.

Although considerably more could be added on contemporary views regarding the egocentric and sociocentric components of self-identity, the key idea here is that both internal and external factors contribute to one's self-identity. This notion is of central importance when considering how self-identity may be affected by disability and how rehabilitation might assist people in reconstructing their identities after injury or illness. This perspective also gives further weight to the application of a biopsychosocial framework (Engel 1977) in rehabilitation, which would require acknowledgement of the interaction between pathological, psychological, and sociopolitical factors in the expression of identity change.

Self-Identity as Development

When considering change in self-identity, Erikson's model of psychological development is also worth considering briefly. Erikson (1950) used the term *ego-identity* and defined eight stages of development that are progressed through, the mastery of each stage being crucial before an individual can move onto the next. Each stage is characterized by specific challenges and tensions.

In essence, what Erikson's model describes is "growing up." All of these stages of course have bearing on self-identity in one way or another, but one particular point of interest is stage 5: identity versus role confusion. Erikson argued that the period of time from adolescence to young adulthood is crucial for the developing identity, as this is the period when people begin to gain a sense of themselves as unique individuals and when they make choices about the sort of person they wish to become. In Erikson's view, formation of a strong ego-identity at this stage is essential before moving onto forming healthy intimate relationships with others (in his sixth stage of psychological development). Indeed, Erikson (1968) coined the phrase "identity crisis" to describe failure to develop ego-identity during adolescence.

From a rehabilitation perspective, illnesses or injuries occurring at this time of life may be particularly salient, as the resultant disability can delay the usual opportunities and experiences that people might otherwise draw on in their formation of ego-identity. Simply having an extended period of time out from one's previous day-to-day life due to being in hospital or residential rehabilitation can contribute to these kinds of interruptions quite apart from the specific effects of the illness or injury. This is of course particularly pertinent to the types of serious injuries that are commonly associated with these age groups (e.g., spinal cord injuries and traumatic brain injury). However, Erikson's model also shows that identity development continues throughout one's lifespan, with different challenges being faced at the different life stages. The key point here is that self-identity is not a static entity, but rather it is constantly remodeled and shaped throughout one's life.

Self-Identity as Narrative

How an individual achieves a sense of self is a complex phenomenon. There are many different "layers" to our sense of self, starting with the basic essential features that may be said to constitute a self (i.e., an ability to reflect consciously on your own existence), which are then added to via more elaborate processes (e.g., memory or language) before that person achieves all the features that are normally included when thinking of one's self-identity. Gallagher (2000) has simplified these into components of the "minimal self," devoid of temporal extension, and "narrative self," which has continuity through time.

The minimal self refers to a minimum set of criteria that are necessary for an organism to be deemed "*a* self." The addition of an autobiographical layer, strongly founded on memory, allows temporal extension and the development of a richer or "extended" form of consciousness, which is termed the narrative self.

Narratives and narrative identity refer to the process of remembering and retelling the stories of our lives. It is through this process of linking past with present that a sense of continuity through time can be established. Narratives are closely related to autobiographical memory but are more than just recall; memories are encoded then recalled, told, and reinterpreted before being re-encoded. This is a process that can help people make sense of the world and produce some coherence to apparently random, chaotic events.

Narratives are closely associated with sense of self, and a disruption in one's life narrative can undermine one's self-identity. There is debate within philosophical circles about the position of narratives in the construction of self-identity, whether these represent how life is actually lived and meaning construed, or whether they are a structure imposed on all events as we recall them. Some, for instance, have argued that narratives *are* our identities (McAdams 1999; Schechtman 1996). Narratives are formulated not as a passive story to follow as if an actor in a play, but a framework by which we make sense of our self-identities in the world. Narratives consist of not one single thread but multiple story lines that interweave; some are major themes, others a more minor part; some are long and persist throughout the course of our lives and others come and go. The narrative self refers to this process that links past with present with future, creating a coherent sense of self through time. "There is no such thing," according to Bruner (2002), "as an intuitively obvious and essential self to know, one that just sits there ready to be portrayed in words. Rather, we constantly construct and reconstruct ourselves to meet the needs of the situations we encounter, and we do so with the guidance of our memories of the past and our hopes and fears for the future" (p. 64). Opponents of narrative identity, however, have argued that life is too unpredictable to conform to such a structure and that not everyone thinks in a narrative way. Others, it is argued, are more "episodic" in their experience of self-identity, viewing themselves as discrete selves during particular episodes of their lives, rather than having one continuous narrative that binds their sense of self into a single whole (Strawson 2004).

Self-Identity as a Composite of Multiple Selves

One notion, widely applied in social psychology and used in psychotherapeutic approaches when dealing with identity loss, is that healthy individuals have a multiplicity of potential selves or identities. Despite the experience we might have of being a single, unified self, it is argued that self-identity is instead a dynamic construction based on an underlying multiplicity of representations. Most theoretical perspectives agree that the experience of multiple self-identities is good for social and psychological well-being (Ryan and Deci 2012). This is by virtue of the flexibility of responses that they provide to any given event; one can respond in different but equally appropriate and self-consistent ways in a variety of situations, for example, present a different "face" at work in the role of doctor to that at home in the role of mother.

Markus and Nurius (1986) introduced the notion of *possible selves*, giving a temporal dimension to the representation of self-identity and providing goals, motivation, and directions for future behavior. Possible selves define how people can think about their potential and their future. Markus and Nurius (1986) postulated that we all carry ideal, hoped-for, and feared selves. We can have several hoped-for selves (e.g., the successful self, the rich self, the fit self) and several feared selves (e.g., the alone self, the incompetent self, the depressed self). These possible selves are cognitive manifestations of hopes, fears, motivations, and goals in that they are metaphors embodying those aspirations.

Some people may have relatively few identities at their disposal, whereas others may be more complex and have a larger repertoire to call upon. The relative importance of particular identities to one's overall sense of self-identity can vary as well. So for some people, loss of a particularly key identity (e.g., being the family breadwinner or being the boss of several dozen employees) can be devastating for a person's overall sense of self, whereas loss of a more minor component might not have such severe consequences. Having multiple and varied identities can "buffer" some of the effects of illness or disability on self-identity by allowing a greater range of responses when some are no longer available (Linville 1987). Self-identity can therefore be construed as dynamic and fluid, changing as circumstances dictate. This idea (that within each person is a selection of possible selves) has been utilized therapeutically with patients who have brain injury, where losses (physical or cognitive) may render the current self-identity no longer viable (Heller et al. 2006; Ylvisaker et al. 2008). In this context, part of rehabilitation might be to explore options regarding the types of person the individual with brain injury might wish to become in the future, and identity-salient goal setting becomes central to person-centered practice. Reciprocally, this suggests that pursuit of certain rehabilitation goals may not be ultimately successful if the individual in question is unable to link these outcomes with a desirable future self.

Threats to Self-Identity

As discussed at the beginning of the chapter, there are many ways that self-identity can be threatened. In this section, we consider three broad types of threats to self-identity that have been associated with injury or chronic illness:

1. Ontological assault
2. Disruptions to body/self relationships
3. Consequences of specific cognitive impairments

Ontological Assault

Loss or change of self-identity has been reported following a diverse range of life experiences: spinal cord injury (Hammell 2007), postnatal depression (Abrams and Curran 2011), and domestic violence (Oke 2008), to give just a few examples. Perhaps common across these different types of phenomena, and contributing to changes in self-identity following them, is the notion of experiencing an alteration to one's worldview: a change in perception regarding what constitutes reality or existence, leading to a change in how one views oneself as a person. Illness and disability experiences in this context have been described as an "ontological assault" (Kleinman 1988) in that they challenge a person's basic understandings and assumptions about life and the world. (Also see Chapters 3, 10, and 14.)

Both the concepts of boundary experiences (Yalom 1980) and liminality (popularized by Turner 1969) have been applied to the process of identity change following illness states, and both relate to this notion of ontological assault. "Boundary experiences" are described as events that lead to the psychological experience of being forced to confront existential issues that we, as individuals, usually would prefer to avoid: "freedom, death, isolation and meaninglessness" (Patterson and Staton 2009, p. 152). *Liminality* was originally used in anthropology to refer to a phase of social development that is undergone as the "rites of passage" as one moves from childhood to adulthood. This stage of life, when an individual is at the threshold between one state and the next, is characterized by ambiguity of identity, place, and status in society (Turner 1969). Latterly, a state of "permanent liminality" has been used to describe the experience of disablement, where people with impairments are treated by society as being neither one thing nor another: "neither sick nor well, neither dead nor fully alive ... state of being is clouded and indeterminate"

(Murphy et al. 1988, p. 238) At their core, both of these concepts contain the notion of a disruption to the normal flow of existence—traumatic events that force one to reevaluate and confront unwelcome truths such as the fragility of one's own existence.

Another notion related to the experience of disruptions to one's worldview is that of biographical disruption (Bury 1982). In the context of acquired chronic illness, the term *biographical disruption* has been used to refer to the undermining of a person's sense of coherence with their past life story, which renders their future uncertain and hard to predict. Bury (1982) introduced the concept of biographical disruption in his longitudinal study on patients following a diagnosis of rheumatoid arthritis. He identified a series of profound changes that occur in a person's life view when faced with a chronic, disabling condition. Notions of pain and suffering and even death, which had been on the peripheries of one's existence, become a reality with the experience of illness. This, Bury (1982) argued, results in recognition of a change in one's social relationships and the possibility of growing dependency on others. It also results in a reexamination of expectations and plans for the future. Bury's (1982) participants were predominantly younger adults who had previously considered ill health to be "in the future," so the diagnosis of arthritis caused a shift in their whole life trajectory, rendering their futures uncertain.

Many illness stories involve talk about a loss of autonomy, and refer to how the medical system can be disempowering. Foucault (1973) used the term the *medical gaze* to describe the depersonalization that occurs as people are reduced to a collection of symptoms and pathologies. There is an inherent power imbalance as more weight is given to objective medical knowledge over the subjective experiential knowledge of the individual. In the acute stages of illness, this is inevitable as control over a person's body is given over to the medical team tasked with a person's recovery. There comes a point during recovery, however, where this autonomy must be actively reclaimed, at which time the "patient" can begin their journey toward becoming a "person" again with control over her body and her life (Frank 2013). Cloute et al. (2008) report on the disempowerment created by a lack of recollection of acute events in their TBI participants. Even years later, this gap in their knowledge of their own lives maintains their "passive position" as a patient. When considering the impact of incomplete recovery on a person's autonomy, it is helpful to draw a distinction between *decisional autonomy* (the ability to make decisions without external restraint or coercion) and *executional autonomy* (the ability or freedom to act on the basis of decisional autonomy). It is decisional autonomy that is more closely linked to the ability to shape one's life and one's self-identity (Cardol et al. 2002). There is also a growing recognition of autonomy as a social relational process; it only makes sense in the context of human society. Various types of social interactions are vital components in the decisions we make, the goals we set, and the life path we choose to pursue (Levack et al. 2014b). Thus, active reclaiming of full autonomy may be limited by cognitive or physical impairments as well as by social and environmental factors.

Disruptions to Body/Self Relationships

The relationship between the body and mind is a complex one. The word *mind* is itself a loaded and controversial term. Some consider the mind to be a separate "thing" from the body, perhaps equating the mind with a soul or spirit that exists apart from the body, yet somehow communicating with and influencing the action of it. This viewpoint is called mind–body dualism or Cartesian dualism after René Descartes, who is attributed with first articulating this perspective (Demertzi et al. 2009). (See also Chapter 5.) Materialists, however, would strongly dispute the notion that there is any nonphysical substance separate from the body that produces our conscious experiences, such as that of self-identity, that instead we are the product entirely of biology (Melnyk 2012). And there is a plethora of other alternative theories and explanations of conscious experience and its relationship to the body—too many to describe here. Regardless of this debate, however, one thing that is unique about having a "mind" is having a capacity to reflect on one's own conscious experience and to internally influence those experiences. So for example, we can make ourselves feel happy or sad to some degree by choosing to think happy or sad thoughts, and this in turn can affect how we behave. The conscious identification and selection of certain thought processes is central to fields such as cognitive behavioral therapy (Beck 1979).

What is pertinent to a discussion of rehabilitation of self-identity, however, is what happens when people lose this sense of strong connection between their conscious experiences and their physical bodies. This can be expressed in terms of both problems with receiving messages from the body (i.e., in the case of disrupted or misinterpreted sensory signals) and with sending messages to the body (i.e., in the case of problems with controlling bodily movements). When sufficiently extreme, both these problems can result in reports of people experiencing a disrupted sense of self.

Most people do not pay much conscious attention to their bodily states on a day-to-day basis; the role of the body in maintaining homeostasis of the vast majority of its physiological functions occurs well below the level of conscious awareness. However, when stability of this homeostasis is questioned, attention must be refocused and meanings ascribed to changes (Bury 1982). Thus, the onset of illness or disability necessitates a reexamination of this relationship; the body that in health has "dis"appeared from our conscious experience in disease "dys"appears and calls for attention (Leder 1990). Being ill or disabled often carries with it a need to attend to the body that was not present before. There is a need to be acutely aware of changes in one's internal milieu, as they may provide a signal of relapse or progression of disease or disability. There may also be a requirement to pay particular attention to basic bodily functions, such as those of the bladder and bowels. The body, instead of being a vehicle for expression, to aid individuals in achievements and fulfilling their desires, may then be experienced instead as "a hindrance" (Jumisko et al. 2005, p. 45) or even as "the enemy" or "an alien force" (Charmaz 1995, p. 662). Gadow (1980) applied Hegel's description of a master–slave dialectic to this phenomenon as follows: (a) the self is experienced as free subjectivity—the body its vehicle and instrument, serving the will of the self as does the slave its master; and (b) the inversion of that relation: the body rebels, refuses to function, and through the asserted independence, the former master—the self—becomes the slave.

Frank used the term "body-relatedness" to describe the degree of intertwining of conscious experience and body that people can experience (Frank 2013). He described this as a continuum from an "associated" body type at one end of the spectrum to "dissociated" at the other. A person with a dissociated body type can view changes in the body as separate and distinct from his or her self-identity, whereas for more associated types, bodily changes necessitate a redefinition of self-identity, with the two (body and self-identity) being inextricably linked.

The nature of one's body–self relationship can change with the onset or development of disability, "undermining the unity between body and self and forc[ing] identity changes" (Charmaz 1995, p. 657). Charmaz (1995) examined the process of adaptation to a new bodily situation in terms of changing self-identity goals and self-identity trade-offs. This describes a process of analyzing which identities or roles are more important. So, for instance, if it is no longer possible to maintain high work standards and good familial relationships, a question can arise regarding which one is more important and which can be sacrificed or adjusted. Successful adaptation to a change in circumstances entails the development of new and deeper meanings, through which people "transcend their bodies as they surrender control" (Charmaz 1995, p. 675). Charmaz's concept of surrender refers not to a process of mental separation of self from body but a process of acceptance and personal growth: "they suffer bodily losses but gain themselves" (Charmaz 1995, p. 675). This has been echoed in other studies where adaptation to chronic impairment has been viewed as acknowledging the "ultimate moral supremacy of the inner self" (Gelech and Desjardins 2010, p. 70).

The body also has an intimate relationship with one's social identity. Goffman (1963) discussed the role of "stigma" and a "spoiled identity" associated with physical imperfections. People with illness or disability risk becoming socially identified and exclusively defined by their impaired bodies (Bury 1988). This has been termed a "master status" as the stigmatized component of one's identity completely dominates all other statuses (Becker 1963, p. 33), reducing "all one's achievements and attributes to a single, stigmatized identity" (Hammell 2006, p. 113). There is a tension here between visible and invisible impairments. Visible impairments can alert others to potential problems and the need to accommodate them or they can create uncomfortable assumptions: for example, "everyone still just sees a twisted body and assumes I have a twisted mind to match" (Padilla 2003, p. 418). Zitzelsberger's participants also comment on how the visible differences can dominate social perceptions: "I was visible as someone different … they would not go and think I could be visible as a woman that could have a relationship, as a woman that could be a friend … as a woman that could be seen in the workplace, as a woman that could be a mother one day" (Zitzelsberger 2005, p. 394). Invisible impairments (with a lack of physical evidence of any problem), such as executive dysfunction after traumatic brain injury, present a different type of challenge, often resulting in a perceived lack of empathy from health professionals and the wider community (Chamberlain 2006).

Cognitive Impairment

However the nature of self-identity is construed, the substance of the brain is certainly involved in its production. Thus, an injury or illness involving the brain is felt to be particularly threatening when considered from the perspective of one's self-identity. Change in self-identity has been reported for instance in the context of traumatic brain injury (Levack et al. 2010; Nochi 1998a), stroke (Salter et al. 2008), and dementia (Caddell and Clare 2010). Furthermore, therapeutic interventions involving deep brain stimulation (e.g., for Parkinson's disease) have been reported to produce experiences of immediate and profound changes in self-identity on occasions, which has caused some to question this procedure on ethical grounds (Schechtman 2010; Witt et al. 2013).

Brain injury can affect cognitive functions associated with the sense of self in many ways. These are discussed next.

Injury to Crucial "Self" Regions

There is evidence of modular organization of function within the brain. Different regions are specialized for distinct cognitive functions, which then require higher-order brain areas to interpret and integrate their outputs. Thinking of a neural correlate to our minimal sense of self, Damasio (2003) has highlighted the role of "interoception" in the formation of self-identity. Damasio (2003) described interoception as being a sixth sense, "most critical for the self" (p. 256), that gives us an internal representation of our bodies, providing the link between our thoughts and our bodies, and enabling us to interpret our feelings and current body states. These functions are attributed to the insula region of the brain (Damasio 2003). From a more extended, narrative identity perspective, it has been proposed that crucial integrative functions for formation of self-identity occur in the left-hemisphere interpreter region (Gazzaniga 2005). Thus, damage to these specific brain regions could produce an alteration in the experience of sense of self.

Others have argued for a more diffuse representation of self-identity in the brain, viewing the sense of self as a "psycho-social dynamic processing system" (Leary and Tangney 2012, p. 36). From this perspective, injury to any of a multitude of brain regions or the connections between them, as may be seen in diffuse axonal injury, could produce a change in one's sense of self.

Memory Impairment

One specific cognitive function that has received much attention when considering changes in sense of self is that of memory. In 1690, Locke postulated that it was continuity of consciousness, derived from our memories, that created our self-identity (Locke 1690). It follows from this that any disruption to our memories could also be associated with a disruption to our sense of self.

It has been suggested that brain injuries involving temporal lobe structures, especially the hippocampus and amygdala, often lead to difficulty reestablishing a sense of self as these areas are recognized as contributing to affective remembering (amygdala) and memory integration (hippocampus) (Perna and Errico 2004). Changes in these neuroanatomical substrates have also been implicated in psychiatric conditions in which a coherent self-identity is lost or damaged, such as schizophrenia (Wright et al. 2000).

Autobiographical memory has semantic and episodic components both of which support identity formation. Episodic memories are of our day-to-day experience and give us our rich autobiographical narratives, and a sense of being able to mentally "travel back in time" (Tulving 1985, p. 387) to a previous event. This ability to relive the past is termed *autonoetic consciousness*. Autobiographical episodic memory is crucial to the formation of coherent self-narratives and gives a sense of personal continuity, a sense of being the *same* individual, persisting through life events.

Semantic memory encodes more factual information, including abstracted information about oneself from autobiographical experience. This includes things such as persisting traits and beliefs, personal likes and dislikes, and knowledge about how one might behave in certain situations (i.e., predictability of responses). This information from semantic memory forms the basis for our self-knowledge, which can then be used to generate our "conceptual self" (Conway 2005).

Memory impairments are particularly problematic for people with traumatic brain injury. Problems with memory after traumatic brain injury may just be for events around the time of injury (post-traumatic amnesia) but may also extend to periods of time before (retrograde amnesia) and after (anterograde amnesia) the insult. Furthermore, in the case of traumatic brain injury, there is evidence of disruption to autobiographical episodic memory irrespective of general cognitive impairment (Piolino et al. 2006). However, even with severe loss of episodic memory, it is possible for individuals to maintain a sense of self based solely on semantic memory (Rathbone et al. 2009), although the experience of this sense of self may be qualitatively different.

Loss of Linguistic Skills

The ability to construct coherent self-narratives is tied up both with autobiographical memory and linguistic capabilities. Language is the primary means by which we can express our identities to others and highlight our individuality. A study of people with aphasia has shown that this impairment can result in a profound effect on one's interpersonal relationships and social life, which (from a social identity perspective) could translate into a change in a person's self-identity, although this is difficult to assess directly given the limitations of communication impairments (Parr 2007).

Executive Impairment and Loss of Interpersonal Skills

Personality change is a well-recognized consequence of brain injury (Yeates et al. 2008). This term refers to a wide-range of alterations in a person's temperament, characteristic traits, and interpersonal skills. As mentioned above, the term *personality change* is more reflective of others' (third-party) views of oneself rather than relating to self-identity or self-concept per se. However, the interrelationships we have with others have a significant bearing on how we come to see ourselves. Other people's opinions are instrumental in forming our own self-identities (Cooley 1902; Mead 1934). Thus, cognitive impairments that contribute to difficulties or changes in these relationships can ultimately have a marked influence on a person's sense of self.

Loss of Agency/Loss of Decision-Making Capacity

Cognitive impairments resulting in emotional lability, impulsivity, or apathy can also lead to a feeling that one is no longer in control of one's own actions (i.e., a loss of sense of agency). One common theme in qualitative research on the experience of life after brain injury is that of having a "loss of sense of personal control over one's body" (Levack et al. 2010, p. 994). This is more than just a feeling of unfamiliarity with a changed body—it also includes the feeling of having a body that is untrustworthy or "could betray [one] indiscriminately" (Jumisko et al. 2005, p. 44). Also associated with some people's experience of a brain injury is a sense of loss of control over emotions and behavioral reactions (Jumisko et al. 2005). Nochi (1998b) discussed an additional tension arising here: between actively incorporating the brain injury into explanations of unwanted behavior, which involves a degree of relinquishing of responsibility for some behaviors, versus fighting against the brain injury explanation for these changes due to a desire to retain a sense of agency and the feeling of being in control of oneself.

Autonomy can also be compromised, particularly in controlled residential or institutional living circumstances (Levack et al. 2010). Cognitive injuries may mean that capacity for decision making is impaired and individuals are judged no longer competent to make decisions about crucial aspects of their lives (e.g., relating to medical treatment, living accommodation, etc.). Such loss of decisional autonomy can erode an individual's sense of being a complete person (Johansson and Tham 2006), although this effect is likely influenced by other factors such as an individual's personal values and the sociocultural context.

Loss of Self-Awareness

Self-awareness has been defined as "the capacity to perceive the 'self' in relatively 'objective' terms, while maintaining a sense of subjectivity" (Prigatano and Schacter 1991, p. 13). Impairment of self-awareness following brain injury can lead to discrepancies between self-reported abilities and the views of others. The relationship between self-awareness and self-identity is a complex one, and a lack of self-awareness could account for some reports of personality change reported by families but denied by patients (Thomas et al. 2014). However, if a change in self-identity is defined by a subjective report of change, then this presupposes sufficient self-awareness to consciously reflect on one's sense of self.

Interventions to Assist with Reconstruction of Self-Identity

Recognizing change in self-identity as an issue, and potentially a problem, for some people after acquired disability raises the question of whether or not self-identity should be a target for rehabilitation interventions. In fact, a growing number of therapeutic strategies already exist that have been developed specifically either to confront change in self-identity or to facilitate recovery following injury or illness via routes that involve attending to self-identity.

It has been observed that, in general, current approaches to rehabilitation tend to focus on treatment of losses (Collicutt McGrath 2008). This approach reinforces a "self-as-damaged-goods" persona (Kovarsky et al. 2007). It has been suggested that a shift to a strengths-based approach, recognizing assets and strengths (Collicutt McGrath 2008) and emphasizing areas of continuity rather than areas of change (Hammell 2006), might be a more positive approach to rehabilitation. Studies looking at people who have adjusted well following brain injury, for instance, highlight acceptance, adaptation, and meaningful occupation as key factors for recovery (Hoogerdijk et al. 2011; Nochi 2000). In studies of physical disability, acceptance of disability as nondepreciating has been linked to an ability to shift one's values—particularly broadening the scope of what is considered important and reducing the emphasis on the value of arguably superficial characteristics such as physique (Wright 1983, p. 159).

Incorporating Self-Identity into Goal Planning

One possible way of attending to the reconstruction of self-identity in rehabilitation is to place it at the center of the goal-setting process. Research on goal setting for traumatic brain injury in particular has begun to look at approaches to goal selection that start not from examining what a person with impairments is currently unable to do but from the question of what kind of person he or she might like to become in the future (McPherson et al. 2009; Ylvisaker et al. 2008). This work has been derived from theories of self-regulation, and in particular, from Carver and Scheier's control-process model of self-regulation (Siegert et al. 2004). In this model, Carver and Scheier proposed that all human behavior is goal directed, but that our goals are organized in a somewhat hierarchal order, with our highest order goal being a "system concept," i.e., that which represents our idealized self (Siegert et al. 2004). Humans then self-regulate thoughts and behavior, including those related to subordinate goals, in order to progress as much as possible toward this system concept.

Ylvisaker et al. (2008) have piloted a new form of goal setting based on this and other concepts of self-regulation, with the examination and discussion of self-identity as a key step in the goal-setting process. Ylvisaker et al. (2008) have labeled this approach "metaphoric identity mapping." In order to elicit information about a person's idealized self (in the context of cognitive impairments), this metaphoric identity mapping first involves identification of an idolized individual (which may be a celebrity, family member, or even a fictional character). The idolized individual is then discussed in terms of the life choices, traits, and attitudes that they embody and toward which the person with brain injury aspires. Steps are then followed, leading to the selection of a goal that would provide a path toward greater achievement of this "system concept."

Early evidence from pilot studies involving this approach have suggested that an everyday metaphor such as a hero figure (to represent a system concept) can be an effective way to converse about complex, abstract ideas with people who have brain injury, and that it can possibly facilitate a different type of engagement in rehabilitation programs to the benefit of those individuals (McPherson et al. 2009; Ylvisaker et al. 2008). Interestingly, however, this pilot work has also suggested that metaphoric identity mapping represents a significant departure from the usual goal-setting practices employed by rehabilitation professionals (McPherson et al. 2009; Ylvisaker et al. 2008). This approach may not be suitable or practical with everyone in all clinical contexts, but it does represent a way of incorporating self-identity as a central component in all rehabilitation interventions.

Narrative Therapy and the Life Thread Model

Narrative therapy is another possible approach that has been applied to self-identity reconstruction following acquired disability. Although, as discussed earlier, the role of narratives in the formation of self-identity is much debated, there is growing support for the use of narratives in identity reconstruction (Carless and Douglas 2008; Morris 2004). The telling of one's illness story has been presented as a therapeutic endeavor in itself. One needs to only look at the number of autobiographical accounts of illness and trauma to see that there is an innate need to share tales of adversity with others. "Stories" according to Frank (2013) "have to repair the damage that illness has done to the ill person's sense of where she is in life ... Stories are a way of redrawing maps and finding new destinations" (p. 53). The audience in this regard is not necessarily important—story telling can be a way of reaffirming one's own presence as the "audience," as well as sharing one's story with others.

In narrative therapy, the nature of interactions with health professionals is relevant. There is a difference between "taking the patient's history" and "hearing the person's story" (Frank 2013, p. 58). The emphasis on the former approach is very much on identifying symptoms or problems as targets for interventions; however, there may be equally as much therapeutic mileage in the latter, with the focus on personal meanings. Through this narrative retelling of events, a sense of coherence may be restored. Based on the role of narratives in identity reconstruction, Ellis-Hill et al. (2008) introduced the "life-thread model" as a way of conceptualizing the inclusion of discursive psychological and social processes alongside physical processes in the delivery of rehabilitation. Similarly, Morris (2004) has advocated for story telling as a key intervention for the reconstruction of self-identity after traumatic brain injury.

Another area where there is recent interest is in the exploration and nurturing of positive personal development arising from experiencing and recovering from traumatic events. Post-traumatic growth is described as an ability to "derive meaning from and experience positive psychological changes even following the most adverse circumstances surrounding illness and trauma" (Ownsworth and Fleming 2011). This work includes examining which factors enable people to turn a traumatic event into a stimulus for positive change. Until recently, it was hypothesized that people with brain injury might be limited in their ability to experience post-traumatic growth due to cognitive impairments. However, studies are emerging that show this is not necessarily the case, and that this group can demonstrate a remarkable degree of growth (Collicutt McGrath 2011; Collicutt McGrath and Linley 2006; Evans 2011). Some studies are just beginning to look at which personal and social factors may be associated with the ability to adjust positively to disability and how post-traumatic growth may be enhanced by rehabilitation strategies (Jones et al. 2011; Nochi 2000).

Identity Discrepancies and the Y-Shaped Model

Gracey et al. (2009) have highlighted the role of discrepancies as being at the heart of identity threats following brain injury. They identified three key types of discrepancies that can impact on self-identity:

1. *Social discrepancies*: differences between how a person views himself or herself and how other members of society view him or her (e.g., as might arise from social stigma)
2. *Interpersonal discrepancies*: differences between how a person views himself or herself (including deficits and abilities) and how his or her family or health professional view him or her
3. *Personal discrepancies*: differences between how people believed they viewed themselves before injury in comparison to how they view themselves after injury

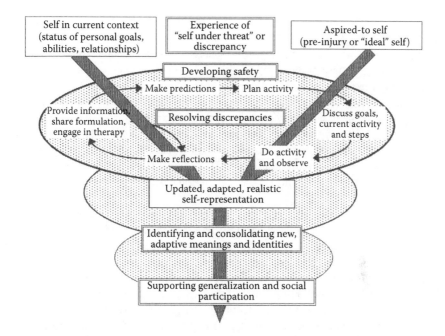

FIGURE 9.1 The converging lines at the top of the Y indicate reduction of discrepancy. The cycle of behavioral experiments is mapped out and the ellipses represent the continued use of behavioral experiments or experiential learning initially to resolve discrepancy and then to support identity development and psychological growth. (Reprinted from Gracey, F. et al., *Neuropsychological Rehabilitation* 19(6): 867–90, 2009. Adapted from Wilson, B. A. et al., *Neuropsychological Rehabilitation: Theory, Models, Therapy and Outcome.* Cambridge: Cambridge University Press, 2009. With permission.)

Gracey et al. (2009) presented a "Y-shaped model" of a possible process of change open to people with altered self-identity (see Figure 9.1). In this Y-shaped model, the top half of the Y represents the discrepant views of self-identity. The target of rehabilitation is to align these views—represented by the merging of the lines into the stem of the Y. They postulated that changes following brain injury are interpreted as a "threat to self," and avoidant coping strategies that may help in the short term actually serve to preserve the discrepancies, i.e., an individual continues to live with two self-identities: the "old me" and the "new me." From this model, Gracey et al. (2009) recommended that initial interventions should aim at reducing identity threats and providing a "safe environment" in which individuals with brain injury can explore their new self-identity. This can then be followed by strategies to enhance awareness and understanding of limitations, and finally approaches to foster adaptive resolution and psychological growth.

Meaningful Occupation

Meaningful occupation has also been seen as a key means by which we express our self-identity to others (and to ourselves), and thus another means by which we might positively influence undesirable changes in self-identity. Occupation refers to the "the everyday activities that people do as individuals, in families and with communities to occupy time and bring meaning and purpose to life" (World Federation of Occupational Therapists 2012). Christiansen (1999) introduced the notion of occupation as the principle means through which people develop and express their identities. He claimed that competence in task performance contributed to identity formation, with development of an acceptable self-identity increasing one's sense of coherence and well-being. Performance deficits are thus, in Christiansen's (1999) view, inherently identity threatening. This leads to the view that rehabilitation interventions become identity-building when therapists provide environments that help people explore possible selves, achieve success in important tasks, and validate the self-identities they have worked hard to achieve in the past.

The value of occupation is also highlighted in studies of people with traumatic brain injury. Changing work roles has been reported as influencing one's sense of self (Johansson and Tham 2006), and productivity has been identified as a key life-defining feature (Cloute et al. 2008). Autobiographical accounts have also emphasized the interconnectedness of occupation, narrative reconstruction, and sense of self-identity (Price-Lackey and Cashman 1996; Smith 2008). "Meaning" and "doing" are intrinsically linked in the process of reconstructing self-identity (Gracey and Ownsworth 2008).

Physical exercise programs also have potential to assist in self-identity reconstruction. Physical therapy can help people connect with their new postdisability bodies and to learn about and connect with new physical sensations and changing abilities. Physical exercise can also exert an influence via the related constructs of self-esteem and physical self-concept (a subcomponent of general self-concept) (Sonstroem and Morgan 1989). Certainly a number of studies have supported the notion of changes in these domains following interventions. For instance, Fines and Nichols (1994) reported improvement in self-concept as measured with the Tennessee Self Concept Scale after a 12-week kayaking program for people with brain injury. Likewise, Driver et al. (2003) found that improvements in physical self-concept and self-esteem resulted from an eight-week aquatic and resistance training program. Participation in group activities has also been promoted because of the influence these can exert on the development of new roles and self-identities, such as that of being a team captain or team member (Cunningham et al. 2012).

Timing of Self-Identity Interventions

Considering rehabilitation from the perspective of change in self-identity also raises the question of when the ideal time is to provide interventions after injury or illness. Current guidelines for rehabilitation after brain injury emphasize the need to provide interventions as soon as possible after a cognitive injury to maximize potential gains at a time of heightened neuronal plasticity. This often leads to a service model where input is maximized in the first three months following injury and tails off thereafter. However, this perspective focuses primarily on physical recovery and takes no account of an individual's readiness for intervention (O'Callaghan et al. 2012), which may be hampered in the early days of recovery by problems with self-awareness (Fleming and

Strong 1999; Sherer et al. 1998) or by simply having insufficient time to come to terms with the implications of newly acquired disability. It seems likely that the reconstruction of self-identity after injury is not going to occur in the same time period as the intensive post-acute hospital-based rehabilitation, but first requires a period of reflection and community living before being actively pursued (O'Callaghan et al. 2012). Papadimitriou's (2008) study of the process of becoming "enwheeled" for patients with spinal cord injuries makes the point that identity (re)construction is a *situated* process. In the accessible environment of the rehab setting, the embodied experience of becoming a wheelchair user is liberating, helping to rebuild a sense of being an independent "active do-er." However, this contrasts with the experiences that wheelchair users report when confronted by the inaccessible "ableist" world, and this dichotomy further complicates rebuilding one's sense of self. People may need time to adapt and develop their new sense of self before they can be expected to fully participate in a community-based goal-orientated rehabilitation program. Consequently, there is currently an unanswered question regarding the timing and location of interventions to address issues of self-identity after the development of long-term impairments or chronic illnesses.

Measurement of Change in Self-Identity

As suggested earlier, there are a number of ways in which rehabilitation professionals might help with the restoration of self-identity after injury or illness. However, further progress toward developing an evidence-based approach to rehabilitation of self-identity is hampered by the lack of an obvious method for measuring change in what might be considered a "healthy" self-identity. Having a way to quantifying problems with self-identity would also be useful for developing a clearer understanding of how self-identity impacts on other health outcomes of interest, such as depression and quality of life.

We do have at our disposal a few validated and widely used tools to assess self-concept. These include the Tennessee Self-concept Scale (Fitts and Warren 1996) and (for brain injury) the Head Injury Semantic Differential Scale (Tyerman and Humphrey 1984). These scales have been applied to the study of change in self-identity following traumatic brain injury and are based on comparing a person's retrospective assessment of his or her self-concept prior to the injury with his or her current self-concept, to produce a measure of "perceived change in self-concept." However, as described earlier, self-concept is only one part of the construct of self-identity. Furthermore, the use of retrospective self-reports has been heavily criticized as introducing systematic bias, even in populations of people without brain injury (Schwarz 2007). There are also problems inherent in using generic instruments in specific patient groups as they may not cover the pertinent areas of function (Gracey et al. 2008; see Chapter 11). This raises the question of whether other methods could be developed to evaluate the status of a person's self-identity as a construct related to well-being, but independent of other variables such as mood.

Work by our team has begun in the area of traumatic brain injury in particular to explore the possibility of developing a measure of sense of self for use after injury. This work began with a metasynthesis of qualitative research, which identified that loss and reconstruction of self-identity were common themes in the lived experience of recovery following brain injury, and therefore an outcome of significant relevance to people with traumatic brain injury (Levack et al. 2010). Following this, a grounded theory study was conducted to examine in more detail the perceptions of people with traumatic brain injury regarding how their injury had (or had not) influenced their sense of self (Levack et al. 2014a). This study provided an initial structure for development and grouping of possible measurement items that related to the experience of change in self-identity after injury. Central to this collection of measurement items was the core theme of desiring to be, or having lost a sense of being, an integrated, valued person, and underpinning this core theme were three subthemes: (1) having a coherent, satisfying and complete sense of oneself; (2) respect, validation, and acceptance by others; and (3) having a valued place in the world.

In parallel with this work, we conducted a concept analysis to examine how the phenomenon of change in self-identity after traumatic brain injury had been previously conceptualized in the literature, across several disciplines including philosophy, psychology, occupational therapy, sociology, and so forth (Thomas et al. 2014). The purpose of this concept analysis was to provide greater clarity regarding the use of the term *self-identity* in literature on traumatic brain injury; to differentiate it from other terms such as *self-awareness*, *self-concept*, *self-esteem*, and *self-knowledge*; and, ultimately, to provide a foundation for better understanding and articulating what construct might be being quantified in a measure of sense of self tailored for people with brain injury. Further research is currently being conducted to test the psychometric properties of the measure that has emerged from this work.

Conclusions

Change in self-identity is rising in prominence on the rehabilitation agenda and can be a significant issue for people living with a wide range of impairments from a variety of etiologies. A reported change in self-identity can be considered a common endpoint for a range of neurological, psychological, and social changes that can accompany any disabling illness. One of the greatest challenges facing rehabilitation is how we can reliably identify problems with self-identity and incorporate identity reconstruction into the rehabilitation program. Nevertheless, considering self-identity as a rehabilitation issue may open new doors for assisting people to adapt positively to life changes that occur with disability and illness.

References

Abrams, L. S., and Curran, L. (2011). Maternal identity negotiations among low-income women with symptoms of postpartum depression. *Qualitative Health Research*, *21*(3), 373–385.

Beck, A. T. (1979). An overview. In A. T. Beck, A. J. Rush, B. F. Shaw, and G. Emery (Eds.), *Cognitive Therapy of Depression* (pp. 1–34). New York: Guilford Press.

Becker, H. (1963). *Outsiders: Studies in Social Deviance*. New York: Free Press.

Bruner, J. (2002). *Making Stories: Law, Literature, Life*. Cambridge, MA: Harvard University Press.

Bury, M. (1982). Chronic illness as biographical disruption. *Sociology of Health and Illness,* *4*(2), 167–183.

Bury, M. (1988). Meanings at risk: The experience of arthritis. In R. Anderson, and M. Bury (Eds.), *Living with Chronic Illness* (pp. 89–116). London: Unwin Hyman.

Caddell, L. S., and Clare, L. (2010). The impact of dementia on self and identity: A systematic review. *Clinical Psychology Review, 30*(1), 113–126.

Caddell, L. S., and Clare, L. (2013). Studying the self in people with dementia: How might we proceed? *Dementia, 12*(2), 192–209.

Cardol, M., De Jong, B. A., and Ward, C. D. (2002). On autonomy and participation in rehabilitation. *Disability and Rehabilitation, 24*(18), 970–974.

Carless, D., and Douglas, K. (2008). Narrative, identity and mental health: How men with serious mental illness restory their lives through sport and exercise. *Psychology of Sport and Exercise, 9*(5), 576–594.

Chamberlain, D. J. (2006). The experience of surviving traumatic brain injury. *Journal of Advanced Nursing, 54*(4), 407–417.

Charmaz, K. (1995). The body, identity and self: Adapting to impairment. *The Sociological Quarterly, 36*(4), 657–680.

Christiansen, C. H. (1999). Defining lives: Occupation as identity: An essay on competence, coherence, and the creation of meaning. *American Journal of Occupational Therapy, 53*(6), 547–558.

Cloute, K., Mitchell, A., and Yates, P. (2008). Traumatic brain injury and the construction of identity: A discursive approach. *Neuropsychological Rehabilitation, 18*(5–6), 651–670.

Collicutt McGrath, J. (2008). Recovery from brain injury and positive rehabilitation practice. In S. Joseph, and P. A. Linley (Eds.), *Trauma, Recovery, and Growth: Positive Psychological Perspectives on Post-traumatic Stress* (pp. 259–274). Hoboken, NJ: John Wiley & Sons.

Collicutt McGrath, J. (2011). Post-traumatic growth and spirituality after brain injury. *Brain Impairment: Special Issue, 12*(2), 82–92.

Collicutt McGrath, J., and Linley, P. A. (2006). Post-traumatic growth in acquired brain injury: A preliminary small scale study. *Brain Injury, 20*(7), 767–773.

Conway, M. A. (2005). Memory and the self. *Journal of Memory and Language, 53*(4), 594–628.

Cooley, C. H. (1902). *Human Nature and the Social Order.* New York: C. Scribner's Sons.

Cunningham, C., Wensley, R., Blacker, D., Bache, J., and Stonier, C. (2012). Occupational therapy to facilitate physical activity and enhance quality of life for individuals with complex neurodisability. *British Journal of Occupational Therapy, 75*(2), 106–110.

Damasio, A. R. (2003). Feelings of emotion and the self. *Annals of the New York Academy of Sciences, 1001,* 253–261.

Demertzi, A., Liew, C., Ledoux, D., Bruno, M.-A., Sharpe, M., Laureys, S., and Zeman, A. (2009). Dualism persists in the science of mind. *Annals of the New York Academy of Sciences, 1157,* 1–9.

Driver, S., O'Connor, J., Lox, C., and Rees, K. (2003). Effect of an aquatics program on psycho/social experiences of individuals with brain injuries: A pilot study. *Journal of Cognitive Rehabilitation, 21*(1), 22–31.

Ellis-Hill, C., Payne, S., and Ward, C. (2008). Using stroke to explore the Life Thread Model: An alternative approach to understanding rehabilitation following an acquired disability. *Disability and Rehabilitation, 30*(2), 150–159.

Engel, G. (1977). The need for a new medical model: A challenge for biomedicine. *Science, 196*(4286), 129–136.

Erikson, E. (1950). *Childhood and Society.* New York: WW Norton and Company.

Erikson, E. (1968). *Identity: Youth and Crisis.* New York: WW Norton and Company.

Evans, J. J. (2011). Positive psychology and brain injury rehabilitation. *Brain Impairment: Special Issue, 12*(2), 117–127.

Fines, L., and Nichols, D. (1994). An evaluation of a twelve week recreational kayak program: Effects on self-concept, leisure satisfaction and leisure attitude of adults with traumatic brain injuries. *Journal of Cognitive Rehabilitation, 12*(5), 10–15.

Fitts, W., and Warren, W. (1996). *Tennessee Self-Concept Scale: TSCS: 2.* Los Angeles: Western Psychological Services.

Fleming, J. M., and Strong, J. (1999). A longitudinal study of self-awareness: Functional deficits underestimated by persons with brain injury. *Occupational Therapy Journal of Research, 19*(1), 3–17.

Foucault, M. (1973). *The Birth of the Clinic* (A. Sheridan, trans. and ed.). London: Tavistock.

Frank, A. W. (2013). *The Wounded Storyteller: Body, Illness, and Ethics (Google eBook).* Chicago: University of Chicago Press.

Gadow, S. (1980). Body and self: A dialectic. *The Journal of Medicine and Philosophy, 5*(3), 172–185.

Gallagher, S. (2000). Philosophical conceptions of the self: Implications for cognitive science. *Trends in Cognitive Sciences, 4*(1), 14–21.

Gazzaniga, M.S. (2005). *The Ethical Brain: The Science of Our Moral Dilemmas.* New York: Dana Press. (Kindle ed., downloaded from http://www.Amazon.co.uk.)

Gelech, J. M., and Desjardins, M. (2010). I am many: The reconstruction of self following acquired brain injury. *Qualitative Health Research, 21*(1), 62–74.

Goffman, E. (1963). *Stigma: Notes on the Management of Spoiled Identity.* New York: Simon and Schuster.

Gracey, F., Evans, J. J., and Malley, D. (2009). Capturing process and outcome in complex rehabilitation interventions: A "Y-shaped" model. *Neuropsychological Rehabilitation, 19*(6), 867–890.

Gracey, F., and Ownsworth, T. L. (2008). Editorial. *Neuropsychological Rehabilitation, 18*(5–6), 522–526.

Gracey, F., Palmer, S., Rous, B., Psaila, K., Shaw, K., O'Dell, J., Cope, J., and Mohamed, S. (2008). "Feeling part of things": Personal construction of self after brain injury. *Neuropsychological Rehabilitation, 18*(5–6), 627–650.

Hammell, K. W. (2006). *Perspectives on Disability and Rehabilitation: Contesting Assumptions, Challenging Practice.* Edinburgh: Elsevier Health Sciences.

Hammell, K. W. (2007). Quality of life after spinal cord injury: A meta-synthesis of qualitative findings. *Spinal Cord, 45*(2), 124–139.

Harris, G. (1989). Concepts of individual, self, and person in description and analysis. *American Anthropologist, 91*(3), 599–612.

Heller, W., Mukherjee, D., Levin, R. L. R., and Reis, J. P. J. (2006). Characters in contexts: Identity and personality processes that influence individual and family adjustment to brain injury. *Journal of Rehabilitation, 72*(2), 44–49.

Hoogerdijk, B., Runge, U., and Haugboelle, J. (2011). The adaptation process after traumatic brain injury: An individual and ongoing occupational struggle to gain a new identity. *Scandinavian Journal of Occupational Therapy, 18*(2), 122–132.

James, W. (1890). *Principles of Psychology: Briefer Course.* New York: Holt.

Jetten, J., Haslam, S. A., and Haslam, C. (2012). The case for a social identity analysis of health and well-being. In J. Jetten, C. Haslam, and S. A. Haslam (Eds.), *The Social Cure: Identity, Health and Well-being.* Hove: Psychology Press. (Kindle ed., downloaded from http://www.Amazon.co.uk.)

Johansson, U., and Tham, K. (2006). The meaning of work after traumatic brain injury. *American Journal of Occupational Therapy, 60*(1), 60–69.

Jones, J. M., Haslam, S. A., Jetten, J., Williams, W. H., Morris, R., and Saroyan, S. (2011). That which doesn't kill us can make us stronger (and more satisfied with life): The contribution of personal and social changes to well-being after acquired brain injury. *Psychology and Health, 26*(3), 353–369.

Jumisko, E., Lexell, J., and Söderberg, S. (2005). The meaning of living with traumatic brain injury in people with moderate or severe brain injury. *Journal of Neuroscience Nursing, 37*(1), 42–49.

Kleinman, A. (1988). *The Illness Narratives: Suffering, Healing and the Human Condition.* New York: Basic Books.

Kovarsky, D., Shaw, A., and Adingono-Smith, M. (2007). The construction of identity during group therapy among adults with traumatic brain injury. *Communication and Medicine, 4*(1), 53–66.

LaFontaine, J. (1985). Person and individual: Some anthropological reflections. In M. Carrithers, S. Collins, and S. Lukes (Eds.), *The Category of the Person: Anthropology, Philosophy, History* (pp. 123–140). Cambridge: Cambridge University Press.

Leary, M. R., and Tangney, J. P. (2012). *Handbook of Self and Identity* (2nd ed.). New York: Guilford Press.

Leder, D. (1990). *The Absent Body.* Chicago, IL: University of Chicago Press.

Levack, W. M. M., Boland, P., Taylor, W. J., Siegert, R. J., Kayes, N. M., Fadyl, J. K., and McPherson, K. M. (2014a). Establishing a person-centred framework of self-identity after traumatic brain injury: A grounded theory study to inform measure development. *BMJ Open, 4*, e004630.

Levack, W. M. M., Kayes, N. M., and Fadyl, J. K. (2010). Experience of recovery and outcome following traumatic brain injury: A metasynthesis of qualitative research. *Disability and Rehabilitation, 32*(12), 986–999.

Levack, W. M. M., Siegert, R. J., and Pickering, N. (2014b). Ethics and goal setting. In R. J. Siegert, and W. M. M. Levack (Eds.), *Rehabilitation Goal Setting: Theory, Practice, and Evidence.* London: Taylor & Francis Group.

Linville, P. W. (1987). Self-complexity as a cognitive buffer against stress-related illness and depression. *Journal of Personality and Social Psychology, 52*(4), 663–676.

Locke, J. (1690). *An Essay Concerning Human Understanding.* Oxford: Oxford University Press.

Markus, H., and Nurius, P. (1986). Possible selves. *American Psychologist, 41*(9), 954–969.

McAdams, D. P. (1999). Personal narratives and the life story. In L. Pervin, and O. John (Eds.), *Handbook of Personality: Theory and Research* (2nd ed., pp. 478–500). New York: Guilford Press.

McPherson, K. M., Kayes, N., and Weatherall, M. (2009). A pilot study of self-regulation informed goal setting in people with traumatic brain injury. *Clinical Rehabilitation, 23*(4), 296–309.

Mead, G. (1934). *Mind, Self, and Society from the Standpoint of a Social Behaviorist* (C. Morris, ed.). Chicago, IL: University of Chicago Press.

Melnyk, A. (2012). Materialism. *Wiley Interdisciplinary Reviews: Cognitive Science, 3*(3), 281–292.

Milne, A. A. (1992). *The Complete Winnie-the-Pooh*. London: Dean.

Morris, S. D. (2004). Rebuilding identity through narrative following traumatic brain injury. *Journal of Cognitive Rehabilitation, 22*(2), 15–21.

Murphy, R. F., Scheer, J., Murphy, Y., and Mack, R. (1988). Physical disability and social liminality: A study in the rituals of adversity. *Social Science and Medicine, 26*(2), 235–242.

Nochi, M. (1998a). "Loss of self" in the narratives of people with traumatic brain injuries: A qualitative analysis. *Social Science and Medicine, 46*(7), 869–878.

Nochi, M. (1998b). Struggling with the labeled self: People with traumatic brain injuries in social settings. *Qualitative Health Research, 8*(5), 665–681.

Nochi, M. (2000). Reconstructing self-narratives in coping with traumatic brain injury. *Social Science and Medicine, 51*(12), 1795–1804.

O'Callaghan, A., McAllister, L., and Wilson, L. (2012). Insight vs readiness: Factors affecting engagement in therapy from the perspectives of adults with TBI and their significant others. *Brain Injury, 26*(13–14), 1599–1610.

Oke, M. (2008). Remaking self after domestic violence: Mongolian and Australian women's narratives of recovery. *Australian and New Zealand Journal of Family Therapy, 29*(3), 148–155.

Owens, T. J. (2003). Self and identity. In J. Delamater (Ed.), *Handbook of Social Psychology* (pp. 205–232). New York: Kluwer Academic/Plenum Publishers.

Ownsworth, T. L., and Fleming, J. M. (2011). Growth through loss after brain injury. *Brain Impairment: Special Issue, 12*(2), 79–81.

Oyserman, D., and Lee, S. W. S. (2008). Does culture influence what and how we think? Effects of priming individualism and collectivism. *Psychological Bulletin, 134*(2), 311–342.

Padilla, R. (2003). Clara: A phenomenology of disability. *The American Journal of Occupational Therapy, 57*(4), 413–423.

Papadimitriou, C. (2008). Becoming en-wheeled: The situated accomplishment of re-embodiment as a wheelchair user after spinal cord injury. *Disability and Society, 23*(7), 691–704.

Parr, S. (2007). Living with severe aphasia: Tracking social exclusion. *Aphasiology, 21*(1), 98–123.

Patterson, F. L., and Staton, A. R. (2009). Adult-acquired traumatic brain injury: Existential implications and clinical considerations. *Journal of Mental Health Counseling, 31*(2), 149–163.

Perna, R. B., and Errico, A. E. (2004). Neurological substrates of personal identity: Implications for neurorehabilitation. *Journal of Cognitive Rehabilitation, 22*(1), 10–12.

Piolino, P., Desgranges, B., Manning, L., North, P., Jokic, C., and Eustache, F. (2006). Autobiographical memory, the sense of recollection and executive functions after severe traumatic brain injury. *Cortex, 42*(2), 176–195.

Price-Lackey, P., and Cashman, J. (1996). Jenny's story: Reinventing oneself through occupation and narrative configuration. *American Journal of Occupational Therapy, 50*(4), 306–314.

Prigatano, G., and Schacter, D. (Eds.). (1991). *Awareness of Deficit after Brain Injury: Clinical and Theoretical Issues.* New York: Oxford University Press.

Rathbone, C. J., Moulin, C. J., and Conway, M. A. (2009). Autobiographical memory and amnesia: Using conceptual knowledge to ground the self. *Neurocase, 15*(5), 405–418.

Ricoeur, P. (1995). *Oneself as Another.* Chicago, IL: University of Chicago Press.

Ryan, R., and Deci, E. (2012). Multiple identities within a single self. In M. Leary, and J. Tangney (Eds.), *Handbook of Self and Identity* (2nd ed., pp. 225–246). New York: Guilford Press.

Salter, K., Hellings, C., Foley, N., and Teasell, R. (2008). The experience of living with stroke: A qualitative meta-synthesis. *Journal of Rehabilitation Medicine, 40*(8), 595–602.

Schechtman, M. (1996). *The Constitution of Selves.* Ithaca, NY: Cornell University Press.

Schechtman, M. (2010). Philosophical reflections on narrative and deep brain stimulation. *The Journal of Clinical Ethics, 21*(2), 133–139.

Schwarz, N. (2007). Retrospective and concurrent self-reports: The rationale for real-time data capture. In A. Stone, S. Shiffman, A. Atienza, and L. Nebeling (Eds.), *The Science of Real-time Data Capture: Self-reports in Health Research* (pp. 11–27). New York: Oxford University Press.

Sherer, M., Bergloff, P., Levin, E., High, W., Oden, K., and Nick, T. G. (1998). Impaired awareness and employment outcome after traumatic brain injury. *Journal of Head Trauma Rehabilitation, 13*(5), 52–61.

Siegert, R. J., McPherson, K. M., and Taylor, W. J. (2004). Toward a cognitive-affective model of goal-setting in rehabilitation: Is self-regulation theory a key step? *Disability and Rehabilitation, 26*(20), 1175–1183.

Smith, C. (2008). Performing my recovery: A play of chaos, restitution, and quest after traumatic brain injury. *Forum: Qualitative Social Research, 9*(2), Art. 30.

Sonstroem, R., and Morgan, W. (1989). Exercise and self-esteem: Rationale and model. *Medicine and Science in Sport and Exercise, 21*(3), 329–337.

Strawson, G. (2004). Against narrativity. *Ratio, 17*(4), 428–452.

Tajfel, H., and Turner, J. C. (1979). An integrative theory of intergroup conflict. In W. G. Austin, and S. Worchel (Eds.), *The Social Psychology of Intergroup Relations* (pp. 33–47). Monterey, CA: Brooks/Cole.

Thomas, E. J., Levack, W. M. M., and Taylor, W. J. (2014). Self-reflective meaning making in troubled times: Change in self-identity after traumatic brain injury. *Qualitative Health Research, 24*(8), 1033–1047.

Tulving, E. (1985). How many memory systems are there? *American Psychologist, 40*(4), 385–398.

Turner, J. C. (1985). Social categorization and the self-concept: A social cognitive theory of group behavior. *Advances in Group Processes: Theory and Research*, 2, 77–122.

Turner, J. C., Hogg, M. A., Oakes, P. J., Reicher, S. D., and Wetherell, M. S. (1987). *Rediscovering the Social Group: A Self-categorization Theory.* Oxford: Blackwell.

Turner, V. (1969). Liminality and communitas. In V. Turner (Ed.), *The Ritual Process: Structure and Anti-structure* (pp. 94–130). Ithaca, NY: Cornell University Press.

Tyerman, A., and Humphrey, M. (1984). Changes in self-concept following severe head injury. *International Journal of Rehabilitation Research*, 7(1), 11–23.

Wilson, B. A., Gracey, F., Evans, J. J., and Bateman, A. (2009). *Neuropsychological Rehabilitation: Theory, Models, Therapy and Outcome.* Cambridge: Cambridge University Press.

Witt, K., Kuhn, J., Timmermann, L., Zurowski, M., and Woopen, C. (2013). Deep brain stimulation and the search for identity. *Neuroethics*, 6(3), 1–13, 499–511.

World Federation of Occupational Therapists. (2012). Definition of occupation, retrieved from http://www.wfot.org/aboutus/aboutoccupationaltherapy/definitionofoccupa tionaltherapy.aspx (accessed September 29, 2014).

Wright, B. (1983). *Physical Disability, a Psychosocial Approach* (2nd ed.). HarperCollins Publishers, New York.

Wright, I. C., Rabe-Hesketh, S., Woodruff, P. W., David, A. S., Murray, R. M., and Bullmore, E. T. (2000). Meta-analysis of regional brain volumes in schizophrenia. *The American Journal of Psychiatry*, 157(1), 16–25.

Yalom, I. D. (1980). *Existential Psychotherapy.* New York: Basic Books.

Yeates, G. N., Gracey, F., and Collicutt McGrath, J. (2008). A biopsychosocial deconstruction of "personality change" following acquired brain injury. *Neuropsychological Rehabilitation*, 18(5–6), 566–589.

Ylvisaker, M., McPherson, K., Kayes, N., and Pellett, E. (2008). Metaphoric identity mapping: Facilitating goal setting and engagement in rehabilitation after traumatic brain injury. *Neuropsychological Rehabilitation*, 18(5–6), 713–741.

Zitzelsberger, H. (2005). (In)visibility: Accounts of embodiment of women with physical disabilities and differences. *Disability and Society*, 20(4), 389–403.

III

Rethinking Rehabilitation Delivery, Research, Teaching, and Policy

10

"This Unfortunate Young Girl…": Rethinking a Necessary Relationship between Disability Studies and Rehabilitation

Contents

Susan
Guenther-Mahipaul

A few years ago, my father handed me a large binder labeled "Susan." I flipped through pages of medical and surgical consultation reports, x-rays, and referrals, full of curiosity. I paused as I gazed upon a report from the then Ontario Crippled Children's Centre that began with "This unfortunate young girl presented with…" Part of me sighed with disbelief, but another part of me accepted it rather quickly. I am a product of rehabilitation. I belong to a group of people who learned as *disabled** children about the pressures from family and healthcare providers to change our bodies—to reach for maximum potential and be as *normal* as possible. Walking, corrective seating and bracing, and managing incontinence all became integral to our existence, equal to or possibly more important than our education. Many of us tended to fall into the language and perceptions that our rehabilitation professionals held about us. Yet I do not believe that I ever viewed myself as an "unfortunate young girl."

Seldom do we meet disabled mentors who show us another side of disability. As a child I learned that being *abled* was good and that people saw my being *disabled* as bad. If health professionals, teachers, and family did not refer to my disability as bad, they often reminded me that I was *special.* Being special stigmatized me, however, because being special meant that I was different but that I should not feel bad about *it.* Through the disability studies literature I have learned that people attribute a meaning and a feeling to disability (Reaume et al. 2009). The meaning of disability is given as follows: different, abnormal, deficient, lacking, inferior, limited, unnatural, etc., and the feeling is that people do not want to *be* disabled. Society juxtaposes normalcy and disability, whereby normalcy legitimizes a good life and the closer a person gets to normal, the closer they are to human.

In this chapter, I argue that although "rehabilitation" offers considerable expertise in the area of disability and chronic illness, it has largely failed to engage with the broader social and cultural context in which disability and disabled people are embedded. My goal is to demonstrate how disability studies provide an important and necessary lens through which to rethink disability and rehabilitation. I do this by discussing prevailing approaches to disability within rehabilitation, and introduce a disability studies perspective as a way forward for rethinking rehabilitation (for further discussion of how disability studies can inform rehabilitation, see Chapters 3 and 4).

The Prevailing Medicalization of Disability in Rehabilitation

As a student occupational therapist, I found my own disability experience at times significantly disconnected from how my peers and I were taught to approach it. I could not connect with many of the theories and models we used to frame our practice. I realized that my everyday disability experiences—for example, the little things to do with managing my needs—were not reflected as part of the models. I noticed that my life did not occur in the silos within which occupational therapists organized my needs. I did not possess the language to voice my views, yet I often played devil's advocate during case-scenario discussions. Since then, I have pursued doctoral studies where I have been drawn to the field of disability studies, its theories and models, and its critical

---

* I use the phrase *disabled* person instead of a *person with a disability* to reflect disability theorists' argument that people are disabled by society and do not live with a "disability" that is medicalized by society (Oliver 1996).

approaches to understanding disability and the role of rehabilitation. I did not expect that I would begin to question my own perceptions on disability.

I am currently in my fifth year of a PhD program and I still experience discomfort with how disability is approached within the classroom and by rehabilitation professors. But I also experience discomfort in how I personally had internalized many of the stigmas and beliefs. It has taken me half a decade to push the boundaries of my discomfort and dramatically shift my perspective on disability. Nevertheless, despite rehabilitation's *deficiencies*, I also appreciate the value of the work that rehabilitation professionals and researchers do every day. As I engage with the theories and writings from disability theorists, I understand why I—and indeed all rehabilitation professionals—need to be critical of our own practices and assumptions, while remaining committed to the important services we provide.

The field of disability studies provides a unique approach to understanding disability and impairment that includes a critique of rehabilitation practices. Perspectives on disability and rehabilitation have been challenged by disability studies theorists, many of whom live with disabilities themselves (e.g., Fougeyrollas et al. 1998; Linton 1998; Longmore 2003; Oliver 1990; Shakespeare 2006a). I believe that their critique of rehabilitation and medicine is warranted. Rehabilitation science has an opportunity to expand how the profession approaches and understands disability. Attitudes, language, and the decisions that medical, rehabilitation, and service providers make on a daily basis all impact disabled people. My own experience helps to illuminate these issues:

> As a 28-year-old I needed a hip replacement and I struggled to find a surgeon who would give me one. I went from walking at home to needing a manual wheelchair full-time because of increasing pain. Eventually an occupational therapist prescribed a power wheelchair with tilt to help with positioning and pain management. Surgeon after surgeon saw my wheelchair and recommended that I limit walking and standing and that then I would "be okay." My symptoms were complicated because of my spina bifida. I simultaneously dealt with a tethered spinal cord, which meant that I was shuffled from neurosurgeon to orthopedic surgeon, each passing the buck to the other to identify the main issues. I deconditioned within months and my hip began to lock while I sat in my wheelchair throughout the day and while sleeping at night.
>
> Of the six orthopedic surgeons I saw, all offered similar *expert* opinions: "You're too young," or "You are not walking so there's no point in replacing your hip." I realized that walking served as a golden ticket to surgery. Before each consultation, I had to complete outcome measures and checklists regarding whether I experienced "great difficulty" or "no difficulty" with, for example, carrying grocery bags in either hand from the car into the kitchen, or whether I could climb stairs without using the railing. I would think to myself, "These are ridiculous questions." I did my grocery shopping using my wheelchair, and I've never been able to walk stairs without using the railing. Surgeon after surgeon told me to use the wheelchair, to not overexert myself, and that, because I was aging with a disability, my limited function was "to be expected."

194   *Rethinking Rehabilitation Delivery, Research, Teaching, and Policy*

Because of my hip, I struggled with driving, living on my own, being in a relationship, and my occupational therapy work became more and more difficult. My
anger with the system grew. I felt angry that no one was willing to help me problem-
solve my situation. No one advocated with me. Therapists told me to conserve my
energy, physicians queried whether I battled depression—which they automatically
linked to my spina bifida rather than the consequences of my hip pain: "I'm sure
it's normal for people like you to have a hard time coping with what's going on," a
social worker told me one day. From my perspective I was coping quite well with
my spina bifida, but I wasn't coping well with the system! My optimistic coping
turned into an alarming defeat. "What does it matter anymore?" I began to wonder.

That year, a friend living with cerebral palsy and a complication of dislocated
hips happened to find a surgeon through the help of his brother, a surgical nurse.
"Susan, you have to go see this guy" he encouraged me, "he thinks like we do about
us." I skeptically secured my own consultation and met Dr. G. a few months later.
"Well you definitely need your hip replaced, no question about it," Dr. G. mumbled
as he studied my x-rays. "When can you come in? I can do it next month."

Seven years later, I enjoy a quality of life that includes walking, using a manual
wheelchair, and being a full-time doctoral student. Recently Dr. G. commented
to one of his fellows during my follow-up appointment, "I have helped people get
back to many things after hip replacement: being grandparents, golfers, travel-
ers. But never did I think that my replacing a hip would lead to a PhD student!"
Dr. G. called me an active person because I worked and led an active lifestyle. His
perspective taught me that not everyone sees me as deficient or broken. "There is
no point for me to compare you to my other clients," Dr. G. explained. However,
he admitted that his colleagues and fellows need to be reminded of this when they
question his views on hip replacement for wheelchair users.

Dr. G. represents a clinician who followed what Hammell (2006c) suggests ought
be the two primary mandates of rehabilitation: (1) working with individuals and personal dimensions of disability/impairment (e.g., to optimize function, increase independence, help people return to their preimpairment function); and (2) addressing the
social dimensions of disability (e.g., to advocate against discrimination and for equal-
ity and opportunity). Despite Hammell's characterization of the importance of both
these roles, the vast majority of rehabilitation practice remains oriented toward only
the first. For many disabled people like myself, rehabilitation is about, as Hammell says,
"learning to live well within impairment in the context of one's environment: a process that might include—but cannot be limited to—enhancement of physical function"
(Hammell 2006c, p. 127). Ameliorating impairments may or may not be part of this
focus. For these reasons, we need a further shift in thinking. Rehabilitation could and
should focus on the lives of disabled people and not (just) their bodies. However, the
majority of rehabilitation professionals find themselves functioning within and maintaining the status quo within a system that continues to perceive disability as personal
tragedy and medicalizes disability through impairment-based interventions (see also
Chapter 3).

The "medical model," also sometimes called the "individual model" of illness, views disability as a pathological condition within the person's body that can be objectively diagnosed, treated (i.e., fixed), and in some cases cured (Barnes and Mercer 2010; Bickenbach et al. 1999; Linton 1998; Oliver 2009; Schultz et al. 2000). Medicalized approaches thus tend to minimize the importance of the material, social, and cultural determinants of disability, which in turn influence the structure and processes of interventions and interactions with disabled people. Associated care practices tend to remain focused on the person first and the environment second (Abberley 2004; Beresford 2004; Hammell 2006a; Oliver 1996; Shakespeare 2006a). In contrast to the medical model, the social model of disability locates disability within the environment and its material, social, and political barriers to social and civic participation. Disability is thus said to result from all the things within the greater society that pose restrictions on disabled people—ranging from individual to institutional prejudices, inaccessible environments, segregated education systems, and limited work accommodations. I describe the social model in more detail below, but the essential difference between the primary locus of disability—body versus environment—provides the point of departure for rethinking rehabilitation practices. First, however, I turn to a discussion of the International Classification of Function, Disability, and Health to further ground an exploration of how disability is conceptualized and understood in rehabilitation.

Critiques of the International Classification of Function, Disability, and Health

The International Classification of Impairments, Disease, and Handicap (ICIDH) first developed in the 1970s (WHO 1980) and later revised into the International Classification of Function, Disability, and Health (ICF) (WHO 2001) represents an example of how the traditional or medicalized view of disability is both challenged and reinforced. Concurrent with the beginnings of the Disability Rights Movement in the 1970s (UPIAS 1976), the World Health Organization (WHO) wanted to classify the outcomes of disease and began to develop an innovative universal nomenclature that could describe and measure the consequences of disease and illness (Hurst 2003; Levasseur et al. 2007). The intent of the WHO was to delineate a common understanding of function, disability, and health and to improve existing health and rehabilitation services. The ICIDH and later the ICF also inadvertently served as the objects on which many of the "conceptual knives of the disability movement were sharpened" (Williams 2001, p. 134). The contested nature of the ICF thus raised further questions and fuelled important debates regarding how to understand disability that has ongoing consequences for policy and rehabilitation practices today.

The ICF attempts to embody the social model in action, defining body functions and structures (impairments), activity limitations and participation restrictions, and contextual factors including environmental and personal factors (WHO 2001). The model is intended to reflect the health of all people—not just those labeled as disabled—and offers a synthesis of medical and social approaches using a biopsychosocial framework (Barnes and Mercer 2010). Rehabilitation researchers use the ICF extensively to conceptualize the dynamic and complex participation of persons with disabilities and their interaction with their environments (Bickenbach et al. 1999; Desrosiers 2005; Levasseur et al. 2007). A strength of the ICF is its recognition of disability as a complex phenomenon that requires dynamic analyses and interventions that span across and link medical and sociopolitical issues. Another important aspect of the ICF framework is its position that disability is not a phenomenon that belongs to a minority group but rather one that affects society as a whole (Shakespeare 2006a). The ICF has created some important shifts in disability healthcare and research; however, it has some key limitations that call into question its promise of change (Hammell 2006b).

The ICF's developers credit the ICF for its ability to measure the effectiveness of rehabilitation services but not the effects of social exclusion (Üstün et al. 2003). The ICF offers a detailed taxonomy to structure data collection (Barnes and Mercer 2010) and classify function. However, by classifying function, it also classifies disability, leading to some unintended consequences (Fougeyrollas et al. 1998; Hammell 2006b). No other minority group has been subject to classification to the same extent as disabled people, who are identified and labeled according to individual bodily and functional differences and allocated health and community services accordingly (Hurst 2003). Although the ICF framework challenges healthcare professionals to acknowledge the impact of the environment on the disability experience, it fails to distinguish between how environments impact an individual's disability and how environments create impairments (Fougeyrollas et al. 1998; Hammell 2006b; Shakespeare 2006a).

Overall, the ICF addresses tensions between multiple perspectives but arguably with limited success. Fougeyrollas and Beauregard (2001) have commented, "We understand that this is a political decision by the WHO that is designed to satisfy the social model. However, it is a cosmetic choice, as we end up with taxonomies that allow each dominant group to maintain its ideological position: An individual's disability is viewed in terms of impairments of body functions, structures, and activity restrictions of the person as a whole" (p. 185). Thus, the ICF, although intended to provide an alternative to individual and social model accounts, also perpetuates a tension between the two (Barnes and Mercer 2010).

The critiques of the ICF demonstrate that there are no shortcuts and/or simplifications to understanding the complex processes of disablement, and attempts to do so can inadvertently reinforce rather than address marginalization. The disabled individual seems to be responsible for social consequences and for how he or she performs and participates in everyday life (Fougeyrollas and Beauregard 2001). Fougeyrollas and Beauregard (2001) point out that "one of the most harmful and violent consequences of this (person-first focus) process occurs when the individual who is different adopts this label and bases his or her identity on all of the disabilities and oppression situations attributed to him or her" (p. 187). Hubbard (2004) acknowledges that the ICF foremost classifies individuals and the use of a client-centered framework focuses on the person and not the environment: "The paternalistic patina, although diminished, prevails" (Hubbard 2004, p. 187). Such critiques call on researchers and clinicians to reflect on and question theories and models because their use directly affects disabled people (Hammell 2006a). As an alternative, the social movement creates radical shifts in how disability is understood, and in doing so challenges healthcare, rehabilitation, and social service providers. In the following, I outline some of the ways that the social model of disability historically shifted perspectives on disability.

The Social Model of Disability: Transitioning Disability Perspectives

The social model of disability represents a shift in thinking that locates disability in society rather than in the bodies of disabled people (Bickenbach et al. 1999; Chatterjee 2005; Hubbard 2004). The Union of the Physically Impaired against Segregation (UPIAS 1976) first defined the social model as follows: "In our view it is society which disables physically impaired people. Disability is something imposed on top of our impairments by the way we are unnecessarily isolated and excluded from full participation in society. Disabled people are therefore an oppressed group in society" (p. 14). Thus, the social model emphasizes societal oppression as the source of the experience of disability (Oliver 1990, 2009), constructs disability as a policy and civil rights issue (Hammell 2006c; Hubbard 2004; Shakespeare 2006a), and sees the causes of disability as resulting from social structure and how society treats and controls disabled people (Altman 2001).

The social model locates impairments in the physical body and minimizes their role in the disablement process. As Oliver (1996, p. 35) has stated, "Impairment is, in fact, nothing less than a description of the physical body." The social model, or at least earlier versions of it (Thomas 2004), is sometimes narrowly interpreted to suggest that medical professionals should play no role in the lives of disabled people. This is perhaps not the intent. What the social model clarifies, however, is that disability for many people is a "long-term social state" that is not necessarily something that needs to be treated medically. Oliver (1996) has expressed:

> Many disabled people experience much medical intervention as, at best, inappropriate, and, at worst, oppressive. This should not be seen as a personal attack on individual doctors, or indeed the medical profession, for they, too, are trapped in a set of social relations with which they are not trained or equipped to deal. (p. 23)

Disability is a complex phenomenon that involves the dynamic interaction between personal and environmental/contextual factors. To medicalize and reduce disability to a bodily impairment creates an obstacle to effectively understand and analyze the complex factors that contribute to disability.

The social model as propagated by the disability movements in North America and the United Kingdom resulted in significant changes that promote(d) positive disability identities and decontextualized disability as a human and civil rights issue. This contributed to changes in policy and legislation and the removal of barriers, whether physical or attitudinal/oppressive (Shakespeare 2006b). Nevertheless, it has been suggested that we need more sophisticated theories and models to understand disability that transcend the social model and do not ignore the effects of impairment or the role of medicine and rehabilitation in helping disabled people to thrive. In the following, I discuss how disability studies can be used to transform rehabilitation practices to make them more relevant to the lives of disabled people.

Disability Studies as a Lens

The social model, at least historically, has underpinned the field of disability studies; however, the field encompasses a broad range of themes, issues, theories, and related social policies (Albrecht et al. 2001). As a disabled person, disability studies provide me an opportunity to reflect on my experiences with the healthcare system, make sense of them, and use that learning in my teaching of current and future rehabilitation professionals. My own narrative helps me to analyze my experiences as situated in complex, dynamic social and personal processes (Williams 2001). I am *not the problem*. My encounter with two physiotherapists, Jenn and Lauren, help illuminate these experiences.*

At the age of 16 my physiotherapist Jenn encouraged my mother and I to start to look for a physiotherapist as I transitioned into adult life. Jenn could only provide services for me until the age of 18 and I credit her for persistently pushing this important transition. I had grown very attached to her. She knew me intimately, she understood my body, and she was keenly aware of my history with my spina bifida and surgical complications that no one else but my parents understood.

My adult physiotherapist, Lauren, came highly recommended by my orthopedic surgeon. She was "brilliant," he said, and he felt strongly that she would know how to problem-solve my physical challenges without the need for surgical interventions. Getting to know Lauren was a culture shock for me. She was less jovial than Jenn, and my therapy sessions often occurred in silence. I learned with time that Lauren was indeed brilliant. She knew how to train my muscles and spine in ways that created significant improvements for me. Lauren added much to my quality of life when it came to strength, endurance, and pain management. But along with her brilliant mind also came a heavy reality about my "situation." I do not think that Lauren felt negatively about how I functioned and coped with my disability, but she perhaps saw my body as a physical flawed entity that would challenge me during my adult years. Her black and white perspective teetered between my challenges with a body that constantly needed to be fixed and maintained and pointing out that I was one of the lucky ones because I possessed a lot of will power to keep fighting.

As I turned 18 I prepared for my first undergraduate year—I was going to live on the university campus. Jenn was excited about my going to university, living on my own, and experiencing the residence life. "You're going to make the best friends you'll ever have," she told me one clinic follow-up day. "You'll love it, Susan. It'll be hard work because you have to promise me that you'll study, but you also have to promise me that you'll party a bit because you're 18 going on 37!" She pored over my course selection calendars with me and recommended programs in which I could specialize. I will always be grateful for her advice regarding taking on a realistic course load. "Whether you take five years to finish your undergraduate degree or four it does not matter," she said. "The more important thing is that you

---

* All names are pseudonyms.

get good grades and that you take the time to put your health first. That's how you'll make it."

In stark contrast to Jenn, Lauren reflected differently about my university journey. At the end of one therapy session, she asked me how things were going and I responded that I was doing well, that I wanted to consider graduate studies down the road. Matter of fact she responded "Don't have babies, your body won't be able to handle it." I vividly recall this conversation still today. "There's always adoption" she added, "but if you really want to pass on your genetics to a child there is also surrogacy." The last thing I recall her advising me was to think hard about not applying for graduate studies. "It's physically demanding even for a person who doesn't have your issues. You'll be lucky if you make it to 40 without any disc problems." I remember thinking, "Why?"

I reflect on my story about Jenn and Lauren because it highlights along with many other narratives of disabled people that our experiences with the rehabilitation system can be negative, and in fact oppressive (for further discussion and a review of related research, see Chapter 3). Disability studies provide a way to unpack these taken-for-granted and matter-of-course notions about disability, impairment, and normalcy in rehabilitation, not only for "clients" but for practitioners. What Lauren never knew was that her attitude seriously affected my self-concept for a very long time, especially my perspectives on my potential to further my career and to perhaps marry and have children. Lauren's focus on my impairments and how I was to age with them were not placed within the larger context of my life. She perceived my disability as nothing but a restriction on my future, i.e., if I was going to go to graduate school I would experience the consequences of my impairment; if I was to not finish graduate school it would be because of my disability, because of my challenges with my body and impairment. She did not think about how I was beginning to dream about a life that would bring me happiness and fulfillment. In her stereotypical perspective of me, Lauren framed all my personal goals in relation to my disability. At the time, almost twenty years ago, I did not react but instead absorbed her judgments. Both Lauren and I did not realize that I had a right to feel insulted and oppressed by her prejudices.

Disability studies raise profound questions about human existence in every social and cultural sphere (Longmore 2003). Empowerment, emancipation, and control for disabled people form the core of the field. Disability studies consider disablement through various theoretical lenses that build from or transcend the social model perspective. These dimensions include "issues such as autonomy, competence, wholeness, independence/dependence, health, physical appearance, aesthetics, community, and notions of progress and perfection" (Linton 1998, p. 118). Disability studies as a discipline bridges the academy and disability communities. It has been described in different ways. I prefer Linton et al.'s (1995) definition of disability studies (as quoted in Longmore 2003, p. 222):

[Disability studies] reframes the study of disability by focusing on it as a social phenomenon, social construct, metaphor, and culture utilizing a minority group model. It examines ideas related to disability in all forms of cultural representations throughout history and examines the policies and practices of all societies to understand the social, rather than the physical or psychological determinants of the experience of disability. Disability Studies both emanates from and supports the Disability Rights Movement which advocates for civil rights and self-determination. The focus shifts the emphasis away from a prevention/treatment/remediation paradigm to a social/cultural/political paradigm. This shift does not signify a denial of the presence of impairments, nor a rejection of the utility of intervention and treatment. Instead, Disability Studies has been developed to disentangle impairments from the myth, ideology, and stigma that influence social interaction and social policy. The scholarship challenges the idea that the economic and social status and the assigned roles of people with disabilities are inevitable outcomes of their condition.

As this description suggests, disability studies theorists do not deny that living with a disability includes the experience of impairment. They argue that the term disability has medical origins, and simply assigning a medical meaning to disability contributes to exclusion and oppression of individuals labeled as disabled. Moreover, these processes are amplified by constructing disability as an individual burden or personal tragedy (Linton 1998). Medicalizing disability keeps it a personal matter, a personal problem that needs to be treated, rather than addressing the social processes that actually restrict or constrict the disabled person's life. This distinction emphasizes that disability is a marker of identity. When medicalized, disabled people get slotted into diagnostic categories such as those embedded in the ICF classification system.

Disability studies help to point out the ways that rehabilitation professionals are potentially and inadvertently "part of the problem" with regard to our clients' disability experiences. Much has been published regarding how rehabilitation professionals have unintentionally contributed to the marginalization of disabled people (e.g., Abberley 2004; Beresford 2004; Hammell 2006b; Krupa 2008). Even within this literature, however, disabled people's experiences have been misrepresented. For example, in her qualitative study of university nursing students with disabilities, Maheady (1999) explored students' experiences as they proceeded through their education and clinical training. Although the study highlighted the strengths of the disabled nursing students (e.g., that their abilities did not differ from their nondisabled peers, and their personal experiences benefited clients), the summary of the study focused largely on negative aspects. For example, Maheady (1999) highlights "important concerns related to potentially unsafe nursing practice" in discussing putative risks to patients posed by disabled nurses (p. 169). She further commented that student accommodations appeared to have negative impacts on their classmates and called for a reexamination of the legal rights of nursing students with disabilities, citing potential threats to patient safety. In a related commentary, Marks (2000) critiqued how Maheady's research methodology, the analysis, and the discussion of findings all centered on a medical-model perspective. She illuminated how the students' rights were not respected, that their experiences had been cruel and bordered on abusive, and that the regular discrimination they experienced during their training appeared to be matter of fact and accepted. Marks concluded her commentary by calling for the nursing profession to address its prejudices toward people with disabilities.

Elsewhere, Abberley (2004) has described the power imbalances between occupational therapists and clients, and discussed how these imbalances are imposed by the requirements placed upon occupational therapists in their work roles. He proposes that occupational therapists are required "to offer evidence of 'success' in order to validate their work" (p. 240). These requirements, he suggests, are passed onto clients who must meet therapist-defined performance criteria and professionally imposed definitions of positive outcome. Although therapists may highly regard client satisfaction in the client–therapist interaction, Abberley (2004) notes that any contradictions in goal setting or clients' education about their new adaptive reality often get resolved in "favour of the therapist" (p. 41). These examples help illustrate how disability studies can be employed to show subtle, complex interactions that can, however unintentionally, create tensions for disabled people in healthcare. Questioning the orientations of rehabilitation professions may serve to foster new relationships between rehabilitation professions and disability scholars who are jointly committed to improving the lives and life circumstances of disabled people.

How Disability Studies Can Inform Rehabilitation

Disability theorists have somewhat harshly critiqued rehabilitation and healthcare professions, which makes infusing disability studies into rehabilitation science challenging (Bickenbach et al. 1999; Hammell 2006a; Hubbard 2004). However, as rehabilitation professionals, we can play a pivotal role in breaking down barriers, promoting advocacy, and promoting disability as a social justice issue. Rehabilitation scholars, for example, argue for an increased focus on principles of the independent living and disability rights movements. They encourage rehabilitation professionals to communicate alternative ways to understand disability, and acknowledge issues of power and expertise within the client–therapist relationship (Hammell 2006c; Hubbard 2004; Magasi 2008a,b; Phelan 2011; Yoshida et al. 1998). In many ways, however, disability studies still seem to exist as a parallel world of which rehabilitation professionals know little (Abberley 2004; Beresford 2004; Hubbard 2004; Linton 1998; Longmore 2003; Oliver 1996). Despite this, I would argue that there are no rigid boundaries between the two fields. Disability studies *can* offer pertinent knowledge to challenge the implicit and explicit assumptions of rehabilitation professionals. Some clinicians naturally understand disability as part of the wider social context and are mindful of how their practices and assumptions impact clients. For me, Jenn represented such a clinician.

I was 13 years old when Jenn, a pediatric physiotherapist, entered my life after I had had several major surgeries on my spine. Jenn knew me through my fourteenth birthday and worked with me almost daily first in inpatient and then outpatient rehab to restore or preserve my preoperative function as best as she could. Before my hospitalization, I walked "independently," but time brought with it changes. Jenn eventually reconsidered her and the surgeon's goal to return me to full time walking and prescribed a wheelchair. We continued to work tirelessly to minimize muscle atrophy in my legs and flexion contractures that were more persistent than our therapy regimen could keep up with.

Jenn saw me as a person, as a teenager with a body that had changed because of impairment, complications, and surgical intervention. My surgeon did not want to accept these changes. He had performed surgeries to prevent destructive functional changes (perhaps even saved my life), and as a result he felt that I should have improved. Jenn recognized her futile attempts to force me to participate in rehabilitation. My rehabilitation sessions were tiring and at times painful. She advocated for my discharge, a recommendation that outraged my surgeon. I learned later that he questioned her judgment and challenged her professional competence. The clinical team argued that I was going to grow weaker and my body would suffer "irreparable" changes. Everything that "they" had gained to this point would be lost. In other words, *I* would be their *failure*. Just before my fifteenth birthday I was discharged home. I returned to grade 10, rekindled friendships, and began to live a productive and active life. I thrived in school.

One year later I felt a spark to return to Jenn and work on my walking. This spark within me, my overwhelming will to persist in my work with Jenn, did not come from my need to walk. Rather, I had developed a peace with my changes, with my disability. It was not about overcoming my disability, about conquering it, or returning to *normal*. Although my parents saw my walking as a freedom from the wheelchair, I felt differently. My wheelchair was an important part of my identity, as it still is today. I know that the ability to walk even within my own disability community is a symbol of status, but this was not the reason I wanted to walk. I just wanted to see where I would go, wheelchair or not. From a rehabilitation lens I learned that the ability to let go of what *could* be and accept the now can be a powerful motivator in change and well-being. I felt that the ability to walk and to have a body that functioned normally was insignificant. At the age of 16 I began on my adult life journey, whether I could walk or not.

Jenn saw me and my impairment experiences within my larger life context. She could sympathize with how my parents approached my needs, which sometimes fell in line with what clinicians recommended and at other times reached beyond the medical context into my social surroundings (e.g., attending university and living on my own). But she also saw many opportunities for me in my adult life, and she participated actively in helping me plan for my future and addressing some of my doubts. I agree with disability theorists like Abberley (2004) that healthcare providers can in their roles and goals with clients "present success as theirs, with failure attributed to their clients" (p. 243). Therapists may believe that they practice in the best interest of clients, which makes challenging these beliefs very difficult (Hammell 2006a; see also Chapter 3). Disability studies offer several unique ways to shift our thinking on disability and rehabilitation.

A disability studies perspective challenges rehabilitation professionals to integrate a focus on social justice within their practices and embrace principles of equal treatment, equal access, and equal outcome (Hubbard 2004). Our day-to-day activities should be about generating a vision that changes the systems within which we work to better serve those who are marginalized and oppressed (Hammell 2006a, 2013; Phelan 2011). In recent years, "client-centered practice" has arguably represented an answer to the tensions between rehabilitation practices and disability studies, yet may actually mislead clinicians (Hammell 2006a; Hubbard 2004; Phelan 2011). Client-centered models still locate disability within the person and his/her body, whereas a disability studies perspective clearly situates disability within society and calls for accountability at a societal and not individual level. A clinician who can reflect on client-centered practice at a societal level is one who is aware of how society constructs disability, and how in his or her practice he or she may unintentionally perpetuate negative disability perspectives (Phelan 2011).

Rehabilitation may be facing a paradigm shift that requires a change at multiple levels, since much of the healthcare system continues to function within a medicalized environment (Hubbard 2004; Oliver 1996). I would suggest that rehabilitation has to redefine its identity; every member needs to reflect on practices that he or she may think come from an interest in supporting his or her clients, and to reflect on potential negative effects of what he or she does. Disability studies offer a potentially powerful role in professional education. Incorporating disability studies into rehabilitation curricula is challenging because of differences in how each field understands and interprets the role of disability within society (Hubbard 2004). The integration of perspectives and voices from disabled people into curricula and teaching provide the greatest opportunities for reflexivity and are pivotal to retaining the richness and integrity of a disability studies lens (Hubbard 2004; Phelan 2011). Currently, healthcare and professional programs have been largely developed from the perspectives of clinicians and situate disabled people in roles of clients, patients, or students (Hubbard 2004; Linton 1998). The direct experience of disabled people coupled with disability theory has the potential to increases students' understanding of why disabling barriers exist and how they can be eradicated (Barnes 2004). As a disabled clinician, it is my exposure to disability theory that has generated the greatest insights into the differences in perspectives and mandates. Perceptions and definitions that place disability and health at either end of each other along a spectrum illustrate the predominant perspectives of disability as dependency, helplessness, lack, and limitation and health as well-being, fitness, and wholesomeness (Reaume et al. 2009). In fact, it is such perspectives and definitions that clinicians can reverse in practice.

Concluding Thoughts: "The Opportunity of Adversity"

My parents and I have had many discussions at the dinner table about my research interests. My father—a logical thinker and problem solver—asked me one day, "Do you think that these disability theorists would have told us that we raised you wrong?" What my parents achieved in raising me well is a notable accomplishment. I feel that they parented me with a combination of a focus on medical aspects of my disability and my accompanying impairments, and an equal focus on my social surroundings. My parents raised me as a cooperative patient and a quiet advocate within a system preoccupied with normalcy. I have had many oppressive experiences within the healthcare system. I feel comfortable to label them as such today. But I also have had great healthcare advocates along my side.

I frequently share with clinicians and students that if a client remembers you ten years from now in a positive way that you have meaningfully influenced their life. As a client myself I feel that making a difference in my life does not mean that I need support to overcome or "rise above" my disability—clinicians can at times equate disability with misfortune and adversity that needs to be "conquered." I intimately understand today that there have been several important opportunities with and perhaps even because of my *adversity*. Clinicians like Jenn saw opportunities for me in life and she helped shape a vision of these opportunities for me and for my family. I therefore credit clinicians like Jenn and Dr. G. in addition to my parents with making the strongest impact on my self-concept as a *person*.

In writing this chapter, I am not suggesting that all rehabilitation professionals currently practice in ways that oppress clients or that oppressive practices are intentional. Yet as a rehabilitation professional myself, I understand why disability scholars have had to reinforce this message and recognize that there are many rehabilitation professionals committed to challenge biomedical thinking, focus on normalcy, and drive to restore function. Although I advocate for a shift in perspectives, I realize that it is not that simple for the rehabilitation professions to change the core underlying theories and philosophies that underpin our practice models and research. Yet that shift is already in motion.

Furthermore, disability studies could also gain from increased partnership with the rehabilitation sciences in order to foster uptake and cross-pollination of ideas. I have not focused on the limitations of the social model within this chapter, but scholars in the disability studies field have increasingly considered "impairment effects" and the importance of their amelioration for disabled people as part of supporting quality of life. A dual consideration of bodily impairment and social limitations that combines the best work from both fields holds tremendous promise in supporting disabled people.

During a national conference for occupational therapists, one of my colleagues gave a keynote address to a filled auditorium. Toward the end of her address, she commented that when she sees a client in a wheelchair she reminds herself that her life is "not that bad." I sat within the crowd, in my wheelchair, and many of my colleagues turned in their seats to look at me with raised eyebrows. Stigma and oppression live around us, and thus around our clients. The language we use, the statements we make, the goals that we set, all reflect how we feel about the concept called disability. Writing about my disability, or using it to understand disability studies and disability theory, has been one of the most challenging scholarly tasks I have undertaken. Thus, I understand that to engage with disability studies literature challenges others as well. As rehabilitation practitioners, educators, and researchers, we have an opportunity to model a disability vision that transcends the current medicalized view of disability, whether we hold this vision within ourselves, our professions, or the greater society. It is precisely this reason why disability studies are so important to our professions.

Acknowledgments

I express my very great appreciation to Dr. Barbara Gibson for her valuable and constructive feedback during the planning and writing of this manuscript. I am particularly grateful for the support given by Christina Minaki who challenged and encouraged me to develop my voice within my narratives. Finally I acknowledge the support of the Canadian Institutes of Health Research for funding my doctoral studies.

References

Abberley, P. (2004). A critique of professional support and intervention. In J. Swain, S. French, C. Barnes and C. Thomas (Eds.), *Disabling Barriers—Enabling Environments* (2nd ed., pp. 239–244). Thousand Oaks, CA: Sage Publications.

Albrecht, G. L., Seelman, K. D. and Bury, M. (2001). *Handbook of Disability Studies.* Thousand Oaks, CA: Sage Publications.

Altman, B. M. (2001). Disability definitions, models, classification schemes, and applications. In G. L. Albrecht, K. D. Seelman and M. Bury (Eds.), *Handbook of Disability Studies* (pp. 97–122). Thousand Oaks, CA: Sage Publications.

Barnes, C. (2004). Disability, disability studies and the academy. In J. Swain, S. French, C. Barnes and C. Thomas (Eds.), *Disabling Barriers—Enabling Environments* (2nd ed., pp. 28–33). Thousand Oaks, CA: Sage Publications.

Barnes, C. and Mercer, G. (2010). Competing models and approaches. In C. Barnes and G. Mercer (Eds.), *Exploring Disability* (2nd ed., pp. 14–42). Malden, MA: Polity Press.

Beresford, P. (2004). Treatment at the hands of professionals. In J. Swain, S. French, C. Barnes and C. Thomas (Eds.), *Disabling Barriers—Enabling Environments* (2nd ed., pp. 245–250). Thousand Oaks, CA: Sage Publications.

Bickenbach, J., Chatterji, S., Badley, E. and Üstün, T. (1999). Models of disablement, universalism and the international classification of impairments, disabilities and handicaps. *Social Science and Medicine, 48*(9), 1173–1187.

Chatterjee, N. (2005). Theory for all and rehabilitation for the few (with money): Who does our theory serve? *Disability and Rehabilitation, 27*(24), 1503–1508.

Desrosiers, J. (2005). Muriel Driver Memorial Lecture: Participation and occupation. *Canadian Journal of Occupational Therapy, 72,* 195–204.

Fougeyrollas, P. and Beauregard, L. (2001). An interactive person-environment social creation. In G. L. Albrecht, K. D. Seelman and M. Bury (Eds.), *Handbook of Disability Studies* (pp. 171–194). Thousand Oaks, CA: Sage Publications.

Fougeyrollas, P., Noreau, L., Bergeron, H., Cloutier, R., Dion, S. and St-Michel, G. (1998). Social consequences of long term impairments and disabilities: Conceptual approach and assessment of handicap. *International Journal of Rehabilitation Research, 21*(2), 127–141.

Hammell, K. (2006a). Contesting assumptions; Challenging practice. In S. Young and C. Jackson (Eds.), *Perspectives on Disability and Rehabilitation: Contesting Assumptions; Challenging Practice* (pp. 187–200). Edinburgh, UK: Churchill Livingstone Elsevier.

Hammell, K. (2006b). Normality and the classification of difference. In S. Young and C. Jackson (Eds.), *Perspectives on Disability and Rehabilitation: Contesting Assumptions; Challenging Practice* (pp. 17–32). Edinburgh, UK: Churchill Livingstone Elsevier.

Hammell, K. (2006c). *Perspectives on Disability and Rehabilitation: Contesting Assumptions; Challenging Practice.* S. Young and C. Jackson (Eds.), Edinburgh, UK: Churchill Livingstone Elsevier.

Hammell, K. (2013). Client-centred practice in occupational therapy: Critical reflections. *Scandinavian Journal of Occupational Therapy, 20,* 174–181.

Hubbard, S. (2004). Disability studies and health care curriculum: The great divide. *Journal of Allied Health, 33*(3), 184–188.

Hurst, R. (2003). The international disability rights movement and the ICF. *Disability and Rehabilitation, 25*(11–12), 572–576.

Krupa, T. (2008). Part of the solution … or part of the problem? Addressing the stigma of mental illness in our midst. *Canadian Journal of Occupational Therapy, 75*(4), 198–207.

Levasseur, M., Desrosiers, J. and Tribble, D. (2007). Comparing the disability creation process and international classification of functioning, disability and health models. *Canadian Journal of Occupational Therapy, 74*(2), 233–242.

Linton, S. (1998). *Claiming Disability: Knowledge and Identity.* New York: New York University Press.

Linton, S., Mello, S. and O'Neill, J. (1995). Disability studies: Expanding the parameters of diversity. *The Radical Teacher, 47*, 4–10.

Longmore, P. K. (2003). *Why I Burned My Book and Other Essays on Disability.* Philadelphia, PA: Temple University Press.

Magasi, S. (2008a). Disability studies in practice: A work in progress. *Topics in Stroke Rehabilitation, 15*(6), 611–617.

Magasi, S. (2008b). Infusing disability studies into the rehabilitation sciences. *Topics in Stroke Rehabilitation, 15*(3), 283–287.

Maheady, D. (1999). Jumping through hoops, walking on egg shells: The experiences of nursing students with disabilities. *Journal of Nursing Education, 38*(4), 162–170.

Marks, B. A. (2000). Commentary: Jumping through hoops and walking on egg shells or discrimination, hazing, and abuse of students with disabilities? *Journal of Nursing Education, 39*(5), 205–210.

Oliver, M. (1990). *The Politics of Disablement.* London: The Macmillan Press.

Oliver, M. (1996). *Understanding Disability: From Theory to Practice.* New York: Palgrave.

Oliver, M. (2009). The social model in context. In T. Titchkosky and R. Michalko (Eds.), *Rethinking Normalcy: A Disability Studies Reader* (pp. 19–30). Toronto, ON: Canadian Scholars Press.

Phelan, S. K. (2011). Constructions of disability: A call for critical reflexivity in occupational therapy. *Canadian Journal of Occupational Therapy, 78*(3), 164–172.

Reaume, G., Titchkosky, T. and Michalko, R. (Eds.). (2009). *Rethinking Normalcy: A Disability Studies Reader.* Toronto, ON: Canadian Scholars Press.

Schultz, I. Z., Crook, J., Fraser, K. and Joy, P. W. (2000). Models of diagnosis and rehabilitation in musculoskeletal pain-related occupational disability. *Journal of Occupational Rehabilitation, 10*(4), 271–293.

Shakespeare, T. (2006a). *Disability Rights and Wrongs.* New York: Taylor & Francis.

Shakespeare, T. (2006b). The social model of disability. In L. J. Davis (Ed.), *The Disability Studies Reader* (2nd ed., pp. 197–204). New York: Routledge.

Thomas, C. (2004). Disability and impairment. In J. Swain, S. French, C. Barnes and C. Thomas (Eds.), *Disabling Barriers Enabling Environments* (2nd ed., pp. 21–27). Thousand Oaks, CA: Sage Publications.

UPIAS. (1976). *Fundamental Principles of Disability.* London: Union of the Physically Impaired Against Segregation.

Üstün, T. B., Chatterji, S., Bickenbach, J., Kostanjsek, N. and Schneider, M. (2003). The International Classification of Functioning, Disability and Health: A new tool for understanding disability and health. *Disability and Rehabilitation, 25*(11–12), 565–571.

WHO. (1980). *International Classification of Impairments, Disabilities, and Handicaps: A Manual of Classification Relating to the Consequences of Disease.* Geneva: World Health Organization.

WHO. (2001). *International Classification of Functioning, Disability and Health (ICF).* Retrieved July 2012 from http://apps.who.int/classifications/icfbrowser/.

Williams, G. (2001). Theorizing disability. In G. L. Albrecht, K. D. Seelman and M. Bury (Eds.), *Handbook of Disability Studies* (pp. 123–144). Thousand Oaks, CA: Sage Publications.

Yoshida, K., Willi, V., Parker, I., Self, H., Carpenter, S. and Pfeiffer, D. (1998). Disability partnerships in research and teaching in Canada and the United States. *Physiotherapy Canada, 50*(3), 198–205.

11

Rethinking Measurement in Rehabilitation

Paula Kersten

Åsa Lundgren-Nilsson

Charles Sèbiyo Batcho

Introduction

The clinical assessment of patients involves interpreting information from various sources, such as patients' reports and therapists' examinations of function, so that clinical decisions related to patients' needs and the appropriateness and nature of their therapy can be determined (Laver Fawcett 2007). In many countries, this information also includes data from outcome measures. Indeed, in some countries or in some services, the use of outcome measures is compulsory and related to reimbursement. For example, rehabilitation centers in the United States use the Functional Independence Measure (FIM) as part of their prospective payment system, along with length of stay and discharge destination (Uniform Data System for Medical Rehabilitation 2014). Increasingly, organizations such as the Commission on Accreditation of Rehabilitation Facilities (CARF) promote the quality, value, and optimal outcomes of services through a consultative accreditation process and continuous improvement services that center on enhancing the lives of persons served (CARF International 2014). This organization works around the world. Measuring outcome of rehabilitation is important for such evaluations. In addition, there is evidence to suggest that the use of outcome measures in clinical practice assists

---

Box 11.1 Plateau Determines Discharge

Sarah and Peter are both taking part in an outpatient rehabilitation program. They both suffered from a stroke a few months earlier and are improving after three weeks of a rehabilitation program. The therapists regularly score their improvements in physical functioning on the Functional Independence Measure (FIM™). After six weeks, Sarah is discharged as the FIM data suggest she has plateaued. Peter's scores are still improving so he continues with rehabilitation. Sarah is very unhappy with this decision as she believes she is still making functional improvements.

---

in "diagnosing" the presence and severity of patients' problems, communication with patients and the team, treatment and discharge planning, the evaluation and improvement of processes of care or treatment, benchmarking against other services, and informing funding priorities and health policies (ACC 2009; Chartered Society of Physiotherapy 2011; Kayes and McPherson 2010; Laver Fawcett 2007; Tyson et al. 2010). But, as our hypothetical case in Box 11.1 suggests, dealing with the interpretation of outcomes data that are not in accordance with what patients think can be a contentious issue—one familiar to many therapists reading this book. The other struggle for many clinicians is how to choose measures that are appropriate in the clinical arena as opposed to the research arena; consequently, clinicians in different fields of rehabilitation report a wide range of validated and nonvalidated outcome measures that are used in their practice (Alviar et al. 2011; Duncan et al. 2000; Haigh et al. 2001; Jette et al. 2009; Korner-Bitensky et al. 2011; Sivan et al. 2011; Skeat and Perry 2008).

In this chapter, we will discuss the use and misuse of outcome measures in clinical practice. We will argue that at times the patient may well be right (such as the fact that s/he is still improving), but the outcomes data are used incorrectly. We will draw on clinical and service examples to illustrate key points, discuss the consequences of poor measurement, and then propose a new future for measurement of outcome.

For the purpose of this chapter, we will define an outcome as, "The observed or measured consequence of an action or occurrence. In a therapeutic process, the outcome is the end result of the therapeutic intervention" (Laver Fawcett 2007, p.10).

Measurement is often defined as the process of assigning numbers to represent quantities of a trait, attribute, or characteristic or to classify objects (Nunnally and Bernstein 1994). However, in this chapter, we will argue that the assignment of numbers does not necessarily result in measurement.

Key Concepts in Outcome Measurement

Outcome measures in rehabilitation tend to fall into two types:

1. There are measures that allow the assessment of phenomena that are measurable with physical equipment and that provide so-called *ratio* or *interval* data (e.g., height, weight, and strength). The difference between a ratio and an interval scale is that the former has a true zero (e.g., height), whereas interval scales do not have a true zero (e.g., the Celsius temperature scale). The really nice thing about ratio and interval data is that we know with absolute certainty that a one-unit difference on the scale equals a one-unit difference elsewhere on the same scale. For example, we can say with 100% certainty that the difference between 102 and 112 m is equal to the difference between 253 and 263 m. This also means that changes in scores on such measures can be compared between patients and change within patients is easily interpretable. The development of these measures should not be underestimated. For example, the establishment of the metric system took seven years during the French Revolution in the 1790s (Alder 2002). Most outcomes in rehabilitation aren't measurable on interval or ratio scales.

2. There are increasing numbers of measures that evaluate the views or observations of patients or clinicians using questionnaire-based outcome measures. Essentially, these measures are a collection of items (tasks or statements) that represent the underlying phenomena (concept or trait) we are interested in and that are not directly observable (i.e., they are latent). For example, we cannot directly observe how much pain, mobility, or community participation someone has. The items in these measures are the best proxies of the latent trait of interest as it is not possible to measure absolutely everything that might be relevant. These items or tasks are scored using response categories to which numbers of increasing magnitude are attached (resulting in so-called *ordinal* measures). For example, each task (item) on the FIM has seven response categories (1 to 7) (Granger et al. 1993); a score of 1 denotes total assistance (i.e., the person does less than 25% of the task), a score of 2 reflects that a person does at least 25% of the task but not as much as 50%, and so forth. The numbers that are assigned to these response options are in increasing order but otherwise arbitrary. While we can say that a four-point increase on an FIM item suggests that the person has improved, we cannot say with certainty the s/he has improved twice as much as someone who improved by only two points on the same FIM item (Lundgren-Nilsson and Tennant 2011). Similarly, a patient that makes a one-point improvement on one FIM item may not necessarily have made the same amount of functional improvement as another patient who improved with one point on another FIM item. Consequently, adding the scores of such measures into a total score is not appropriate (but done all the time).

The remainder of the chapter will focus on outcome measures that use categories.

Outcomes of Interest in Rehabilitation

Rehabilitation is a very complex intervention, supporting disabled people with complex needs and helping them to gain skills they need to attain independence and self-determination (Gutenbrunner et al. 2006; World Health Organization 2001). Given this complexity, it is not surprising that measurements of outcomes in rehabilitation are given lots of attention, although no consensus about the best approach to measurement is available. Over the years, outcomes of interest in rehabilitation have changed, with an early focus on improvements in impairment to a more contemporary view that patients' perceptions of outcome are important. For example, constructs such as function, participation, and health-related quality of life are now key in determining the extent of rehabilitation outcome. As health services continue to improve and people's expectations continue to change, what we will wish to measure as a meaningful outcome may well continue to change too (many chapters, including Chapter 1, reflect further on what outcomes matter). Key to the development of such new measures is robust theory. Theory is not simply someone's hunch but a model that is derived through a variety of means such as qualitative research (e.g., through focus groups, interviews, or observations), review of the literature, and expert opinion (experts should include people with the condition of interest). From this, theoretical understanding items for a measure can be generated, an item reduction process can take place, and the measure will be tested psychometrically (against its theoretical predictions and with the underlying model/concept/trait in mind) (Streiner and Norman 2008). This will be further explored next.

Outcomes Measurement Development and Use—The State of Play

A discussion on the need for quality assurance and outcomes started in the 1970s (Brook et al. 1976), and at the time, the need for national registers was also stressed. Nowadays, it is a clear requirement that the health-care sector provides not only evidence-based care but also cost-effective care within limited resources (Richardson and McKie 2005, 2007). This includes consideration of changing the patterns of use of hospital resources, including falls in the average length of stay, and shifting the focus of care from hospital to home or nearby community centers. Health care is also moving toward more individualized treatment, and the participation of the patient in the decision process has gained much prominence (Nolte and McKee 2008). To support these challenges, increasing emphasis has been placed upon the use of standardized patient reported outcomes instruments (PROs), including the patient's subjective experiences of the health care provided. While linking outcome to funding and quality of care appears fair, we have concerns over the way tools are used since they are ordinal measures.

To explain the basis of our concerns, we first need to look at approaches to measurement development. At present, there are two dominant approaches to the development of outcome measures: classical test theory (CTT) and modern test theory. Within the modern test theory, there are two approaches, that of the Rasch model and item response theory (IRT) (Andrich 2011). We will discuss CTT and modern test theory in the following.

Classical Test Theory

In CTT, it is assumed that the observed scores on items are the sum of the true score (which we can't directly measure) and measurement error. However, as Cano et al. (2010) state, neither the true score nor the measurement error can be determined, and the approach is therefore flawed. Importantly, the best conclusion that can be made following satisfactory tests of validity and reliability using CTT is that an outcome measure is an ordinal scale. Box 11.2 illustrates a key problem with ordinal scales.

Box 11.2   Ordinal Outcome Measures— Same Score, Different Level of Functioning

---

Melody and Sunita are two women with multiple sclerosis who both score a 9 on the Barthel Index (a 20-item scale, ranging from 0 to 20). The only difference in their scores on this measure arises from two items:

1. *Bowels: Melody has occasional accidents (once per week), scoring 1 on this item. Sunita has no accidents and scores 2 on this item.*
2. *Transfers: Melody requires minor help with transfers (verbal/easy physical help from one person), scoring 2 on this item. Sunita requires major physical help (from one or two people) but can sit and scores 1 on this item.*

Simply summing the scores on the Barthel Index for Melody and Sunita will lead to the conclusion that their level of dependency is the same. However, we don't know if the level of difficulty for these two items is the same, or whether the amount of physical improvement needed to move one point on each of these items is equivalent. In other words, the "whole" or the "total" does not equal the sum of the parts of the scale (Stucki et al. 1996). The fact that the Barthel Index is an ordinal scale has been shown using robust methods over 15 years ago (Tennant et al. 1996) but it remains in use in clinical practice and trials alike. The example given above clearly emphasizes that ordinal data should not be summed and that comparisons between individuals and within individuals are flawed.

Now let's look at the consequences of using ordinal scales when measuring change. It has been shown that a one point change on a given part of the measure may actually resemble a much larger change than another one point change elsewhere on the measure. For example, a change in raw scores that occur over the bottom or upper end of an outcome measure actually resembles a much larger change on the underlying interval metric, whereas raw score gains in the middle of the measure are more minimal (Grimby et al. 2012; Kersten et al. 2010). This can lead to a wrong interpretation of patients' change. Take for example a fictional scale that measures function and on which scores range from 0 to 30 (Figure 11.1). Patient A, who exhibits a four-point change in the middle of the (ordinal) scale (e.g., score improves from 10 to 14) would be considered as responding better to rehabilitation compared to patient B whose raw score improved by only two points over the upper end of the same scale (e.g., score improves from 28 to 30). In other words, just looking at the scores would result in the conclusion that patient A has made a greater improvement. However, when we use analytical techniques that allow the conversion of ordinal data to interval-level data, we come to a different conclusion: patient B has made much greater improvements than patient A. This finding is a common feature of ordinal scales, and ignoring this can lead to biased clinical judgments. It follows then that the concept of a minimal (clinically) important difference (MID or MCID) derived from ordinal data is erroneous, since such statistics are simply change scores.

So without using any mathematics or statistics, we have shown that ordinal data are very problematic. For those readers who wish to explore the more technical underpinnings of these arguments, we refer them to suitable texts by others (Küçükdeveci et al. 2011; Merbitz et al. 1989; Svensson 2001; Wright and Linacre 1989). Of course, we are not the first to highlight problems with ordinal data. Over a decade ago, Ben Wright (1999) wrote: "As long as primitive counts and raw scores are routinely mistaken for measures by our colleagues in Social, Educational and Health research, there is no hope of their professional activities ever developing into a reliable or useful science.

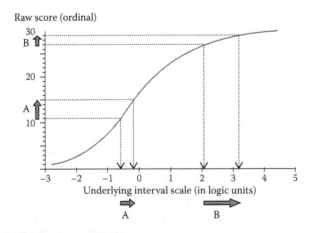

FIGURE 11.1 Item response curve.

We owe it to them, and to ourselves, to teach them how to construct measures which work as well as the ubiquitous physical measures by which they manage their every-day living, so that they can do a better job in making sense out of the profusions of data which they collect so enthusiastically." However, ordinal data continue to be used in reimbursement schemes such as the prospective payment system in the United States (Uniform Data System for Medical Rehabilitation 2014), benchmarking schemes such as the Australasian Rehabilitation Outcomes Centre (AROC 2014), and national databases such as the UK Rehabilitation Outcomes Collaborative database Turner-Stokes et al. (2012).

Modern Test Theory—The Rasch Model

The Rasch model has been explained in methodological books since the 1960s (such as Andrich 1978; Bond and Fox 2001; Rasch 1980 [1960]) as well as more educational papers relevant to rehabilitation over the past 20 years or so (e.g., Fisher 1993; Hobart and Cano 2009; Küçükdeveci et al. 2011; Tennant and Conaghan 2007; Tennant et al. 1996, 2004a; Tesio 2003; Tesio et al. 1997; Wright and Linacre 1989). Underpinning this theory are mathematical models that specify the conditions under which equal interval measurements can be estimated from outcome measurement data. We won't be looking at the actual formula here, as we want to keep you reading, but those interested in this should consult key texts (Andrich 1978; Rasch 1980 [1960]; Wright and Stone 1979). Essentially, the Rasch model is a probabilistic model. It is based on the premise that the probability of a correct or positive response to an item is a mathematical function of person and item parameters. The item parameter, or item difficulty, determines the probability that someone with a given level of ability (e.g., level of upper limb activity) will give a positive or otherwise response to an item. Thus, someone with low levels of upper limb activity is more likely to give a negative answer to an item than someone with high levels of upper limb activity. Measures should include items (derived from theory) that range from easy to endorse or pass to difficult to endorse or pass (as stated in the theory of the trait being measured). One of the fundamental requirements of the Rasch model is that the comparison of two people is independent of which items from the total set of items in the measure they completed or were scored on (Rasch 1980 [1960]). Item parameters (item difficulties) and person parameters (person abilities) can be placed along an *interval* logarithmic scale. Item parameters are estimated indepen-dent of the distribution of abilities in the particular group of persons for whom the items are appropriate (Bond and Fox 2001). Similarly, person parameters are estimated independently of the distribution of their responses to the measure's items. This concept is called *specific objectivity* and is a necessary requirement for the unidimensionality of a scale. Unidimensionality refers to a single trait being measured. As Thurstone (1931) said, "The measurement of any object or entity describes only one attribute of the object measured. This is a universal characteristic of all measurement" (p. 257).

Rasch (1980 [1960]) demonstrated that these requirements of a measure can be sum-marized by a formula that specifies the probabilistic expectations of items and persons. Specifically, he proved that the probability of a correct (or positive) response to an item (i.e., for it to be passed or endorsed) is a logistic function of the difference between the person and the item parameter. Thus, the ratings from outcomes data are analyzed according to the Rasch model to check whether they meet a priori specifications (Andrich 2011). For this reason, the Rasch model is said to be located within an experimental measurement paradigm. Rasch analysis is ideally used to develop and test new measures, and there are some good examples (Batcho et al. 2012; McFadden et al. 2012; Penta et al. 2001). The approach is also very helpful in establishing if legacy measures fit the Rasch model. Once it is shown that the data fit the Rasch model, it can be said that this is a robust outcome measure and interval level data can be derived. Figure 11.2 provides a clinical example.

The top panel in Figure 11.2 is the threshold map of the ACTIVLIM-Stroke. It indicates patients' expected responses for each item as a function of ability. The bottom panel represents the relationship between ordinal scores and the linear measures expressed in logits (log odds probability units). The figure helps to illustrate the score of a patient by examining how his or her actual answers to each of the questions in the measure fit with expected answers. For example, on admission to rehabilitation (marked in the figure as Test 1), this patient is likely to give the response "impossible" to the items *standing for a long time* through to *carrying a heavy load*. On discharge (Test 2), the patient is likely to

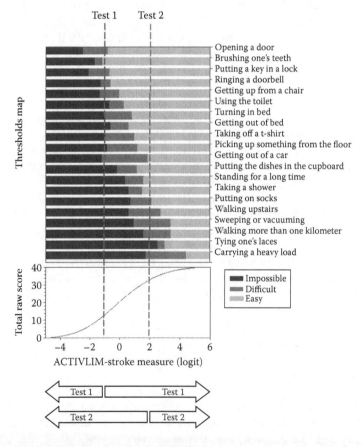

FIGURE 11.2 Example of threshold map and conversion from ordinal to linear scores.

give the response "impossible" to virtually no items. This visual representation of the patient's functioning allows us to determine what has really changed in the patient's everyday life and if unexpected answers contribute to the final observed score.

Rasch analysis also helps to examine the invariance of each item in a measure to test if scores on the item are independent of other personal factors. In other words, the scores on any given item should only be resulting from a given amount of the latent trait under investigation, as opposed to personal characteristics such as age, gender, culture, diagnosis, and so on. Scale invariance is a key point and deserves to be considered while interpreting clinical change. Let's illustrate this by looking at invariance of culture of a measure of activities of daily living (ADL) (Box 11.3).

BOX 11.3 CULTURAL VARIANCE OF ITEMS

Two patients with stroke, Mark and Feng, are in the same rehabilitation unit in Auckland and have the same level of ADL functioning. Mark, originally from Auckland, reports no difficulty with eating. Feng, who emigrated from China recently, on the other hand reports that eating is very difficult.

Given that the level of ADL functioning for Mark and Feng are the same, it pays to explore why they respond so differently to this particular item. Mark ordinarily uses a knife and fork to eat his meals, while Feng uses chopsticks. No wonder the item attracts a biased response. Let's consider a patient of 77 years old, who moved from the Benin Republic (West Africa), to live in Belgium where he had a stroke. Evaluating his functional recovery using outcome measures that are based on cultural values of western nations may not have the same meaning in the patient's country of origin where he is supposed to go back to after rehabilitation.

So, if a scale dedicated to evaluate ADL limitations (as an example) is not invariant regarding key personal factors, the interpretation of scores between different patients may be problematic as we do not know if a difference in favor of a patient is due to the actual level of function or to other personal factors. Similarly, in multicenter studies that involve different countries (or different cultural groups within a country), the cultural validity of outcomes has to be assessed. Invariance should of course also be evaluated for other key variables that are predicted to result in bias. These typically (but not exclusively) include age groupings, gender, culture, and diagnostic groups. Rasch analysis allows this investigation to be done by checking the actual observed scores and expected scores for each item, given the person's overall level of functioning. If variance is observed, techniques can be used that allow the data to be used but accounting for the observed variance. There are many examples in the literature (Arnould et al. 2012; Krageloh et al. 2012; Lundgren-Nilsson et al. 2005; Tennant et al. 2004b).

Once an outcome measure has been shown to fit the Rasch model, it is possible to produce tables that convert raw ordinal scores to their corresponding interval level scores. For example, we have produced conversion tables for a measure of consultation and relational empathy (Kersten et al. 2012), a measure of physical and social integration following stroke (Kersten et al. 2010), and a measure of activity limitations following stroke (Batcho et al. 2012). Clinicians can simply sum the scores from the outcome measure and use the conversion table to look up the interval equivalent. Evaluating a patient's change or comparing patients then becomes an accurate exercise as clinicians are no longer dealing with ordinal data. This also has huge potential for researchers. A well-built outcome measure that fits the Rasch model can be used to develop a computer adaptive test (CAT), which we will discuss later in this chapter.

Modern Test Theory—Item Response Theory

IRT is placed within a statistical modeling paradigm (Andrich 2011). A key difference with the theory of Rasch measurement is that IRT aims to find the best model that fits the available data (Fayers 2007), whereas Rasch analysis aims to test if data from an outcome measure fit the stringent requirements of the Rasch model only. In other words, the two approaches (Rasch and IRT) come from different paradigms (Andrich 2004). It is beyond the scope of the chapter to discuss this but for the interested reader we refer them to some key texts (Andrich 2004; Fayers 2007).

Computer Adaptive Testing

Patient reported outcome measures are often developed for evaluations of populations and not focused on individual patients. In order to meet patients' needs, instruments adjustable to their situation might be advantageous. For example, a measure of physical function that can be used in the acute setting, the inpatient rehabilitation setting, and the community setting would allow a comparison of the patient's functioning over their entire rehabilitation journey (La Porta et al. 2011). This could be achieved by developing item banks that would cover the whole range of the underlying unidimensional construct under investigation. However, due to the large amount of items, these can become impossible to use at the individual level. To tailor an individual test, but keeping the possibility to compare persons, a computerized adaptive test (CAT), a form of computer-based test, can be used that adapts to the patient's ability level. Here individual patients are only exposed to a few items from the bank. For example, an item bank could be created for physical functioning and consist of items from several of scales (both generic and diagnosis specific), co-calibrated on the same interval metric. This would make it possible to compare patients between diagnoses and different levels of physical functioning asking only a few questions of the patient. Exchange rates could also be given between existing measures. CATs are said to be efficient at rating a single dimension with precision (Fayers 2007).

How might a CAT work? Essentially, the basic computer-adaptive testing method is an iterative algorithm. Let's use an example of physical functioning:

a. At the start of the test, we don't know the level of someone's physical functioning. The individual is therefore offered the item that lies in the middle of the scale in terms of its difficulty (as previously determined with Rasch analysis).
b. The individual answers the item correctly/positively or incorrectly/negatively.
c. Depending on the individual's answer, the next item selected from the pool of available items will either be a harder one or an easier one, and so forth.
d. The estimate of the person's ability (the person parameter) is constantly updated during the test, and the tests stop when the person ability has been estimated with a pre-set acceptable level of error.

Using CAT approaches, people can receive different tests that are better targeted at their level. This makes the test more specific and acceptable to the person. In addition, CATs are shorter and therefore less burdensome on the respondent. As CAT data are collected using a computer or handheld electronic device, the data are stored electronically. In health outcome measurement, there is a rapid increase in the use of item banks and CAT (Heinemann et al. 2010; Hou et al. 2011; Siemons et al. 2012; Simms et al. 2011; Tennant et al. 2004a; Vogels et al. 2011), and they will most certainly be part of the future approaches to measuring outcome in clinical practice.

Conclusions and Thoughts for the Future

We believe the future for measuring outcomes of rehabilitation is bright if robust methodologies, underpinned by sound and proven theories described in this chapter, are used along with technologies. The rapid development of information technologies available to support the health-care systems around the world has the potential to improve both the information flow and associated decision making for patients, health-care professionals, providers, and researchers alike:

1. For clinical use, a CAT combined with modern technologies could be integrated with the patient's normal life. Information can be accessed by applying basic technical solutions available to routinely collect clinical information via the web and or mobile phones. Thus, through mobile applications and online platforms, patients' data can be collected and monitored and clinicians can interact with patients online. An important effect of this technical development is the patients' increased use of the new technology to access their own health-related information and the establishment of patient networks.
2. Electronically available data can provide unprecedented opportunities to continuously monitor all patients and focus on those patients with signals indicating that there are problems or where the patient reports this through the system. Health-care professionals will have important information about their patients' progress and also their rehabilitation potential, as pooled data will be available from other patients. This will greatly assist their ability to plan the rehabilitation program as well as discharge.
3. Health-care organization will be able to use robust data for benchmarking and quality improvement work. Benchmarking is already happening in many places,

such as the AROC benchmarking system. However, with more robust approaches to outcome measures used in such initiatives, they will have more accurate data upon which to compare services.

4. Researchers will be able to use these data to compare their outcomes of trials with those from other research centers across the world, while being able to still use some of their own country-specific measures. This is achievable because item banks can be created, which share common questions but facilitate also the use of unique country-specific items.

Box 11.4 provides an example of our proposed shared future.

Box 11.4   Example from Sweden of a Shared Future

In Sweden, the history of national quality control registers started in the last decade (Garpenby and Carlsson 1994), with the primary aim to generate information that can improve health care for patients. The goal of the registers is to become tools for continuous quality improvement and to support high and consistent quality of care throughout Sweden, ultimately to ensure patient benefit in the form of the best possible care.

During the last decade, there has been immense development in computer technology, the World Wide Web, and new communications systems. In this scenario, databases for quality control can be very useful to support high and consistent quality of care and the possibilities to gather data. The data can be entered online and immediately be available with feedback to the person, which theoretically should enhance willingness to supply data.

The vision for the quality registries today in Sweden is to create an overall knowledge system that is actively used on all levels for continuous learning, quality improvement, and management of health-care services.

Clinical Messages

Having read this chapter, the practicing clinician may well be left with uncertainty as to how to choose a measure—and then what to do with data from an outcome measure. First and foremost, the clinician should be clear what outcomes s/he aims to impact on through the therapeutic intervention. After all, there is nothing worse than measuring something one has no means to change. This should be followed by a search of the outcome measurement literature for a suitable measure. More often than not, this will reveal several candidate measures. There are some useful guidelines for reviewing the evidence of the reliability, validity, and responsiveness of such measures. We recommend the use of a recently developed tool, the COSMIN checklist (consensus-based standards for the selection of health status measurement instruments), for this purpose (Mokkink et al. 2010; Terwee 2011). However, we would add to this checklist and suggest that measures that have not been evaluated with the Rasch model should only be considered to be ordinal and be treated as such. In other words, such measures should not be used to evaluate change on latent variables. By contrast, measures that fit the Rasch model and that provide a conversion from ordinal to interval data can be adopted readily and used to do examine change over time. If such evidence is not available, the clinician can use the outcome measure, so long as it fits the requirements of CTT and is only evaluated using nonparametric statistics (Grimby et al. 2012; Küçükdeveci et al. 2011; Merbitz et al. 1989; Svensson 2001; Wright and Linacre 1989). It is of course also important to compare the findings from outcomes data with the perceptions of patients.

Conclusions

This chapter has illustrated the serious limitations of ordinal scales and argued that misuse of such scales can lead to incorrect conclusions, to the detriment of our patients and services. We have argued that the use of modern approaches to measurement, and in particular the use of the Rasch model, will result in robust/fundamental measurement. This is a theory-driven approach, well tried and tested, and one for which there are ample educational papers, introductory textbooks, and courses. Given the evidence underpinning this theory, it is time to stop misusing data from ordinal scales in clinical practice, time to educate our young health-care students the right techniques, and time to move toward data sharing to ensure quality improvements in health care and rehabilitation.

The changes we propose will involve a shift from a production-driven mindset toward a collaborative service-oriented paradigm. A prerequisite for a successful change is the ability for information to be seamless across organizations, and also, in some way, directly involving the patient. While technologies to make that happen are already available, critical ethical and legal issues need to be addressed. This will require a new ethical, legal, and practical model, flexible enough to accommodate rapidly changing technological development. This is only the beginning of something of a shift in paradigm.

References

ACC. 2009. *Guide to Outcome Measure Reporting*. Wellington: Accident Compensation Corporation.

Alder, K. 2002. *The Measure of All Things. The Seven-Year-Odyssey That Transformed the World.* London: Abacus.

Alviar, M. J., J. Olver, C. Brand, J. Tropea, T. Hale, M. Pirpiris, and F. Khan. 2011. Do patient-reported outcome measures in hip and knee arthroplasty rehabilitation have robust measurement attributes? A systematic review. *Journal of Rehabilitation Medicine* 43 (7):572–583.

Andrich, D. 1978. A rating formulation for ordered response categories. *Psychometrika* 43:561–573.

Andrich, D. 2004. Controversy and the Rasch model: A characteristic of incompatible paradigms? *Medical Care* 42 (1 Suppl):I7–I16.

Andrich, D. 2011. Rating scales and Rasch measurement. *Expert Review of Pharmacoeconomics and Outcomes Research* 11 (5):571–585.

Arnould, C., L. Vandervelde, C. S. Batcho, M. Penta, and J. L. Thonnard. 2012. Can manual ability be measured with a generic ABILHAND scale? A cross-sectional study conducted on six diagnostic groups. *BMJ Open* 2:e001807. doi: 10.1136/bmjopen-2012-001807.

Australasian Rehabilitation Outcomes Centre (AROC). 2014. Available at http://ahsri.uow.edu.au/aroc/index.html. University of Wallongong.

Batcho, C. S., A. Tennant, and J. L. Thonnard. 2012. ACTIVLIM-Stroke: A crosscultural Rasch-built scale of activity limitations in patients with stroke. *Stroke* 43 (3):815–823.

Bond, T. G., and C. M. Fox. 2001. *Applying the Rasch Model. Fundamental Measurement in the Human Sciences.* London: Lawrence Erlbaum Associates.

Brook, R. H., K. N. Williams, and A. D. Avery. 1976. Quality assurance today and tomorrow: Forecast for the future. *Annals of Internal Medicine* 85 (6):809–817.

Cano, S., A. F. Klassen, A. Scott, A. Thoma, D. Feeny, and A. Pusic. 2010. Health outcome and economic measurement in breast cancer surgery: Challenges and opportunities. *Expert Review of Pharmacoeconomics and Outcomes Research* 10 (5):583–594.

CARF International. 2014. Available at http://www.cart.org/home/.

Chartered Society of Physiotherapy. 2011. Measuring for quality improvement in physiotherapy. Available at http://www.csp.org.uk/director/policymakers/Keyissues/measuringforquality.cfm.

Duncan, P. W., H. S. Jorgensen, and D. T. Wade. 2000. Outcome measures in acute stroke trials: A systematic review and some recommendations to improve practice. *Stroke* 31 (6):1429–1438.

Fayers, P. M. 2007. Applying item response theory and computer adaptive testing: The challenges for health outcomes assessment. *Quality of Life Research* 16 (Suppl 1):187–194.

Fisher, A. G. 1993. The assessment of IADL motor skills: An application of many-faceted Rasch analysis. *American Journal of Occupational Therapy* 47 (4):319–329.

Garpenby, P., and P. Carlsson. 1994. The role of national quality registers in the Swedish health service. *Health Policy* 29 (3):183–195.

Granger, C. V., B. B. Hamilton, J. M. Linacre, A. W. Heinemann, and B. D. Wright. 1993. Performance profiles of the functional independence measure. *American Journal of Physical Medicine and Rehabilitation* 72 (2):84–89.

Grimby, G., A. Tennant, and L. Tesio. 2012. The use of raw scores from ordinal scales: Time to end malpractice? *Journal of Rehabilitation Medicine* 44 (2):97–98.

Gutenbrunner, C., A. B. Ward, and M. A. Chamberlain. 2006. White book on physical and rehabilitation medicine in Europe. In *Europa Medicophysica*. Section of Physical and Rehabilitation Medicine and European Board of Physical and Rehabilitation Medicine, Union Européenne des Médecins Spécialistes (UEMS) and Académie Européenne de Médecine de Réadaptation pp. 287–332.

Haigh, R., A. Tennant, F. Biering-Sørensen, G. Grimby, Č. Marinček, S. Phillips, H. Ring, L. Tesio, and J.-L. Thonnard. 2001. The use of outcome measures in physical medicine and rehabilitation within Europe. *Journal of Rehabilitation Medicine* 33 (6):273–278.

Heinemann, A. W., D. Tulsky, M. Dijkers, M. Brown, S. Magasi, W. Gordon, and H. DeMark. 2010. Issues in participation measurement in research and clinical applications. *Archives of Physical Medicine & Rehabilitation* 91 (9):S72–S76.

Hobart, J., and S. Cano. 2009. Improving the evaluation of therapeutic interventions in multiple sclerosis: The role of new psychometric methods. *Health Technology Assessment* 13 (12):1–177.

Hou, W.-H., J.-H. Chen, Y.-H. Wang, C.-H. Wang, J.-H. Lin, I. P. Hsueh, Y.-C. Ou, and C.-L. Hsieh. 2011. Development of a set of functional hierarchical balance short forms for patients with stroke. *Archives of Physical Medicine and Rehabilitation* 92 (7):1119–1125.

Jette, D. U., J. Halbert, C. Iverson, E. Miceli, and P. Shah. 2009. Use of standardized outcome measures in physical therapist practice: Perceptions and applications. *Physical Therapy* 89 (2):125–135.

Kayes, N. M., and K. M. McPherson. 2010. Measuring what matters: Does "objectivity" mean good science? *Disability and Rehabilitation* 32 (12):1011–1019.

Kersten, P., A. Ashburn, S. George, and J. Low. 2010. The subjective index for physical and social outcome (SIPSO) in stroke: Investigation of its subscale structure. *BMC Neurology* 10:26.

Kersten, P., P. J. White, and A. Tennant. 2010. The visual analogue WOMAC 3.0 scale—Internal validity and responsiveness of the VAS version. *BMC Musculoskeletal Disorders* 11:80.

Kersten, P., P. J. White, and A. Tennant. 2012. The consultation and relational empathy measure: An investigation of its scaling structure. *Disability and Rehabilitation* 34 (6):503–509.

Korner-Bitensky, N., S. Barrett-Bernstein, G. Bibas, and V. Poulin. 2011. National survey of Canadian occupational therapists' assessment and treatment of cognitive impairment post-stroke. *Australian Occupational Therapy Journal* 58 (4):241–250.

Krageloh, C. U., P. Kersten, D. R. Billington, P. H.-C. Hsu, D. Shepherd, J. Landon, and X. J. Feng. 2012. Validation of the WHOQOL-BREF quality of life questionnaire for general use in New Zealand: Confirmatory factor analysis and Rasch analysis. *Quality of Life Research* 22 (6):1451–1457.

Küçükdeveci, A. A., A. Tennant, G. Grimby, and F. Franchignoni. 2011. Strategies for assessment and outcome measurement in physical and rehabilitation medicine: An educational review. *Journal of Rehabilitation Medicine* 43 (8):661–672.

La Porta, F., M. Franceschini, S. Caselli, P. Cavallini, S. Susassi, and A. Tennant. 2011. Unified Balance Scale: An activity-based, bed to community, and aetiology-independent measure of balance calibrated with Rasch analysis. *Journal of Rehabilitation Medicine* 43 (5):435–444.

Laver Fawcett, A. 2007. *Principles of Assessment and Outcome Measurement for Occupational Therapists and Physiotherapists: Theory, Skills and Application.* Chichester: John Wiley & Sons.

Lundgren-Nilsson, Å., G. Grimby, H. Ring, L. Tesio, G. Lawton, A. Slade, M. Penta, M. Tripolski, F. Biering-Sørensen, J. Carter, C. Marincek, S. Phillips, A. Simone, and A. Tennant. 2005. Cross-cultural validity of functional independence measure items in stroke: A study using Rasch analysis. *Journal of Rehabilitation Medicine* 37 (1):23–31.

Lundgren-Nilsson, Å., and A. Tennant. 2011. Past and present issues in Rasch analysis: The functional independence measure (FIM™) revisited. *Journal of Rehabilitation Medicine* 43 (10):884–891.

McFadden, E., M. C. Horton, H. L. Ford, G. Gilworth, M. McFadden, and A. Tennant. 2012. Screening for the risk of job loss in multiple sclerosis (MS): Development of an MS-specific Work Instability Scale (MS-WIS). *Multiple Sclerosis* 18 (6):862–870.

Merbitz, C., J. Morris, and J. C. Grip. 1989. Ordinal scales and foundations of misinference. *Archives of Physical Medicine and Rehabilitation* 70 (4):308–312.

Mokkink, L. B., C. B. Terwee, D. L. Knol, P. W. Stratford, J. Alonso, D. L. Patrick, L. M. Bouter, and H. C. W. de Vet. 2010. The COSMIN checklist for evaluating the methodological quality of studies on measurement properties: A clarification of its content. *BMC Medical Research Methodology* 10:22.

Nolte, E., and M. McKee, eds. 2008. *European Observatory on Health Systems and Policy Series. Caring for People with Chronic Conditions. A Health System Perspective*. Maidenhead: Open University Press.

Nunnally, J. C., and I. H. Bernstein. 1994. *Psychometric Theory*, 3rd ed. New York: McGraw-Hill.

Penta, M., L. Tesio, C. Arnould, A. Zancan, and J. Thonnard. 2001. The ABILHAND questionnaire as a measure of manual ability in chronic stroke patients: Rasch-based validation and relationship to upper limb impairment. *Stroke* 32 (7):1627–1634.

Rasch, G. 1960/1980. *Probabilistic Models for Some Intelligence and Attainment Tests*, revised and expanded ed. Chicago: University of Chicago Press.

Richardson, J., and J. McKie. 2005. Empiricism, ethics and orthodox economic theory: What is the appropriate basis for decision-making in the health sector? *Soc Sci Med* 60 (2):265–275.

Richardson, J., and J. McKie. 2007. Economic evaluation of services for a National Health scheme: The case for a fairness-based framework. *Journal of Health Economics* 26 (4):785–799.

Siemons, L., P. M. Ten Klooster, E. Taal, C. A. W. Glas, and M. A. F. J. Van de Laar. 2012. Modern psychometrics applied in rheumatology—A systematic review. *BMC Musculoskeletal Disorders* 13:216.

Simms, L. J., L. R. Goldberg, J. E. Roberts, D. Watson, J. Welte, and J. H. Rotterman. 2011. Computerized adaptive assessment of personality disorder: Introducing the CAT-PD project. *Journal of Personality Assessment* 93 (4):380–389.

Sivan, M., R. J. O'Connor, S. Makower, M. Levesley, and B. Bhakta. 2011. Systematic review of outcome measures used in the evaluation of robot-assisted upper limb exercise in stroke. *Journal of Rehabilitation Medicine* 43 (3):181–189.

Skeat, J., and A. Perry. 2008. Exploring the implementation and use of outcome measurement in practice: A qualitative study. *International Journal of Language & Communication Disorders* 43 (2):110–125.

Streiner, D. L., and G. R. Norman. 2008. *Health Measurement Scales: A Practical Guide to Their Development and Use*. Oxford: Oxford University Press.

Stucki, G., L. Daltroy, J. N. Katz, M. Johannesson, and M. H. Liang. 1996. Interpretation of change scores in ordinal clinical scales and health status measures: The whole may not equal the sum of the parts. *Journal of Clinical Epidemiology* 49 (7):711–717.

Svensson, E. 2001. Guidelines to statistical evaluation of data from rating scales and questionnaires. *Journal of Rehabilitation Medicine* 33 (1):47–48.

Tennant, A., and P. G. Conaghan. 2007. The Rasch measurement model in rheumatology: What is it and why use it? When should it be applied, and what should one look for in a Rasch paper? *Arthritis Care and Research* 57 (8):1358–1362.

Tennant, A., J. M. L. Geddes, and M. A. Chamberlain. 1996. The Barthel Index: An ordinal score or interval level measure? *Clinical Rehabilitation* 10 (4):301–308.

Tennant, A., S. P. McKenna, and P. Hagell. 2004a. Application of Rasch analysis in the development and application of quality of life instruments. *Value in Health* 7 (Suppl 1):S22–S26.

Tennant, A., M. Penta, L. Tesio, G. Grimby, J. L. Thonnard, A. Slade, G. Lawton, A. Simone, J. Carter, A. Lundgren-Nilsson, M. Tripolski, H. Ring, F. Biering-Sørensen, C. Marincek, H. Burger, and S. Phillips. 2004b. Assessing and adjusting for cross-cultural validity of impairment and activity limitation scales through differential item functioning within the framework of the Rasch model: The PRO-ESOR project. *Medical Care* 42 (1 Suppl):I37–I48.

Terwee, C.B. 2011. *The COSMIN Checklist.* Amsterdam: EMGO Institute for Health and Care Research.

Tesio, L. 2003. Measuring behaviours and perceptions: Rasch analysis as a tool for rehabilitation research. *Journal of Rehabilitation Medicine* 35 (3):105–115.

Tesio, L., C. V. Granger, and R. C. Fiedler. 1997. A unidimensional pain/disability measure for low-back pain syndromes. *Pain* 69 (3):269–278.

Thurstone, L. L. 1931. Measurement of social attitudes. *Journal of Abnormal and Social Psychology* 26:249–269.

Turner-Stokes, L. et al. 2012. Healthcare tariffs for specialist inpatient neurorchabilitation services: Rationale and development of a UK casemix and casting methodology. *Clinical Rehabilitation* 26 (3):264–279.

Tyson, S., J. Greenhalgh, A. F. Long, and R. Flynn. 2010. The use of measurement tools in clinical practice: An observational study of neurorehabilitation. *Clinical Rehabilitation* 24 (1):74–81.

Uniform Data System for Medical Rehabilitation. 2014. Available at http://www.udsmr.org/WebModules/Pro/Pro_About.aspx. University of Buffalo.

Vogels, A., G. Jacobusse, and S. Reijneveld. 2011. An accurate and efficient identification of children with psychosocial problems by means of computerized adaptive testing. *BMC Medical Research Methodology* 11 (1):111.

World Health Organization. 2001. *International Classification of Functioning, Disability and Health: ICF.* Geneva: World Health Organization.

Wright, B. D. 1999. Common sense for measurement. *Rasch Measurement Transactions* 13:704–705.

Wright, B. D., and J. M. Linacre. 1989. Observations are always ordinal; measurements, however, must be interval. *Archives of Physical Medicine and Rehabilitation* 70 (12):857–860.

Wright, B. D., and M. H. Stone. 1979. *Best Test Design.* Chicago: Mesa Press.

12

Te Waka Oranga: Bringing Indigenous Knowledge Forward

Hinemoa Elder

Introduction

Reconsidering the concept of rehabilitation has unique significance for indigenous peoples. Indigenous concepts of recovery and the restoration of balance have developed separately to those of the Western scientific tradition, so considering rehabilitation via an indigenous lens alongside considering what might constitute indigenous approaches can illuminate what this might mean in practice—its strengths and weaknesses. Overall, indigenous concepts privilege unique, culturally determined aspects that can enhance outcomes for the indigenous peoples' healing and well-being. The need for a distinct indigenous critique of the term rehabilitation is also important for a number of other reasons. The United Nations Declaration on the Rights of Indigenous Peoples emphasizes the importance of indigenous peoples' rights to access their own knowledge systems (United Nations 2008) in the context of illness or injury. In addition, disparities are well documented for indigenous peoples in the prevalence of injury and illness associated with rehabilitation services and responses by these services (Harwood 2010; New Zealand Guidelines Group 2006; Oakley-Brown et al. 2006). Attempting to meet what might constitute the rehabilitation needs of indigenous peoples is the responsibility of practitioners from all cultural backgrounds. There are two key reasons for this. First, the indigenous workforce is not sufficient to meet these needs, and second, all practitioners have a duty of care, and in some countries a statutory requirement (New Zealand Health Practitioner Competence Assurance Act 2003), to be culturally competent to work with indigenous peoples.

In this chapter, I present a pragmatic review of the literature pertaining to the use of indigenous knowledge across activities that might constitute "rehabilitation." Following this, *whakaaro Māori* (Māori thinking—that of the indigenous people of New Zealand) with reference to the concept of rehabilitation is described with a specific exemplar. *Whakaaro Māori* here means conceptualization that comes from *Te Ao Māori* (the Māori world view). For example, the *whakataukī* or proverbial saying "*He tao rākau, e taea te karo. He tao kī, e kore e taea*"—which can be translated as "The thrust of a weapon can be parried, a lashing of the tongue cannot,"—illustrates that Māori thinking about the meaning and impact of words is the opposite of the English saying "sticks and stones may break my bones but words will never hurt me." Māori research is highlighted because of the author's ethnicity, knowledge, research, and practice.

One fundamental problem in reconsidering rehabilitation is that of relativism (Hoeman 2008; Macklin 1999). Considering indigenous concepts relative to the concept of rehabilitation locates the idea and practice of rehabilitation itself as the central reference point. This risks further marginalization of indigenous evidence. Ensuring that indigenous concepts are not subsumed by the concept of rehabilitation is argued for here.

Overall, the contention of this chapter is that the use of indigenous methods that strengthen indigenous identity is preventative of illness and injury as well as a protective factor for well-being. The need for indigenous research methods to examine the ways in which these approaches are successful is also underlined. To assist with the understanding of Māori words and concepts, a glossary is included at the conclusion of this chapter.

International Indigenous Literature

While there is no universally accepted definition of indigenous peoples, the report by Special rapporteur Jose R. Martinez Cobo (1983) to the United Nations Economic and Social Council completed in 1983 provides a working definition:

> Indigenous communities, peoples and nations are those which, having a historical continuity with pre-invasion and pre-colonial societies that developed on their territories, consider themselves distinct from other sectors of the societies now prevailing on those territories, or parts of them. They form at present non-dominant sectors of society and are determined to preserve, develop and transmit to future generations their ancestral territories, and their ethnic identity, as the basis of their continued existence as peoples, in accordance with their own cultural patterns, social institutions and legal system. This historical continuity may consist of the continuation, for an extended period reaching into the present of one or more of the following factors:
>
>> Occupation of ancestral lands, or at least of part of them
>> Common ancestry with the original occupants of these lands
>> Culture in general, or in specific manifestations (such as religion, living under a tribal system, membership of an indigenous community, dress, means of livelihood, lifestyle, etc.)

Language (whether used as the only language, as mother-tongue, as the habitual means of communication at home or in the family, or as the main, preferred, habitual, general or normal language)

Residence in certain parts of the country, or in certain regions of the world

On an individual basis, an indigenous person is one who belongs to these indigenous populations through self-identification as indigenous (group consciousness) and is recognized and accepted by these populations as one of its members (acceptance by the group). This preserves for these communities the sovereign right and power to decide who belongs to them, without external interference

It is also relevant to consider Article 33 of the United Nations Declaration on the Rights of Indigenous Peoples, which states that Indigenous peoples have the right to determine their own identity or membership in accordance with their customs and traditions. This does not impair the right of indigenous individuals to obtain citizenship of the States in which they live. Indigenous peoples have the right to determine the structures and to select the membership of their institutions in accordance with their own procedures

With this context in mind, searches using terms *indigenous, Māori, aborigine, aboriginal, Torres Strait Islander, native, American Indian, Alaskan Native,* and *Basque* were used. I found a significant indigenous literature where the term rehabilitation is used or where what can be broadly considered rehabilitative practices are described (Coyhis and Simonelli 2008; DeVerteuill and Wilson 2010; Hartmann and Gone 2012; Kahn et al. 1988; Shaw et al. 2011; Spicer 2001; Watts 2001). This body of literature crosses several disciplines. These include the fields of alcohol and other drug use rehabilitation and recovery, psychiatry, offender programs, mental health services, and response to different types of injury. Descriptions of exactly how indigenous knowledge is applied and research methods that appropriately investigate the impact are few (Chenhall 2008; Dell et al. 2011). Practice suggests that this type of information is not commonly published in peer-reviewed journals; rather it is held in local protocols and reports to funding bodies.

Surprisingly, the term rehabilitation does not appear to have been previously interrogated from an indigenous perspective. No papers were found critiquing the term. From an indigenous perspective, this word and its underlying constructs locate the pathology centrally, as this is the aspect that drives the very need for rehabilitation. There is an absence within this premise of the potential presence of cultural healing resources. Practice-based evidence suggests the term reinforces the idea of a state of being that is insufficient. In addition, the concept of rehabilitation assumes that a return to an earlier, healthier state is the goal. For indigenous peoples, this approach could be perceived as invalidating of current and ancestral resources as it focuses on what is wrong and is silent to cultural strengths (Smith 1999). For some it could also reinforce the idea that earlier healthier states of being are unattainable given the long history of intergenerational disharmony secondary to the impact of colonization (Jackson 2007). One further difficulty is that the concept of rehabilitation is commonly linked to independence as a core goal. For example, The World Health Organization states on their website, "rehabilitation provides disabled people with the tools they need to attain independence" (WHO 2014). There are two aspects that invite question here. First, indigenous communities continue to privilege interdependence (Elder 2012a; Watts 2001). Second, the identity of "disabled people" is not one that resonates with indigenous communities (Nikora et al. 2004) as well as being critiqued by others (Patston 2007; see also Chapter 10).

Some indigenous approaches use the term rehabilitation to describe their use of culturally defined practices, ways of being, and ways of knowing (Coyhis and Simonelli 2008; Hartmann and Gone 2012; Shaw et al. 2011). These approaches commonly call on ancestral mores that inform interwoven roles and responsibilities as both process and content material for the healing process (Gone 2009). In the context of writing about the challenges of developing a trauma system for indigenous peoples, Plani and Carson (2008, p. 47) underline the importance of indigenous communities "owning" the system. They recommend these communities "be encouraged to design their own culturally appropriate rehabilitation programs" (Plani and Carson 2008, p. 50). The word rehabilitation in these settings seems to serve as a code word to alert funders, referrers, and patients alike to a process where peoples' health state is modified in some way. Perhaps there is also the desire to present indigenous approaches within an evidence-based frame thereby ensuring funding from which they might otherwise be excluded (Hoeman 2008, p. 72).

Use of indigenous knowledge has been identified as a key indicator of improving outcomes in a range of settings (Coyhis and Simonelli 2008; Hartmann and Gone 2012; Shaw et al. 2011; Spicer 2001; Watts 2001). The importance of making this knowledge available is emphasized across urban (Hartmann and Gone 2012; Wright et al. 2011) and remote contexts (Schmidt 2000; Shaw et al. 2011). A number of tensions have been described in urban settings. These include how to incorporate traditional healing methods in the context of the harsh realities of poverty in city life, finding appropriate tribal matches for kinship support, the maintenance of cultural integrity and support of elders who provide this kinship, and enthusiasm for traditional healing versus concerns about confidentiality and trust (Hartmann and Gone 2012). Attempts to build access to indigenous knowledge can be delivered through employing indigenous cultural workers (Stathis et al. 2006). Indigenous health workers' personal health behaviors and credibility are also noted as having a significant influence on their capacity to support positive health outcomes (Kahn et al. 1988; Thompson and Roberston 2010).

Culturally embedded notions of rehabilitation emphasize as critical the recovery of healthy cultural identity (Coyhis and Simonelli 2008; Spicer 2001). There has been some debate about this issue (Brady 1995; Spicer 2001). It has been suggested that some practitioners of indigenous healing support the use of substances to alter the state of consciousness in some traditional ceremonies. Taking an indigenous cultural approach may limit what is shared with other cultural groups and this may limit access to approaches from diverse cultures that might be beneficial (Brady 1995). These concerns have been rebutted by Spicer (2001), finding that revivals of cultural traditions acted both as antidotes to substance abuse and as powerful therapeutic tools. The caveat being that, because of study design issues, it has been difficult to make strong inferences about the outcomes of programs (Spicer 2001).

The central tension identified here is between indigenous evidence and the evidence-based practice paradigm. The legitimacy of indigenous evidence, which does not fit within the evidence-based practice paradigm, is therefore questioned, which leads to difficulties in maintaining funding for established programs (Shaw et al. 2011). The absence of control groups and lack of outcome measures (not seen as absolute requirements by some indigenous methodologies) make inferences of indigenous programs difficult (Spicer 2001). On the other hand, the evidence-based practice approaches, where cultural adaptation is used, has been interpreted as "forced acculturation" (Trickett 2011, p. 64). There is an urgent need for indigenous research that rigorously addresses these issues, and while it must be firstly responsive to indigenous communities, it can also respond to Western scientific critique.

This body of research highlights the many barriers to bringing indigenous knowledge forward in the dominant sociopolitical and research contexts. These include geographic and economic factors (Schmidt 2000), knowledge paradigm factors, and attitudinal factors (DeVerteuill and Wilson 2010). Despite these elements, there is also evidence of locally meaningful ways in which indigenous cultural restoration is used in response to illness and injury with some benefit to individuals and their communities (Chenhall 2008; Coyhis and Simonelli 2008; Dell et al. 2011).

The significant remaining gap in the literature is an analysis of the concept of rehabilitation from an indigenous perspective as well as, and importantly, an indigenous conceptualization without reference to the English term *rehabilitation*. This is now addressed from a Māori perspective.

Māori Literature

There is a paucity of specific literature about rehabilitation from a Māori perspective (Harwood 2010). This is of grave concern when considering the numbers of Māori in groups that have alcohol and other drug-use problems, mental illness, and injury (New Zealand Guidelines Group 2006; Oakley-Brown et al. 2006). The aim of my ongoing research program is to examine the impact on outcomes of the use of Māori concepts of healing and recovery.

Two key areas of research and practice that apply the concept of rehabilitation are reviewed here in relation to Māori—traumatic brain injury (TBI) and psychiatric illness. What this applied literature review shows is that, despite significant overrepresentation of the indigenous population in both areas, very few culturally responsive rehabilitation approaches have been developed. One post-stroke controlled trial about self-directed rehabilitation used recordings of personal stories of recovery against adversity for participants to review at their leisure (Harwood et al. 2012). The authors did not emphasize this method as using the cultural tradition of storytelling, however, they found improved outcomes in both Māori and Pacific peoples with stroke.

Māori

Māori are the indigenous people of Aotearoa, New Zealand. The Population Census in 2013 found almost 15% of the New Zealand population (approximately 660,000 people) reported being of Māori descent. Māori continue to be a young population with approximately 35% of Māori under 15 years old (Statistics New Zealand, n.d.).

Tangata Whenuatanga (Māori culture) is manifested via *Te Reo Māori* (the Māori language) and *Te Ao Māori* (the Māori world view) with attendant *tikanga* (protocols). While it is acknowledged that in contemporary Aotearoa, New Zealand, Māori identity occurs in a wide range of ways, there is validity in considering a general Māori experience (Durie 2001). The concept of Māori approaches and Māori responsive services is also recognized across government policy (Ministry of Health 2008). For the purposes of this discussion, a definition of culture is needed. While the concept of culture is complex, *The New Oxford English Dictionary* (1998) provides the following useful framework: "the customs, arts or social institutions and achievements of a particular nation, people, or other social group." This definition forms the parameters of what is described in this article.

Traumatic Brain Injury

TBI is a significant health problem in Aotearoa, New Zealand. Between 22,000 and 33,000 people have a TBI each year, compared to between 7000 and 8000 who have a stroke (Feigin et al. 2012). The costs of these injuries are difficult to quantify. However, direct costs of stroke and TBI combined are reported as over NZ$200 million annually (Feigin et al. 2012). Approximate costs of TBI alone have been quoted as NZ$100 million per year (New Zealand Guidelines Group 2006).

New Zealand data are consistent with international findings; children and young people are high-risk populations for TBI. Almost 20% of young people will have sustained a TBI before they turn 14 years old (McKinlay et al. 2009). Being male, coming from a family with four or more adverse life events, and where punitive parenting is used have been identified as risk factors for childhood TBI (McKinlay et al. 2009).

Incidence rates of TBI in Māori were significantly higher for all age bands of *tamariki* (young people) compared to non-Māori (Barker-Collo et al. 2009). Recent data have shown that violence-related TBI rates in Māori are three times that of non-Māori (Feigin et al. 2012). Given the age structure of the Māori population with a mean age 23 (compared to 37 years of age in non-Māori), this overrepresentation is a cause for grave concern (Statistics NZ, n.d).

Nonaccidental TBI (NATBI) is one important area of concern. Māori infants have been shown to have one of the highest rates of subdural hematoma in the world, with most being of likely nonaccidental origin (Kelly and Hayes 2004). The annual incidence rates are estimated between 32.5 and 38.5/100,000 compared to non-Māori rates of 14.7 to 19.6/100,000 (Kelly and Farrant 2008). In another NATBI study, 77% of a sample of infants admitted to hospital with suspected NATBI were Māori (Kelly et al. 2009).

Evidence of disparities in Māori access to support for serious injury, including TBI, is well documented (Jansen et al. 2008). Accident Compensation Corporation (ACC) data show that while Māori have 1.5–2.5 the rates of all injuries compared to non-Māori, they are also less likely to access appropriate rehabilitation (Jansen et al. 2008; New Zealand Guidelines Group 2006). One report also describes a disproportionately high rate of serious injury in Māori (Jansen et al. 2008).

The topic of TBI from a Māori perspective needs to be carefully considered, mindful that in *Te Ao Māori* (the Māori world view) "*he tapu te upoko*" means the head is sacred (Moko-Mead 2003). Utilization of *Rangahau Kaupapa Māori* (Māori led, Māori determined) methods is required so that the process of investigation remains culturally safe. Cultural safety is an approach that has led to profound transformation in the delivery of health services that has led by Māori nurses since the early 1990s. The emphasis of this approach is on the patient's and their families' experience needing to be culturally safe in order to ensure positive outcomes (Browne et al. 2009; Ramsden and Spoonley 1993). The paucity of this type of research investigating Māori TBI points to a fundamental omission in the way in which Māori TBI and rehabilitation are considered (Elder 2012a).

Māori and Psychiatric Illness

There is significant overrepresentation of Māori in populations of those with mental illness, both in community samples (Arroll et al. 2009; Baxter et al. 2006; Oakley-Brown et al. 2006) and in inpatient settings (Gaines et al. 2003). For this reason, it is important to consider Māori psychiatric rehabilitation. There is no national set of guidelines for Māori psychiatric rehabilitation. Limited published evidence exists of services developing their own models of care using Māori practices and values (Elder et al. 2009).

Māori constructs of psychological distress and accident are commonly linked to some form of transgression of *tikanga* (protocols) (Durie 2001). Little research investment has been made into what Māori think about classification systems of mental disorder, their validity, or into traditional Māori ways of grouping psychological and behavioral aberrance or injury. Specific *iwi* (tribal) knowledge, shared in *wānanga* (traditional learning retreats), is evidence of a precolonial knowledge base about the brain, mind, spiritual experiences, and their disturbance (Wikaira 2008).

Strengthening cultural identity has been found protective in suicide prevention (Coupe 2005) and improving treatment outcomes for Māori (Durie 2001). However, there are no best practice guidelines about how Māori cultural identity might be incorporated into individual and *whānau* (extended family) treatment plans, how this could be measured, and how these concepts could be operationalized to assist in rehabilitation and prevention.

What Might Rehabilitation Mean for Māori?

The dominant interpretation of rehabilitation can be summarized as a treatment journey directed at maximizing the functional independence of a person by decreasing impairment, handicap, and disability (Arffa 2006; Gordon et al. 2006; Ylvisaker 1998). Or in the language of the international classification of functioning, disability and health (ICF), rehabilitation is activity that addresses changes to body structures/functions, activity limitations, and participation restrictions (ref). However, little or no reference is made in definitions of rehabilitation to a cultural worldview regarding responses to cultural function that may be impacted by injury or disease processes (Gordon et al. 2006; Stucki 2005). This absence of cultural considerations has been elegantly highlighted by others (Uomoto 2005). Harwood (2010) provides a detailed description of what rehabilitation could do to respond to the needs of indigenous people with particular focus on Māori; however, the concept of rehabilitation itself remains without critique.

The differential response from minority cultural groups to the rehabilitation goal of "community integration" has been highlighted in recent literature (Sander et al. 2010).

A recognized limitation of current rehabilitation programs and standardized measures of programs is a lack of patient and family priority customization. The importance of creative activities such as dance, music, and extended social networks in some cultures is absent from these generic measures (Sander et al. 2010).

The dominance of focus on independence as a rehabilitation goal in the literature and the practices of rehabilitation professionals create a significant challenge from a Māori perspective. *Te Ao Māori* places value on *inter*dependence as well as independence, predicated on *whakapapa* (genealogical links). These values emphasize interconnection and embody central aspects of *Te Ao Māori* (Marsden 2003; Moko-Mead 2003; Royal 2002; Smith 2003). Interestingly, a TBI state of science paper (Gordon et al. 2006) emphasized the importance of therapeutic interdependency by stating, "establishment of an effective therapeutic relationship is associated with better traumatic brain injury rehabilitation outcomes" (p. 358). Exploration of how the values of interdependence and independence for *whānau* and clinicians might improve TBI rehabilitation outcomes therefore has merit. No Māori research was found addressing this issue to date.

Similarly, inclusion of real-world *cultural* activities could be regarded as an important part of Māori TBI rehabilitation, given the importance of providing both an enriched environment of therapeutic interventions and one that reflects real-world experiences (Beaulieu 2002). One study using *kapahaka* (traditional song and movement) as a form of sensory modulation for mental-health patients provides promising findings that invite further research (Hollands et al. 2014).

Examples of *Kaupapa Māori* Rehabilitation in Action

Three specific examples illustrate the implementation of *Kaupapa Māori* (Māori specific) practices in what could be termed traditional clinical rehabilitation services. Two of these come from the mental-health sector and one from adult TBI rehabilitation. One of the inherent challenges for these services is to attempt to measure the impact of *Kaupapa Māori* interventions in the context of the range of other modalities of treatment offered. Tanewhakapiripiri is the name of the *Kaupapa Māori* psychiatric rehabilitation unit that opened at the Mason Clinic Regional Forensic Service in Auckland in 2006. Rationale for the rehabilitation approach taken is well described (Tapsell 2007). Development of an outcome measure relating to this *Kaupapa Māori* rehabilitation has been a core feature of the service (N. Wiki, personal communication, 2010). Tamahere Hospital and Healing Centre is another example of a *Kaupapa Māori* rehabilitation facility in the Waikato region also utilizing *tikanga Māori* to improve outcomes for forensic psychiatric *whānau* (R. Wirihana, personal communication, 2011). "'Whatever it takes, *Ū ki te whānau ora*,' Home Rehabilitation and Support Services Ltd," is a *Kaupapa Māori* provider that ensures a *Kaupapa Māori* model of care is provided for *whānau* (and indeed more widely) since early 2009 (http://www.whateverittakes.co.nz/). This service provides a range of supports for *whānau* including inhome live-in care. Building Māori cultural competency within the clinical team members who interface with "Whatever it takes" staff and clients is reported to be a challenging area of work (C. Tinana-Williams, personal communication, 2011). While there is no standard measure of Māori cultural competency for clinical staff, "Whatever it takes" staff provide cultural competency training for clinical personnel. However, they have found that cultural competencies are regarded as a luxury rather than a requirement (C. Tinana-Williams, personal communication, 2011). However, none of these services have published outcomes of their interventional approaches. In my experience, publication of such outcomes is difficult for services in general. A lack of a specific budget line for such work and lack of research expertise within services are two barriers. In some services, the data could be regarded as commercially sensitive.

Policy Imperatives

Policies form another contextual layer that expose the interface between Māori and the concept of rehabilitation. The New Zealand Disability Strategy, developed in consultation with disabled people and the disability sector, outlines fifteen objectives (Minister for Disability Issues 2001). Objective 11 is specific to Māori and includes this statement:

> Build the capacity of disabled Māori through equitable allocation of resources within the context of Māori developmental frameworks, establish more disability support services designed and provided by Māori for Māori, ensure mainstream providers of disability services are accessible to and culturally appropriate for disabled Māori and their *whānau*, train more Māori disability service provider professionals and increase the advisory capacity of Māori. (Minister for Disability Issues 2001, p. 25)

The document also states that it is important to,

> ensure that Government funded or sponsored *marae*-based initiatives meet the access requirements of disabled people (and encourage all other *marae*-based initiatives to also meet those requirements). (Minister for Disability Issues 2001, p. 25)

While not directly locating disability services on *marae* (a traditional meeting house), the strategy recognizes *marae* as places are important to Māori and in meeting the needs of Māori with disabilities. This suggests that *marae* could also be considered one appropriate context for activities that come from *Te Ao Māori* (the Māori world) perspectives of healing and restoring balance.

Informed analysis of the New Zealand Disability Strategy and proposed culturally effective framework for Māori provide useful consideration (Nikora et al. 2004; Ratima et al. 1995). Lack of acknowledgement of *tikanga Māori* is reported as one aspect that influences the barriers of access for Māori to disability support services including rehabilitation services (Nikora et al. 2004; Ratima et al. 1995).

Te Kōkiri, the mental health and addiction plan in effect from 2006 to 2015, outlines the need for building quality health services, including rehabilitation services, based on research to inform innovation in health and disability care delivery based in the New Zealand context (Ministry of Health 2006). However, these documents do not detail an analysis of the concept of rehabilitation from a Māori perspective. While on one hand, government departments and agencies have identified addressing the poor health status of Māori as of the highest priority (Accident Compensation Corporation 2005; Ministry of Health 2005), the paucity of investment in research that is clinically applicable to Māori rehabilitation means that the ability to improve knowledge transfer in these areas is significantly limited.

Indigenous TBI Literature

What can be learned about rehabilitation from other indigenous cultures? The indigenous TBI literature is largely North American with a small number of relevant Australian studies also identified. The North American literature uses the terms *American Indian* and *Alaskan Natives* (AI/AN), and *native* and non-native, and is adult-centric (Adekoya and Wallace 2002; Blackmer and Marshall 1999; Keightley et al. 2009, 2011; Langlois et al. 2003; Nelson et al. 2007; Rutland-Brown et al. 2005). As found elsewhere, male TBI is more than twice as common as in the female population in North America (Adekoya and Wallace 2002; Langlois et al. 2003). The authors note the likely underreporting of TBI because of use of tribal facilities and remote locations (Adekoya and Wallace 2002). TBI rates secondary to assault were highest in African Americans and AI/AN reported as 30/100,000, a rate four times that of whites (Langlois et al. 2003).

Prevention strategies related to motor vehicle accidents (MVAs) and assaults are linked to increased rates of elevated blood alcohol levels and limited use of seatbelts and helmets (Adekoya and Wallace 2002; Rutland-Brown et al. 2005). Imperatives for prevention include high rates of mood and anxiety disorder in the indigenous TBI population compared to the indigenous comparison group without TBI (Nelson et al. 2007). How rehabilitation services respond to these sequelae in these populations remains unreported.

Differences in presentation between native and non-native people admitted for inpatient TBI rehabilitation included high alcohol and other drug use, 83% of the native group compared to 9% of non-native patients; as well as in follow-up support where non-native postdischarge support was 90% compared to 33% for native patients (Blackmer and Marshall 1999). Again, no published work was found that interrogates this presentation complexity from an indigenous perspective. Nor were papers found that question the veracity of nonindigenous compared to indigenous drug and alcohol self-reporting.

Two qualitative studies found that there were significant challenges for aboriginal adults in Canada with TBI in transitioning home from the hospital (Keightley et al. 2009, 2011). Two Australian papers examined cultural influences on motivation for participation in rehabilitation (Saltapidas and Ponsford 2007) and experience and beliefs about TBI and the association with outcome in the same cohort (Saltapidas and Ponsford 2008). Although indigenous people were not included in their sample, the findings resonate here. Significant cultural differences were identified that suggested role was very important in minority cultures and that having a value system different to what was termed *Australian* was associated with worse outcomes. Interestingly, Saltapidas and Ponsford commented that:

> variables related to culture, emotional response, understanding of injury and education were more strongly related to outcome than variables traditionally related to outcome, such as age and injury severity. (p. 11)

Māori Research That Illuminates Concepts Related to Rehabilitation

The word rehabilitation, the values inherent to rehabilitation, and its practice have been identified as problematic for Māori in recent research (Elder 2012b). Illustrative verbatim quotes from that research (followed by the names of the *marae*, traditional meeting houses from where the quote originates) are now presented to show the different ways Māori participants expressed their views related to the concept of rehabilitation (refer to the glossary for explanation of Māori words and concepts).

> [Y]ou can't *whakamaori* (translate into *Te Reo Māori*) "rehab." Because it's already a whole *kaupapa*. It's not really Māori, so you actually have to think away from the notion of that *ki te whakaaro Māori* (in Māori thinking). Things have to be done differently, eh?
>
> One of the things that was really interesting in the disability *kōrero*, with this woman that I met with. She was paraplegic as the result of a car accident, from Rotorua. She was talking about how whenever they went anywhere the brothers, uncles, and that would have to carry her from the car into the *marae* or into the *whare*. What all the rehab people had tried to teach her was to be independent, to be able to do everything on her own. She said one of the biggest things that she learnt was actually because she's Māori she doesn't have to be independent, and getting her brothers and uncles to carry her to the *marae* meant they had learnt about her disability. And they learnt to cope with her and everything as well. What they been taught at the spinal injury unit was totally not relevant.
>
> **Araiteuru Marae**

The important role of *Te Reo Māori* (the Māori language) was underlined.

> [I]f we are going to even consider a rehabilitating thing for our people. And we already know without our reo you're nothing, you just part of this other mono-culture.
>
> **Rauru Marae**

Traditional Māori concepts, such as the *poutama* (a traditional Māori design and concept using a series of steps to embody improvement and aspiration), were used to describe ideal approaches to TBI healing and detail specific interventions for consideration.

> I like that idea something around *poutama*, puts it back into a Māori world. Saying there's always going to be steps and challenges, it reminds me of that *whakatauki*, "*taihoa tonu te hinengaro, he kake ai a Tane-nui-a-rangi ki tikitiki o rangi o te hiringa i te mahara,*" the one thing that helped Tane take that journey through the twelve heavens, up the *poutama*, was the power of his mind, the power of determination. And that was all set about because he had this *whānau* structure pretty much around him, who believed. So that was what the *poutama* was built of? You know like that (makes stepwise gesture). Every step is supposed to be challenges and struggles so you build new characteristics about yourself. But I think things like *waiata, oriori, mirimiri* those kinds of things that we have in our world have to

be an absolute part of his rehabilitation, part of his poutama. Because, that's kind of whether the whānau are into Māori things or not. They might not be into it, but actually he needs it because the brain stuff says that people need touch, they need sound and singing, like music therapy. It's not just a Māori, it's not only Māori healing, but that's how we do it in our world, but it's universal. That those things need to be done. But if you can do it through a Māori world, then not only is he gaining strength in his head injury, he's also gaining new insight into being a Māori.

Araiteuru Marae

The metaphor of a revitalized piece of wood used in traditional carving in the building of a marae was presented as another way to think about rehabilitation from a Māori perspective.

So rehabilitation to me with those ones that have had brain trauma and that, it's about giving them hope, giving them a future, giving them that permanence that they are part of the this world. It's not about judging them. If I was to judge that piece of wood, when it was like full of holes, if you thought like that, well we'd chop a tree down. We don't think like that. I thought, well let's give that piece of timber a *mauri*, a life force, a reason for it to be here. So we carved it. Did the carvers like it? No, they swore at me because their chisels were getting blunted by all the nails. Those holes, and that. So they were swearing at me, then I told them the story about *mauri*, and they stopped swearing at me and started carving again, so you know rehabilitation to me is about *mauri*.

Pukemokimoki Marae

Overall, these selected quotes highlight distinctive culturally determined views that respond to the word rehabilitation and that speak about practices that promote healing for Māori people.

Te Waka Oranga

Te Waka Oranga, derived from my PhD research (Elder 2013a,b), is a way of responding to TBI that differs from the current TBI rehabilitation paradigm. This difference is predicated on the theoretical premise that TBI causes damage to the physiology of the brain as well as to *wairua*, a Māori cultural concept indicating connectivity between people and all aspects of the universe. Wairua is sometimes translated as spiritual well-being. Given this culturally determined injury to *wairua*, the *whānau* hold cultural knowledge salient to healing the *wairua* injury that occurs alongside the physical injury in TBI.

Te Waka Oranga takes the shape of a *waka* (see Figure 12.1). This iconic Māori symbol has been chosen for a number of reasons. The *waka* is a Māori mode of transport that exemplifies movement across the ocean, requiring collective work and purpose to bring the desired destination to those on board. This way of conceptualizing the recovery destination is uniquely Māori and is well recognized by ocean going *waka* practices. This idea of bringing the destination to the *waka*, rather than the *waka* moving to the destination, changes the quality and energy of the endeavor. It privileges the idea that the destination already exists and it can be brought to the fore. The Māori world view invites other world views to come aboard

FIGURE 12.1 *Te Waka Oranga.*

the *waka* in line with Māori values of *anaakitanga* (hospitality to visitors) and *whanau-ngatanga* (building connections). In this way, one side of the *waka* is paddled by the *whānau*, and the other side by the professionals involved. The approach deliberately avoids the use of the word rehabilitation. This choice also respects that rehabilitation is an English word that comes from a different paradigm of knowledge, with its own history and traditions.

This tool could be considered a Māori approach to rehabilitation; however, the relativistic problems are pronounced by linking this indigenous approach to the nonindigenous concept of rehabilitation. The concepts and actions in *Te Waka Oranga* come from *Te Ao Māori* and were not derived in response to questions about rehabilitation. Rather, this work has been developed from responses to *pūrākau* (stories) about children and young people who had sustained a TBI. This tool was designed with children, adolescents, and young adults in mind. The word *mokopuna* (grandchild) is used here for the index patient to honor the voices of participants in the original research. A brief overview of this approach is now described.

The framework assumes that being Māori is a fluid continuum and *whānau* may not "look" or "feel" Māori to others or within themselves. Nevertheless, *whānau* are supported by the *Te Waka Oranga* process to access their *mātauranga* (knowledge systems) held within their own *whakapapa* (genealogy).

Using the framework has two stages. First is *Hoe tahi*, which literally means "paddle as one." This stage establishes the ground rules of *Te Waka Oranga*. The second stage is called *Te Haerenga o Te Waka Oranga*, or "the journey of *Te Waka Oranga*." This involves the launching, sailing, and arriving on *Te Waka Oranga*, the combined work of the *whānau* and clinical team in bringing recovery goals to fruition.

The two stages involve consideration of four navigational elements or tools used to ensure that all the culturally salient aspects are considered. These are *Tangata* (people), *Wā* (time), *Wāhi* (place), and *Wairua* (defined here as a profound and uniquely Māori sense of connection). The *Waka Oranga* connects these elements. *Te Waka Oranga* is located within a Māori world view, thereby attempting to provide a safe space for Māori to invite non-Māori world views, people, and activities that can benefit the *whānau* in the context of TBI. These non-Māori are guests in *Te Ao Māori*. Thus, as guests, the roles, reciprocal connections, and responsibilities are outlined in order to paddle as one.

Using this approach involves completing *taonga* (pictorial representations). These *taonga* provide a paper record of the collective *whānau* and clinician work. The *taonga* may exemplify periods of a day, a week, or a procedure or other episode of care as needed. The completed and partially completed *taonga* can be displayed on the walls of the *whānau* home, on the walls of the hospital room, and as part of the clinical records. Reflecting the *mokopuna*'s world is another important aspect of the *taonga*. They can be decorated in ways that reflect the *mokopuna*'s preferences.

The framework could also be used as *toi whakaari*, or acted out. In this way, chairs can be set up representing the places of the paddlers; *hoe* (paddles) could be used. The index patient, *koroua*, and *kuia* (elders) have a special place on board, representing their central roles in navigation. Participants could use this approach to discuss the use of their skills and feelings to prepare the *waka*, to move the *waka*, and to bring the agreed destination forward.

This first stage of work establishes the mechanisms of paddling as one. A *waka* cannot safely and cleanly move through the water toward its destination unless all are paddling in unison. This stage covers the protocols of encounter as well, establishing where and how these activities occur. It is envisaged that this set of practices is repeated on a regular basis. In this way, respect and trust among the *whānau* and the clinicians are strengthened by contact. These practices are represented on the *taonga* template as the horizontal lashings that bind the *waka* together as well as the positioning of the *kaihoe* (paddlers) on board the *waka*. This stage requires both *whānau* and professionals to agree to guidelines for their working together that reinforce their sense of connection. These practices are the tools for negotiating differences of opinion, managing situations that have not progressed as planned, and discussing difficulties. The codes from the original participant narratives provide a helpful guide. These are "acknowledging each other," "setting the foundation for the *korero* (discussion)," "exploring around the topic," "putting thoughts and feelings into words," "using humor," "sharing experiences," "getting back to the *kaupapa* (subject)," and "questions."

Limitations to *Te Waka Oranga*

There are, as with any proposed model or way of thinking, some possible limitations to *Te Waka Oranga*. One is that some Māori may not feel comfortable working in this way if they do not feel connected to these aspects of their cultural identity. However, if *marae* contact is used as a default measure of activated Māori identity, then those less comfortable with this approach are likely to be a minority of Māori as most Māori visit their *marae* often or very often (Statistics New Zealand and Ministry of Culture and Heritage

2003; Te Puni Kokiri 2010). It remains to be investigated as to whether this approach would also be appropriate for adult patients and their *whānau*. The word *mokopuna* (grandchild) has been used. Given that adults remain grandchildren in terms of their place within *whakapapa* (genealogy), this possibility holds promise. Interestingly, feedback from Pākehā (non-Māori New Zealanders) is that despite reflecting a different world view to the dominant European model they are more familiar with, *Te Waka Oranga* also has relevance for them and for many of their experiences.

Conclusions

Cultural concepts of healing and restoration are notably absent from definitions of rehabilitation; the impact of injury on overall cultural functioning such as in culturally valued roles, responsibilities, skills, and activities is not included in internationally recognized classification systems of function and disability (Stucki 2005; see Chapters 1 and 2 for further discussion of these issues). One contention here is that the absence of culture from definitions of functional impairment secondary to injury is one reason why Māori culture and rehabilitation has not been more actively targeted as an area of clinical or research interest.

There is a profound gap in the literature related to indigenous concepts of injury and response to injury and illness. Given the epidemiological evidence of high prevalence of indigenous illness and injury, there is compelling need for indigenous researchers and those who support their work to determine a research agenda. It is difficult to explain why this absence of substantive research exists. It is possible that other health conditions have taken priority. Improvements in data collection and their publication provide a more robust platform for advocating for indigenous lead research in this area.

Until recently, there has been no literature recognizing the significance of injury to the head in Māori culture. This cultural silence suggests a need for research informed by and building on existing *Rangahau Kaupapa Māori* to begin to understand more about how injury is conceptualized and therefore how it needs to be responded to and prevented. Only then can more practical application of such knowledge be implemented to improve outcomes for Māori. The debate about what this means for rehabilitation remains ongoing. A central challenge will be to implement such approaches in culturally safe ways where fidelity to the approach is maintained. That will require significant infrastructural support.

One example of a Māori approach and response to injury, *Te Waka Oranga*, has been presented. This manifests a different paradigm of collaboration between *whānau* and professionals where there is an assumption of cultural knowledge within *whānau*. Using the *waka* tradition of bringing the recovery destination to the *whānau* is an example of practice that brings indigenous knowledge forward. It is possible that this approach may be of use to other indigenous peoples and cultures and in other conditions affecting the brain.

Glossary

hauora: Māori concept of holistic health
he: A or some
hui: Meeting

iwi: Tribe
kaitiaki: Guardian
karakia: Prayer
kaumatua: Esteemed elder of either gender
kaupapa: Subject, reason
koroua: Grandfather, esteemed elder
kuia: Grandmother, esteemed female elder
Māori: Indigenous people of Aotearoa, New Zealand
marae: Traditional Māori campus of related areas and buildings for meetings
mātauranga: Knowledge, knowledge systems
mokopuna: Grandchild
noho: Stay, sit
noho puku: Self-reflection
ora: Well-being
oranga: Healing, health
Pākehā: Non-Māori non-Pacific New Zealander
Pepeha: Traditional form of address that identifies elements of the natural world that define the person's identity
pūrākau: Story
rōpū: Group
tangata: Human being, person
tapu: Sacred
Te Ao Māori: The Māori world
Te Reo Māori: The Māori language
upoko: Head
wā: Time
wāhi: Place
wairua: Sometimes translated as the spiritual dimension of well-being, profound connection, uniquely Māori
waka: Canoe, vessel, conveyance
wānanga: Traditional fora for learning and discussion
whakapapa: Genealogy
whā: Four
whānau: Extended family
whānau ora: Well-being of the extended family system
whanaungatanga: Process of making relational links
whare: House

References

Accident Compensation Corporation. (2005). *Māori Second Quaterly Report 2004–2005.* Wellington: Accident Compensation Corporation.

Adekoya, N., and Wallace, L. (2002). Traumatic brain injury among American Indians/Alaska Natives—United States, 1992–1996. *Morbidity and Mortality Weekly Report Surveillance Summaries, 51*(14), 303–304.

Arffa, S. (2006). Traumatic brain injury. In C. Coffey, & R. Brumback (Eds.), *Pediatric Neuropsychiatry*. Philadelphia, PA: Lippincott Williams & Wilkins.

Arroll, B., Goodyear-Smith, F., Kerse, N., Hwang, M., Crengle, S., Gunn, J. et al. (2009). The prevalence of depression among Māori patients in Auckland general practice. *Journal of Primary Care, 1*(1), 26–29.

Barker-Collo, S. L., Wilde, N. J., and Feigin, V. L. (2009). Trends in head injury incidence in New Zealand: A hospital-based study from 1997/1998 to 2003/2004. *Neuroepidemiology, 32*, 32–39.

Baxter, J., Kingi, T., Tapsell, R., Durie, M., and McGee, M. (2006). The prevalence of mental disorder among Māori in Te Rau Hinengaro: The New Zealand Mental Health Survey. *Australian and New Zealand Journal of Psychiatry, 39*(5), 401–406.

Beaulieu, C. (2002). Rehabilitation and outcome following pediatric traumatic brain injury. *Surgical Clinics of North America, 82*, 393–408.

Blackmer, J., and Marshall, S. (1999). A comparison of traumatic brain injury in the Saskatchewan native North American and non-native North American populations. *Brain Injury, 13*(8), 627–635.

Brady, M. (1995). Culture in treatment, culture as treatment. A critical appraisal of developments in addictions programs for indigenous North Americans and Australians. *Social Science & Medicine, 41*(11), 1487–1498.

Browne, A. J., Varcoe, C., Smye, V., Reimer-Kirkham, S. R., Lynam, M. J., and Wong, S. (2009). Cultural safety and the challenges of translating critically oriented knowledge into practice. *Nursing Philosophy, 10*, 167–179.

Chenhall, R. (2008). What's in a rehab? Ethnographic evaluation research in indigenous Australian residential alcohol and drug rehabilitation centres. *Anthropology and Medicine, 15*(2), 105s–116s.

Cobo, J. R. M. (1983). *Study of the Problem of Discrimination against Indigenous Populations*. New York: United Nations Economic and Social Council.

Coupe, N. (2005). *Whakamomori. Māori Suicide Prevention*. Massey University, Palmerston North.

Coyhis, D., and Simonelli, R. (2008). The Native American healing experience. *Substance Use & Misuse, 43*, 1927–1949.

Dell, C. A., Seguin, M., Hopkins, C., Tempier, R., Mehl-Madrona, L., Dell, D. et al. (2011). From benzos to berries: Treatment offered at an Aboriginal youth solvent abuse treatment centre relays the importance of culture. *Canadian Journal of Psychiatry, 56*(2), 75–83.

DeVerteuill, G., and Wilson, K. (2010). Reconciling indigenous need with the urban welfare state? Evidence of culturally-appropriate services and spaces for Aboriginals in Winnipeg, Canada. *Geoforum, 41*, 498–507.

Durie, M. (2001). *Mauri Ora. The Dynamics of Māori Health*. Oxford: Oxford University Press.

Elder, H. (2012a). An examination of Māori tamariki (child) and taiohi (adolescent) traumatic brain injury within a global cultural context. *Australasian Psychiatry, 20*(1), 20–23.

Elder, H. (2012b). He tapu te upoko: The head is sacred. Indigenous child and adolescent traumatic brain injury in NZ. A theory and framework. Paper presented at the 20th World Congress International Association Child and Adolescent Psychiatrists and Allied Professionals, Paris, France.

Elder, H. (2013a). Indigenous theory building for Māori children and adolescents with traumatic brain injury and their extended family. *Brain Impairment, 14*(3), 406–414.

Elder, H. (2013b). *Te Waka Oranga*. An indigenous intervention for working with Māori children and adolescents with traumatic brain injury. *Brain Impairment, 14*(3), 415–424.

Elder, H., Milne, M., Witehira, H., Mendez, P., Heslin, A., Cribb-Su'a, A. et al. (2009). Whakaora nga moemoea o nga tupuna—living the dreams of the ancestors. Future planning in a Kaupapa Māori CAMHS team. *Australasian Psychiatry, 17*(Suppl), S104–S107.

Feigin, V. L., Theadom, A., Barker-Collo, S. L., Starkey, N., McPherson, K., Dowell, A. et al. (2012). Incidence of traumatic brain injury in New Zealand: A population-based study. *The Lancet, 12*(1), 53–64.

Gaines, P., Bower, A., Buckingham, W., Eager, K., Burgess, P., and Green, J. (2003). *New Zealand Health and Classification and Outcomes Study: Final Report*. Auckland: Health Research Council.

Gone, J. P. (2009). A community based treatment for Native American historical trauma: Prospects for evidence-based practice. *Journal of Consulting and Clinical Psychology, 77*(4), 751–762.

Gordon, W. A., Zafonte, R., Cantor, J., Brown, M., Lombard, L., Goldsmith, R. et al. (2006). Traumatic brain injury rehabilitation: State of the science. *American Journal of Physical Medicine & Rehabilitation, 85*, 343–382.

Hartmann, W. E., and Gone, J. P. (2012). Incorporating traditional healing into an urban American Indian health organization: A case study of community member perspectives. *Journal of Counselling and Psychology, 59*(4), 542–554.

Harwood, M. (2010). Rehabilitation and indigenous people: The Māori experience. *Disability and Rehabilitation, 32*(12), 972–977.

Harwood, M., Weatherall, M., Talemaitoga, A., Barber, A. P., Gommans, J., Taylor, W. J. et al. (2012). Taking charge after stroke: Promoting self-directed rehabilitation to improve quality of life—A randomized controlled trial. *Clinical Rehabilitation, 26*(6), 493–501.

Hoeman, P. S. (2008). *Rehabilitation Nursing: Prevention, Intervention and Outcomes*. St. Louis, MO: Mosby Elsevier Health Sciences.

Hollands, T., Sutton, D., Wright-St. Clair, V., and Hall, R. (2014). Māori mental health consumers' sensory experiences of kapa haka. *New Zealand Journal of Occupational Therapy*, submitted.

Jackson, M. (2007). Globalisation and the colonising state of mind. In M. Bargh (Ed.), *Resistance: An Indigenous Response to Neoliberalism*. Wellington: Huia.

Jansen, P., Bacal, K., and Crengle, S. (2008). *He ritenga whakaaro: Māori Experiences of Health Services*. Auckland: Mauriora Associates.

Kahn, M. W., Lejero, L., Antone, M., Francisco, D., and Manuel, J. (1988). An indigenous community mental health service on the Tohono O'odham (papago) Indian reservation: Seventeen years later. *Journal of Community Psychology, 16*(3), 369–379.

Keightley, M., Kendall, V., Jang, S.-H., Parker, C., Agnihotri, S., Colantonio, A. et al. (2011). From health care to home community: An Aboriginal community-based ABI transition strategy. *Brain Injury, 25*(2), 142–152.

Keightley, M., Ratnayake, R., Minore, B., Katt, M., Cameron, A., White, R. et al. (2009). Rehabilitation challenges for Aboriginal clients recovering from brain injury: A qualitative study engaging health care practitioners. *Brain Injury, 23*(3), 250–261.

Kelly, P., and Farrant, B. (2008). Shaken baby syndrome in New Zealand: 2000–2002. *Journal of Paediatrics and Child Health, 44*(3), 99–107.

Kelly, P., and Hayes, I. (2004). Infantile subdural haematoma in Auckland, New Zealand: 1988–1998. *The New Zealand Medical Journal, 117*(1201), U1047.

Kelly, P., MacCormick, J., and Strange, R. (2009). Non-accidental head injury in New Zealand: The outcome of referral to statutory authorities. *Child Abuse and Neglect, 33*, 393–401.

Langlois, J., Kegler, S., Butler, J., Gotsch, K., Johnson, R., Reichard, A. et al. (2003). Traumatic brain injury-related hospital discharges. Results from a 14-state surveillance system. *Morbidity and Mortality Weekly Report Surveillance Summaries, 52/SS-4*, 1–18.

Macklin, R. (1999). *Against Relativism—Cultural Diversity and the Search for Ethical Universals in Medicine.* New York: Oxford University Press.

Marsden, M. (2003). The woven universe. Selected writings of Rev. Māori Marsden. Otaki: Estate of Rev. Māori Marsden.

McKinlay, A., Kyonka, E. G. E., Grace, R. C., Horwood, L. J., Fergusson, D. M., and MacFarlane, M. R. (2009). An investigation of the pre-injury risk factors associated with children who experience traumatic brain injury. *Injury Prevention, 13*(16), 1–5.

Minister for Disability Issues. (2001). *The New Zealand Disability Strategy. Making a World of Difference.* Wellington: Ministry of Health.

Ministry of Health. (2005). *CBG Health Research Reducing Inequalities Contingency Funding Evaluation.* Wellington: Ministry of Health.

Ministry of Health. (2006). *Te Kōkiri: The Mental Health and Addiction Plan 2005–2015.* Wellington: Ministry of Health.

Ministry of Health. (2008). *Te Puāwaiwhero: The Second Māori Mental Health and Addiction National Strategic Framework 2008–2015.* Wellington: Ministry of Health.

Moko-Mead, H. (2003). *Tikanga Māori, Living by Māori Values.* Wellington: Huia.

Nelson, L., Rhodes, A., Noona, C., Manson, S., and AI-SUPERPFP Team. (2007). Traumatic brain injury and mental health among two American Indian populations. *Journal of Head Trauma Rehabilitation, 22*(2), 105–112.

New Zealand Guidelines Group. (2006). *Traumatic Brain Injury: Diagnosis, Acute Management and Rehabilitation. Evidence Based Practice Guideline.* Wellington: Accident Compensation Corporation.

The New Oxford English Dictionary. (1998). Oxford: Oxford University Press.

Nikora, L., Karapu, R., Hickey, H., and Awekotuku, N. T. (2004). *Disabled Māori and Disability Support Options.* Hamilton: Māori and Psychology Research Unit, University of Waikato.

Oakley-Brown, M. A., Wells, J. E., and Scott, K. M. (2006). *Te Rau Hinengaro.* Wellington: Ministry of Health.

Patston, P. (2007). Constructive functional diversity: A new paradigm beyond disability and impairment. *Disability and Rehabilitation, 29*, 1625–1633.

Plani, K., and Carson, P. (2008). The challenges of developing a trauma system for indigenous people. *Injury. International Journal for the Care of the Injured, 39S5*, S43–S53.

Ramsden, I. H., and Spoonley, P. (1993). The cultural safety debate in nursing education in Aotearoa. *New Zealand Annual Review of Education, 3*, 161–174.

Ratima, M., Durie, M., Allan, G., Morrison, P., Gillies, A., and Waldon, J. (1995). A culturally effective framework for the delivery of disability support services to Māori. *New Zealand Journal of Disability Studies, 1*, 60–75.

Royal, T. (2002). Indigenous worldviews, a comparative study. A report on research in progress: Te Wānanga-o-Raukawa.

Rutland-Brown, W., Wallace, L., Faull, M., and Langlois, J. (2005). Traumatic brain injury hospitalizations among American Indians/Alaska Natives. *Journal of Trauma Rehabilitation, 20*(3), 205–214.

Saltapidas, H., and Ponsford, J. (2007). The influence of cultural background on motivation for and participation in rehabilitation and outcomes following traumatic brain injury. *Journal of Head Trauma and Rehabilitation, 22*(2), 132–139.

Saltapidas, H., and Ponsford, J. (2008). The influence of cultural background on experiences and beliefs about traumatic brain injury and their association with outcome. *Brain Impairment, 9*(1), 1–13.

Sander, A. M., Clark, A., and Pappadis, M. R. (2010). What is community integration anyway? Defining meaning following traumatic brain injury. *Journal of Head Trauma and Rehabilitation, 25*(2), 121–127.

Schmidt, G. (2000). Barriers to recovery in a First Nations community. *Canadian Journal of Community Mental Health, 19*(2), 75–87.

Shaw, G., Ray, T., and McFarland, B. (2011). The Outstation model of rehabilitation as practiced in central Australia: The case for its recognition and acceptance. *Substance Use & Misuse, 46*, 114–118.

Smith, G. (2003). Kaupapa Māori theory. Theorizing indigenous transformation of education and schooling. Paper presented at the Kaupapa Māori Symposium. NZARE/AARE Joint Conference.

Smith, L. T. (1999). *Decolonizing Methodologies* (9th ed.). Dunedin: University of Otago Press.

Spicer, P. (2001). Culture and the restoration of self among former American Indian drinkers. *Social Science & Medicine, 53*, 227–240.

Stathis, S. L., Letters, P., Doolan, I., and Whittingham, D. (2006). Developing an integrated substance use and mental health service in the specialised setting of a youth detention centre. *Drug & Alcohol Review, 25*(2), 149–155.

Statistics New Zealand. (n.d). QuickStats about Māori. Retrieved January 4, 2012, from http://www.stats.govt.nz/census/2006censushomepage/quickstats/quickstats-about-a-subject/maori.aspx.

Statistics New Zealand, and Ministry of Culture and Heritage. (2003). *A Measure of Culture: Cultural Experiences and Cultural Spending in New Zealand*. Wellington: Statistics New Zealand, Ministry of Culture and Heritage.

Statistics NZ. (n.d). National ethnic population projections, in projections overview. Retrieved October 11, 2011, from http://www.stats.govt.nz/browse_for_stats/population/estimates_and_projections/projections-overview/nat-ethnic-pop-proj.aspx.

Stucki, G. (2005). International classification of functioning, disability and health (ICF): A promising framework and classification for rehabilitation medicine. *American Journal of Physical Medicine and Rehabilitation, 84*, 733–740.

Tapsell, R. (2007). The treatment and rehabilitation of Māori in forensic mental health services. In W. Brookbanks, & A. Simpson (Eds.), *Psychiatry and the Law* (pp. 397–422). Wellington: LexisNexis.

Te Puni Kokiri. (2010). *2009 rangahau i ngā waiaro, ngā uara me ngā whakapono mō te Reo Māori. 2009 Survey of Attitudes, Values and Beliefs Towards the Māori Language.* Wellington: Te Puni Kokiri.

Thompson, M., and Roberston, J. (2010). A review of the barriers preventing indigenous health workers delivering tobacco interventions to their communities. *Australia and New Zealand Journal of Public Health, 35*(1), 47–53.

Trickett, E. J. (2011). From "Water boiling in a Peruvian town" to "Letting them die," community intervention, and the metabolic balance between patience and zeal. *American Journal of Community Psychology, 47*, 58–68.

United Nations. (2008). *United Nations Declaration on the Rights of Indigenous Peoples.* New York: United Nations.

Uomoto, J. M. (2005). Multicultural perspectives. In W. M. High (Ed.), *Rehabilitation for Traumatic Brain Injury* (pp. 247–267). Oxford: Oxford University Press.

Watts, L. (2001). Applying a cultural models approach to American Indian substance dependency research. *American Indian and Alaskan Native Research, 10*(1), 34–50.

WHO. (2014). Available at http://www.who.int/topics/rehabilitation/en/.

Wikaira, S. (2008). Ngāpuhi classification system of the mind. In H. Elder (Ed.). Hokianga. Personal Communication.

Wright, S., Nebelkopf, E., King, J., Maas, M., Patel, C., and Samuel, S. (2011). Holistic system of care, evidence of effectiveness. *Substance Use & Misuse, 46*(11), 1420–1430.

Ylvisaker, M. (1998). *Traumatic Brain Injury, Children and Adolescents* (2nd ed.). Newton, MA: Butterworth Heinemann.

13

Whose Behavior Matters? Rethinking Practitioner Behavior and Its Influence on Rehabilitation Outcomes

Nicola M. Kayes

Suzie Mudge

Felicity A. S. Bright

Kathryn McPherson

Introduction

Historically, we have spent considerable research resources on developing and testing interventions that focus on changing patient beliefs and behavior, their impairments, and/or their wider context. We have sought to understand those patient characteristics that predict rehabilitation outcomes (Cooper et al. 2002; van Almenkerk et al. 2013; van der Hulst et al. 2005; Vieira et al. 2011) and attempted to understand many patient behaviors (and related concepts) such as motivation, readiness to change, adherence, compliance, and personality (Clay and Hopps 2003; Lerner 1997; Maclean and Pound 2000; Sela-Kaufman et al. 2013; Wall et al. 2006). It is, perhaps unsurprisingly, common in rehabilitation practice to ascribe failure to achieve expected results to the patient, using descriptors (or labels) such as unmotivated, not ready, noncompliant, or difficult (Anderson and Funnell 2000; Potter et al. 2003a; van Hal et al. 2013). The impact of patient beliefs and perceptions on patient response to injury or illness and associated outcomes has also long been recognized. This has led to countless theories and models that aim to elucidate the mechanisms by which beliefs and perceptions serve to influence outcome and numerous associated interventions (Cameron and Leventhal 2003; French et al. 2006; Leventhal et al. 1984; Mikhail 1981). Indeed, approaches such as cognitive behavioral therapy, which explicitly draws on the hypothesis that beliefs, emotions, and behavior interact, are now routinely incorporated into practice (Brunner et al. 2013; Kerns et al. 2001; Thomas et al. 2006; Waldron et al. 2013).

By comparison, research pays very little attention to the characteristics of rehabilitation practitioners beyond the importance of technical skill and interprofessional working or teamwork. There has been remarkably little consideration of how the way we work, who we are, or how we think may influence outcomes. We screened the titles and abstracts of all papers published in 2012 in three of the most highly cited rehabilitation journals: *Archives of Physical Medicine and Rehabilitation, Clinical Rehabilitation,* and *Disability and Rehabilitation.* Only 14 of the 658 (2.1%) articles published in 2012 oriented toward practitioner behavior or related processes. Of these, six papers were focused on rehabilitation processes such as quality of care and goal setting where the practitioner was one of a number of aspects discussed. Only eight papers (1.2%) explicitly set out to explore, measure, or discuss aspects relating to rehabilitation practitioners or their behavior, which may influence the rehabilitation process and, potentially, patient outcome.

In contrast, most of us are likely at some point to have observed two practitioners carrying out the same intervention, but yielding quite different results (DiMatteo et al. 1993; Kaplan et al. 1989). Researchers also commonly recognize the potential for practitioners to influence outcomes in noting the potential for therapist effects to confound findings in the design of scientifically robust studies. They seek to manage these effects through randomization and other statistical analyses and adjustments (Walters 2010) and by incorporating a range of strategies to maintain treatment fidelity (Persch and Page 2013). While this indicates clear recognition that practitioners influence outcome (by way of confounding at least), such recognition does not appear to have translated into a great deal of work (at least in the form of publication) that aims to advance our understanding of exactly *how* rehabilitation practitioners influence outcome. Neither has it led to the development and testing of vast amounts of practitioner-oriented interventions.

In this chapter, we explore how practitioner thoughts, feelings, and behavior may directly and indirectly influence patient outcome. We argue that despite growing evidence supporting this case, the practitioner is rarely the focus of attention as a potentiating, or limiting, factor of rehabilitation processes or outcomes. We propose that more explicit attention should be given to practitioner characteristics and behavior as a legitimate explanatory factor both in practice and research and propose some possible strategies to facilitate this. Our approach augmented (and augments) the discussion regarding socio-relational rehabilitation in Chapter 8.

The chapter is presented in two parts. First, we draw on current evidence to substantiate our argument that practitioner factors, in particular their thoughts, feelings, and ways of working, influence rehabilitation outcome in important ways. Second, we propose a number of theoretically informed strategies to harness this largely untapped resource to optimize rehabilitation outcome in research and practice.

The Role That Practitioner Factors Play in Influencing Outcome

The focus on rehabilitation practitioners' role in relation to outcome has commonly been on their personal characteristics (e.g., age, gender) (Demou et al. 2012) or aspects perceived to be related to their level of expertise (e.g., years of experience, education, and training). For example, studies routinely define expert practitioners solely based on years of experience (Stevenson et al. 2005). Practitioners themselves place a great deal of credence on clinical experience and commonly use this to underpin or validate their clinical decision making (Graham et al. 2013; O'Halloran et al. 2012; Stevenson et al. 2005; Turner and Whitfield 1997). The tacit assumption is that patients being treated by more experienced therapists will have better therapeutic outcomes. However, while there is some evidence supporting the relationship between clinical experience and outcome in specific situations and where the intervention is fairly prescriptive (Rose 2013), this is not more generally supported by the literature (Demou et al. 2012; Resnik and Hart 2003; Whitman et al. 2004). Similarly, while there is emphasis placed on level of education and training (Buys and Casteleijn 2007; Parry and Brown 2009; Shields et al. 2013; Sran and Murphy 2009; Van Vuuren and Nel 2010), evidence for the relationship between ongoing education and improved patient outcomes is equivocal (Resnik and Hart 2003; Whitman et al. 2004). Even if the evidence strongly supported the relationship between years of experience or education and outcome, it seems insufficient to simply wait for practitioners to accrue the years of experience deemed necessary for expert status if something else could be done. Rather than "accept" that there will only ever be a small subgroup of educated practitioners who can expect to achieve optimal outcomes, we suggest it is worth exploring other practitioner-related characteristics or aspects, some of which may be more amenable to change. Two growing bodies of work of relevance here include research exploring (a) the role that practitioner's thoughts, feelings, and attitudes may have in influencing outcome; and (b) the practitioner's approach to practice or *way of working*. Each of these and associated evidence are discussed in more detail.

Practitioner's Thoughts, Feelings, and Attitudes

Given that the body of work on patient response is underpinned by the notion that human behavior is driven by a complex interaction of factors, in which personal beliefs, thoughts, and feelings play a fundamental role (see earlier discussion), it seems surprising we have been slow to consider that the same principles may also apply to practitioner behavior. In this section, we discuss empirical evidence that has explored the connection between the way we think and act and its association with rehabilitation outcome.

Some aspects of practitioner beliefs and attitudes have been explicitly considered in the field of evidence-based medicine. Findings indicating that evidence-based practice may result in better treatment outcomes have generated a great deal of interest in understanding factors influencing uptake of research findings by practitioners (Duncan et al. 2002; Hubbard et al. 2012; Salbach et al. 2010). A range of barriers to uptake have been identified such as time constraints, lack of support, insufficient critical appraisal skills, insufficient breadth of research, and consequently a lack of applicability to specific clinical situations (Graham et al. 2013; Nilsagård and Lohse 2010; Salbach et al. 2007). While many of these relate to the practice context, evidence suggests that factors such as *self-efficacy* in accessing and using evidence-based practice and a *positive attitude* toward research may contribute to an increased use of research in clinical decision making (Salbach et al. 2010). Given that practitioners beliefs and perceptions influence integration of research findings into practice, it is arguably a reasonable proposition to suggest that practitioner beliefs and perceptions may play a role in other areas of rehabilitation practice.

Research that explores the theoretical orientation and beliefs of practitioners working in lower back pain injuries supports the argument that beliefs and perceptions can and do influence clinical management, recommendations, and use of education strategies (Bishop et al. 2008; Brown and Richardson 2006; Darlow et al. 2012; Hendrick et al. 2013; Houben et al. 2005). For example, a cross-sectional postal survey with 2000 physiotherapists and 2000 general practitioners presented vignettes of patients with nonspecific lower back pain. They found that practitioners with a behavioral orientation (a belief in "a biopsychosocial model of disease in which pain does not have to be a consequence of tissue damage, and can be influenced by social and psychological factors," Bishop et al. 2008, p. 189) generally acted consistent with practice guidelines including promotion of activity. In contrast, the clinical management and recommendations from those with a biomedical orientation ("where disability and pain are a consequence of a specific pathology ... and treatment is aimed at treating the pathology and alleviating the pain," Bishop et al. 2008, pp. 188–89) were more divergent from practice guidelines. Practitioners were more likely to recommend that the patient remain off work despite the iatrogenic disability such a recommendation is associated with. Similarly, while Houben et al. (2005) primarily set out to examine the measurement properties of the Pain Attitudes and Beliefs Scale for physiotherapists, they found a relationship between a biomedical orientation and a greater likelihood that practitioners would view daily activities as harmful for people with nonspecific lower back pain. Like Bishop et al. (2008), they found that physiotherapists with a biomedical orientation were more likely to recommend patients limit their activity and work. Consistent with these findings, research with nurses working in a multidisciplinary pain team found that beliefs regarding treatment control (i.e., whether they believed pain was controlled internally, by powerful others, or by chance) were related to the type of treatments they would endorse (Brown and Richardson 2006).

Qualitative work exploring perspectives of physiotherapists who treat patients with lower back pain offers further insight into the impact of practitioner beliefs and perceptions (Daykin and Richardson 2004; Josephson et al. 2013; Trede 2000). A notable example is the research by Trede (2000) regarding educational practices of physiotherapists in relation to lower back pain. Trede suggested a key assumption underpinning educational practices for physiotherapists in her study was that practitioner *knowledge* leads to *trust* in the practitioner and that this in turn leads to patient *compliance*. This assumption led physiotherapists to adopt what she suggested to be a therapist-centered approach to patient education, where the practitioner was positioned as the expert with important medical facts to convey. Physiotherapists in this study primarily viewed their role as technicians: "In reality you tend to be more of a technician" (Trede 2000, p. 431). This was at odds with patient perspectives, which suggested that a personalized communication approach would be a more effective educational tool.

While the majority of evidence presented here is from cases of lower back pain, a similar pattern is evident in other rehabilitation populations and contexts. Beliefs, perceptions, attitudes, and interpretations of practitioners have been found to inform key clinical decisions associated with discharge planning (Atwal et al. 2012), treatment provision (Kolehmainen et al. 2008), and end-of-life care (Ruppe et al. 2013); perceptions regarding roles and responsibilities (Oakley et al. 2010); the quality of the relationship between patients and practitioners (Jackson et al. 2012); the adoption of standardized outcome measures (King et al. 2011); and quality of care (Lee et al. 2006).

What is not so well explored is the mechanism by which practitioner beliefs and perceptions influence patient outcomes (a limited "treatment theory" focus perhaps as described in Chapter 1). One theory that has been considered is the tripartite efficacy framework (Jackson et al. 2012). Lent and Lopez (2002) distinguish three types of efficacy beliefs relevant in the context of dyadic relationships including self-efficacy (beliefs in own ability), other-efficacy (belief in significant other's ability), and relation-inferred self-efficacy (RISE—belief in how significant other views them). They theorize that a dynamic interaction across these efficacy beliefs may have important implications for relationship formation and related outcomes. This framework extends commonly understood individualistic models regarding the influence of efficacy beliefs on behavior to consider the dynamic nature of efficacy beliefs in the context of interpersonal relationships. Jackson et al. (2012) tested this theory in patient–practitioner relationships in an exercise program for lower limb musculoskeletal disorders. Of particular interest is that perceptions of a better quality patient–practitioner relationship existed when each member of the dyad was confident in his or her own ability (self-efficacy); when each member was confident in the other person's ability (other-efficacy); *and* when they believed that the other person was also confident in their ability (RISE). What's more, practitioner *other-efficacy* beliefs (i.e., their belief in their patient's self-efficacy) were found to have a highly significant relationship with patient perceptions regarding the quality of the relationship. The theoretical explanation that the tripartite efficacy framework provides is consistent with prior work highlighting the interdependent nature of the dyadic relationship in other health contexts where it has been proposed that individuals in a therapeutic relationship "have the potential to influence each other's cognitions, emotions and behaviours in a reciprocal way" (Kenny et al. 2010, p. 763).

In summary, the evidence provided in this section indicates a growing body of work identifying the role that practitioner beliefs and attitudes have as a covariate of rehabilitation outcome. The tripartite efficacy framework provides a useful theoretical understanding of the interaction between practitioner's thoughts and feelings and how that can serve to impact their behavior within the dyadic therapeutic encounter. Next, we move beyond practitioner thoughts and beliefs to more explicitly consider the role of practitioner behavior, or their way of working, on outcome.

Practitioner's Way of Working— Their Approach to Practice

In using the term "way of working," we refer not to the technical aspects of practice unique to each discipline but rather the way in which we work that supersedes or contextualizes that expertise and goes across disciplinary boundaries. Ways of working might include, for example, an approach to practice, interpersonal skills, or the almost intangible, yet distinct "something" that marks out one practitioner from another with regard to how they connect and engage with a patient. There is mounting evidence over the last decade arguing that *who* and *how* we are with our patients may have a potentiating effect on rehabilitation outcome (DiMatteo et al. 1993; Hall et al. 2010; Jensen et al. 2000; Kaplan et al. 1989; Kayes and McPherson 2012).

Exploration of the ways in which "expert" practitioners work provides some insight into how way of working may influence outcomes. For example, Resnik and Jensen (2003) distinguished "expert" and "average" physiotherapists based on the clinical outcomes of the respective practitioners' patients with lower back pain from a large database of over 24,000 patients. They interviewed subgroups of the expert and average practitioners to develop a theory of expert practice using grounded theory. Expert practitioners were distinguished by a person-centered approach to practice, where therapists expressed views of patients as active participants, and empowerment of patients was a primary goal of their intervention. They described achieving this through clinical reasoning, patient education, and the establishment of a good patient–practitioner relationship. This approach was understood by them to arise from the interplay between clinical reasoning, values, virtues, as well as therapist knowledge. The expert practitioners' way of working in this study was based on their ethic of care and a respect for the individual, which led to a communication style involving active listening (Resnik and Jensen 2003). Resnik and Jensen's (2003) expert practitioners also placed importance on ongoing learning, which together with a recognition of one's own limitations leads to the use of reflection in practice, a strategy well supported by the literature (Schutz 2007; Wilding 2008). Reflection in practice provides an opportunity to refine and improve practice. Expert therapists in this study described a conscious and explicit process, whereas "average" therapists did not and, in some cases, reported difficulty with reflection on practice (Resnik and Jensen 2003).

How a practitioner engages and interacts with patients has long been considered an important component of rehabilitation with the role of rapport in rehabilitation appearing in the literature as early as 1957 (Shontz and Fink 1957). Both patients and practitioners report aspects of the therapeutic relationship such as trust, collaboration, and mutual respect as potentially augmenting treatment effects (Fadyl et al. 2010; Hargreaves 1982; Hills and Kitchen 2007; Leach 2005; Potter et al. 2003a,b; Punwar and Peloquin 2000; Resnik and Jensen 2003; Roberts and Bucksey 2007; Stenmar and Nordholm 1994). The growing body of research exploring the association between aspects of the therapeutic relationship and outcome suggests that this is not just a "nice thing to have," but rather it may be vital to optimize rehabilitation outcomes. A strong therapeutic relationship has been linked with improved outcomes following rehabilitation in brain injury, musculoskeletal and cardiac conditions, cancer survivors, stroke, and spinal cord injury (Evans et al. 2008; Ezrachi et al. 1991; Guidetti et al. 2009; Hall et al. 2010; Prigatano et al. 1994; Strauser et al. 2010; Wright et al. 2014). Hall et al. (2010) synthesized findings from 13 prospective rehabilitation studies, concluding that aspects of the patient–practitioner relationship were linked to treatment adherence, patient satisfaction, and improvements in a range of outcomes including pain, depression, and physical function. In neurorehabilitation, higher ratings of therapeutic relationship have been associated with improvements in a diversity of outcomes including productivity (Klonoff et al. 2001; Schönberger et al. 2006c), cognitive retraining tasks (Klonoff et al. 2007, 2010), driving status (Klonoff et al. 2010), awareness (Schönberger et al. 2006a), compliance (Schönberger et al. 2006a,c), depression (Schönberger et al. 2006b), and patient ratings of collaborative success (Schönberger et al. 2006b).

Patient engagement in rehabilitation is said to be crucial to achieving good outcomes, especially in the face of long-term injury or illness (Kortte et al. 2007; Medley and Powell 2010). We have recently completed a conceptual review of engagement in health care where we proposed engagement to be both a *state* ("engaged in") and a *process* ("engaging with") (Bright et al. 2014). Almost all measures of engagement (Hall et al. 2001; Lequerica et al. 2006; Macgowan 2006; Meaden et al. 2012) focus primarily on the patients' *state* of engagement and on their behavioral indicators of engagement, such as whether they are actively participating, committed, and enthusiastic, or whether they demonstrate a high level of vested interest, energy, and effort. However, shifting focus to engagement as *a process* places an emphasis on reciprocity, or even that responsibility for engagement rests with the practitioner. For example, our conceptual review highlighted a trusting and mutually respectful relationship, creating a therapeutic space where the patient feels listened to and understood, seeing the person behind the impairment and being responsive and flexible to the patient's needs and preferences as hallmarks of the *process* of "engaging with" (Bright et al. 2014). We proposed the following working definition of engagement drawing together findings from this conceptual review:

> Engagement is a co-constructed process and state. It incorporates a process of gradually connecting with a person and/or therapeutic program, which enables the individual to become an active, committed, and invested collaborator in healthcare.

This working definition highlights that both patient and practitioner play an important role in the process of engagement as it occurs through the rehabilitation period. We suggest that there is reason to see practitioner engagement as at least as relevant as patient engagement, if not more so, accentuating the important role that the practitioner's way of working has in the process of engagement as something we can indeed individually (as practitioners) and collectively (as professions) influence.

Summary

In this section, we have argued for a shift away from the dominant focus on patient behavior and called for a more in-depth exploration into the role that practitioner thoughts, beliefs, feelings, and ways of working have in influencing rehabilitation outcome. We have drawn on evidence in two key developing bodies of work to demonstrate that the way we think and the way we work influence patient experience and outcome, over and above our specific disciplinary-based rehabilitation techniques. This is by no means the first time aspects outside of *technical expertise* have been acknowledged to be important. Indeed, our thinking has been stimulated by our prior work exploring perspectives of what constitutes good quality health and social care in a mixed disabled population that highlighted that a context-appropriate balance between *technical competence* and a *human approach* to care is seen as crucial, and in some instances, humanness may be even more important than technical skill (Fadyl et al. 2010). Despite the work that we and many others before us have done highlighting aspects outside of technical expertise to be important, a tangible shift in practice is yet to be demonstrated.

There may be many reasons for this lack of tangible change in practice, including systemic issues serving to maintain status quo (many other chapters in this text propose or identify different aspects of this, including Chapters 6 and 10). However, we argue that the absence of change is at least in part because we still view technical skills and the exercise of clinical expertise as the most legitimate way to spend our time. Interestingly, professionals tend to believe that patients indeed share this view. For example, in Trede's (2000, p. 430) research cited above, despite patient participants distinguishing between *hands-on technique* and *hands-off attitude* and emphasizing the importance of hands-off attitude, physiotherapist participants believed that their patients placed greater value on their technical skill—"They pay the money and you have to actually treat them. You can't sit there and talk to them" (Trede 2000, p. 431). This weighing up (or perhaps trading off) between the technical aspects (the hands-on) and human aspects (the hands-off) of physiotherapy and the tensions that arise for physiotherapists have been cited elsewhere (Fadyl et al. 2010; Mudge et al. 2014). The empirical evidence presented above clearly illustrates a relationship, albeit complex and as yet poorly understood, between practitioner thoughts, feelings, and ways of working and rehabilitation outcome. A greater focus on these aspects in rehabilitation practice and research seems long overdue. In the following, we extend our discussion in light of the evidence presented to propose some strategies for how we could rethink practitioner behavior to optimize rehabilitation outcome.

How Could/Should We Rethink Practitioner Thoughts, Feelings, and Ways of Working to Optimize Rehabilitation Outcome?

Taking the view that we bring many important factors (beyond our technical expertise) to the therapeutic encounter opens up different ways of thinking about rehabilitation practice. Our sense is that we need to reconsider the skills we prioritize and seek to develop and essentially revisit what is viewed as a legitimate way to spend clinical time. The evidence suggests that it almost certainly requires change. We acknowledge that practitioners do not work in isolation but rather within systems and structures that serve to influence the way we think and work. We do not wish to minimize this in any

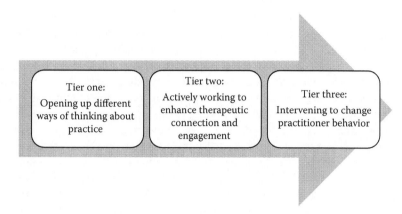

FIGURE 13.1   Three tiers of response to rethinking practitioner thoughts, feelings, and ways of working.

way, because clearly the system in which we work can exert significant influence over key aspects of our work and indeed what "work" is given legitimate status over other "work" (McPherson and Reid 2007). However, for the purpose of this section, we focus predominantly on influencing change at the level of the individual practitioner.

We propose, at the level of the individual practitioner, that behavior change (within practitioners) may occur at three levels (see Figure 13.1). At the most basic level, simply being more aware and acknowledging the possible role that we have on rehabilitation outcome may inadvertently result in changes to practice and rehabilitation outcome. At this level, the strategies we propose focus on tools that may facilitate a critical reflection on practice and open up different ways of thinking. At the second tier of response, we highlight the potential of a more active and targeted shift in practitioner behavior, emphasizing therapeutic connection with patients and facilitating their engagement in rehabilitation. Finally, at the third tier, we acknowledge the complexity of enacting change in clinical behavior, especially when change requires a fundamental shift in the way one works or what one perceives to be the fundamental and legitimate components of practice, or when one's professional identity is perceived to be threatened. As such, at this level, we argue for the merits of more actively intervening with practitioners to facilitate change. Each of these tiers and associated strategies is discussed in more detail below.

Tier One Response: Opening Up Different Ways of Thinking about Practice

Given that in "Practitioner's Thoughts, Feelings, and Attitudes," we identified the relationship between beliefs and attitudes on a range of practitioner behaviors and rehabilitation outcome, it is not hard to imagine how even a minor shift in thinking could influence changes in practitioner behavior. Reflecting on our practice is supported by many (Gibson et al. 2010; McGlynn and Cott 2007; Schutz 2007; Wilding 2008) and can be thought of as the first step in becoming more aware of our own position in relation to the position of others, including the patient, colleagues, profession, and family. One example of a tool to promote clinical reasoning and reflection was developed by Atkinson and Nixon-Cave (2011). The Physical Therapy Clinical Reasoning and Reflection Tool provides a structure comprising the ICF's distinction between body structures and functions, impairments, activity limitations, and participation restrictions in addition to some suggested reflection points that guide the practitioner to examine their clinical reasoning, assessment process, and underlying assumptions. They present a case report of a student who successfully used the tool to aid their clinical reasoning during the treatment planning process. Although the authors acknowledge that further research is warranted to demonstrate the success of this tool, we note it here as an example of how a tool that requires the practitioners to reflect on their decision-making processes and underlying assumptions may aid practitioners to more explicitly consider how their underlying beliefs and attitudes may be influencing their clinical decision making.

Another semistructured approach to reflection is use of autoethnography in practice, an approach that "uses the self as the basis for exploring broader socio-cultural issues" (Bright et al. 2012, p. 998). Hoppes et al. (2007) have utilized autoethnography as a mechanism for promoting reflection throughout occupational therapy training. In this study, students attended a formal introductory lecture on autoethnography; they also participated in regular sharing and discussions with their peers to deepen their insights and understanding of their reflections. They utilized their experiences, reflections, and stories about patients and practitioners to construct an autoethnography of practice. The authors contend that although the students find the course challenging, by the end, they arrive at a deep and profound understanding of occupational therapy practice. There are several examples of autoethnography used by practitioners and researchers to formally reflect on their practice or research (Bright et al. 2012; Hinckley 2005; Mudge et al. 2014). In our own autoethnographic work (Bright et al. 2012; Mudge et al. 2014), we advocate the usefulness of autoethnographic methods in driving change through the systematic exploration of underlying assumptions, questioning our background and ways of working. We have found that the sharing of reflections, thoughts, and emotions with colleagues made the process revealing and, as such, incredibly powerful in terms of practice development (Bright et al. 2012; Mudge et al. 2014).

Considering the epistemological foundations of practice, associated assumptions and how they may constrain or indeed open up new ways of thinking may be another form of reflection on practice. For example, Nicholls and Holmes (2012) argued that a key distinguishing feature of physiotherapy is therapeutic touch, and proposed that the disciplined approach to touch that endures in physiotherapy may constrain other, more humanistic approaches to care (Nicholls and Holmes 2012). Shaw and DeForge (2012) argued for expert physiotherapists as *bricoleurs*. While the language may alienate practitioners, the findings should not. They suggested that, as bricoleurs, expert physiotherapists critically examine and deconstruct knowledge and assumptions and embrace multiple epistemologies versus privileging any one way of knowing, opening the doors to new ways of thinking and working: "Physiotherapists acting as bricoleurs consider the multiple contextual and dynamic factors that impact not only the care they choose to provide, but the perspectives, needs, and preferences of their clients" (Shaw and DeForge 2012, p. 426). All these strategies could be thought of as tools that aim to facilitate practitioners to more explicitly reflect on and question their practice, expose some of the assumptions that underpin the way they work, and make more visible the beliefs and perspectives that influence their clinical reasoning, decision making, and behavior. These strategies sit at tier one of Figure 13.1 because they are somewhat passive, in that they stem from the premise that changing the way we think about practice will in turn change the way we act (arguable in light of the evidence presented earlier in the Practitioner's Thoughts, Feelings, and Attitudes section) versus actively identifying and targeting specific practice behaviors where change could be enacted. In the following, tier two proposes a more active response.

Tier Two Response: Actively Working to Enhance Therapeutic Connection and Engagement

A logical response to the evidence presented in the early part of this chapter is for a more active focus on therapeutic connection and engagement given their relationship to outcome. This is not a novel suggestion, and there have been numerous calls for similar collaborative and person-centered approaches to practice over many years. Unlike many of these calls, however, we view this as a potential "core" component of practice, an explicit approach enacted as a legitimate *intervention* in its own right, as a "human technology" (as we have referred to it previously [Kayes and McPherson 2012]), not simply a recommended way of working or a nice thing to have. In some ways, this is similar to what occupational therapists refer to as "therapeutic use of self" (Punwar and Peloquin 2000). Viewing therapeutic connection and engagement as an intervention differs to routine practice where (a) efforts to strengthen the therapeutic relationship and facilitate engagement appear to be seen as secondary to other supposedly more active ingredients considered to be the domain of disciplinary-based interventions; (b) the development of a good therapeutic relationship may be more due to chance than design; or (c) the skills needed to optimize connection and engagement are considered unteachable, instead developing intuitively through experience over time, eventually marking out the expert from the average practitioner (Kayes and McPherson 2012; Resnik and Jensen 2003).

To equip practitioners with the knowledge and skills to more actively work to enhance therapeutic relationship and engagement, we suggest that two things are needed:

- A strong theoretical foundation to underpin practice in this area
- Clear guidance on how to operationalize theoretical components for practice

While there is a long history of theory and research in the "human or humanistic" aspects of other fields, such as psychotherapy, counseling, and mental health, research specific to the rehabilitation context has been slow to emerge. With regard to theoretical basis, research exploring therapeutic relationship in rehabilitation has almost primarily drawn on Bordin's (1979) theory of working alliance, a theory borrowed in completeness from the psychotherapy literature. In his theory of working alliance, Bordin proposes that the quality of the working alliance is a function of the concordance and agreement on therapeutic tasks and goals and the strength of the emotional bond between the practitioner and the patient. Other candidate theories that may have some relevance in the rehabilitation context include Gelso and Carter's (1985) work as well as that of Pinsof's (1994) integrative systems perspective of therapeutic alliance. In their work, Gelso and Carter distinguish between working alliance, the real relationship (i.e., the relationship that forms between two individuals whether or not that exists within a therapeutic context), and the transference relationship (i.e., redirection of feelings from one person to another). They argue that all exist in a therapeutic relationship (Gelso 2009a,b; Gelso and Carter 1985). What is not clear, however, is the relative importance of each in the context of patient outcome. In contrast to the more dyadic theories of therapeutic relationship proposed by Bordin, Gelso, and Carter, Pinsof argues that therapeutic alliance is not confined to a relationship between two individuals (the patient and the practitioner), but rather is something that exists between and within therapist and patient *systems*. This is particularly relevant in rehabilitation where the therapist is one of a number of people in the *therapists system* exerting influence over the relationship and the rehabilitation process and, likewise, where the *patients system* tends to extend at the very least to significant others and family members. We have argued elsewhere that there is almost certainly a need for more research specific to the rehabilitation context to ensure that practice is well underpinned by a strong conceptual basis (Kayes and McPherson 2012). Nonetheless, while the theories here have to some extent been borrowed from other health settings, they provide a good starting point for critically thinking about the therapeutic relationship in rehabilitation practice.

The therapeutic relationship has been variably defined, but most definitions agree on the key foundations on which a good therapeutic relationship is built including trust, communication, rapport, mutual respect, active collaboration, and empathy. A definition provided by Cole and McLean (2003) captures this well: "a trusting connection between therapist and client through collaboration, communication, therapist empathy, and mutual understanding and respect" (p. 262). There have been several attempts made in the literature to operationalize these core concepts for the purpose of practice, as well as to inform clinical training. A good example of this is in a paper written by Benbassat and Baumal (2004) where they discussed an interviewing style that would promote empathy for medical students. They discussed empathy as one of three basic tenets of professionalism (alongside respect and competence) and proposed that developing empathy is a process of gaining insight, identification with the patient, and compassion. They suggested a person-centered interviewing style to be key to this process and characterize this style as having a willingness to listen, listening without interruption, sustaining respect and interest, making eye contact, allowing the patients to talk freely about aspects of their story that move beyond their symptoms, and inviting elaboration through open-ended questioning.

Our autoethnographic reflection of person-centered practice within the context of a clinical trial (Bright et al. 2012) is consistent with the advice of Benbassat and Baumal. In this autoethnography, we proposed that a primary strategy was "an active process of listening to get to know, to uncover and understand what is meaningful" (Bright et al. 2012, p. 44) and identified four key strategies and techniques for achieving this, including (1) seeing active and mindful listening as a therapeutic tool, a valid intervention in its own right; (2) allowing time; (3) supporting clients to prioritize what is meaningful; and (4) viewing our role differently. As one extract in the clinical researchers' reflections stated:

We now consider, active listening a therapeutic tool rather than a means to an end for assessment and prescription. Listening and talking took on a primary role in our intervention. It replaced assessing and doing techniques that had dominated our earlier ways of working. Because we showed a genuine interest in people's hopes and dreams and did not immediately dismiss them, this helped the client value them, which in turn enabled them to work towards these hopes and dreams. It was so powerful, the act of having that conversation, just being there with an open mind to inquire. (Bright et al. 2012, p. 1001)

It has been proposed that the therapeutic relationship is a fundamental component of client-centered practice (Hammell 2006). However, drawing on the work described earlier, we would also argue that truly working in a client-centered way can positively impact the quality of the therapeutic relationship, that is, perhaps there is to some extent a mutually dependent relationship between the two.

While the evidence expressly looking at therapeutic connection and engagement in the rehabilitation context is perhaps more limited than we would like (something we are actively working to amend at the current time), the literature that does exist, such as that referred to above, offers some guidance on tangible strategies for practitioners to integrate into practice. The response we often hear when we discuss this topic with practitioners is "Oh, we already do that." Hammell questions elsewhere (and indeed in Chapter 3 of this text) this blanket response in her discussion on client-centered practice:

It is easy for the health-care professions to claim to be client-centered (indeed, the professions possess the power to make any claim they choose). This is both politically expedient and beneficial for the professions' public image. However, it is important to engage in a sceptical analysis of whether the professions 'walk the walk' or whether they just 'talk the talk'. (Hammell 2006, p. 150)

We agree that practitioners view the therapeutic relationship and engagement as important. There is a range of literature that supports that practitioners hold this view (Jensen et al. 2000; Leach 2005; Potter et al. 2003a; Roberts and Bucksey 2007; Shontz and Fink 1957; Stenmar and Nordholm 1994). Neither do we contest that practitioners may already routinely work in the way described earlier, such as perhaps the expert practitioners Resnik refers to (Resnik and Jensen 2003). However, we would argue that this happens less frequently than it should, and that all evidence to date suggests it occurs largely by chance versus through the active repositioning of the therapeutic relationship and engagement as fundamental components of the rehabilitation process that require active and explicit management.

Tier Three Response: Intervening to Change Practitioner Behavior

We know that it is challenging to change behavior based on knowledge alone (Michie et al. 2009; van der Ploeg et al. 2007) and practitioner behavior is no exception. Indeed, many practitioners acknowledge difficulties in changing their approach or behavior despite intentions to the contrary (Jones et al. 2013; Mudge et al. 2014; Norris and Kilbride 2014). In response to the challenge of translating research findings into clinical practice, the field of implementation science has recently proliferated (as noted in the "theory" framework in Chapter 1). Implementation science acknowledges the complexity of translating research findings into clinical practice, challenging the notion that dissemination of research findings in publications and presentations at conferences is sufficient to drive change. This new field acknowledges that there are many barriers and points of influence at multiple levels (intervention characteristics, inner and outer settings, individual and organizational characteristics, process) that must be addressed if change is to occur (Damschroder et al. 2009). Without minimizing the impact of organizational barriers to knowledge translation, there are indeed personal barriers that may be addressed, which may lead to change for an individual and their networks. A number of techniques have been described in the literature and we will summarize them here.

Lal and Korner-Bitensky (2013) present a compelling argument for the use of motivational interviewing to change practitioner behavior as a way to facilitate translation of knowledge into practice. Motivational interviewing is a communication technique used by a range of health professionals, usually to help patients identify their own internal and external reasons for making behavioral change associated with improved health (Miller and Rollnick 2002). This review suggests that motivational interviewing could be used as a technique to engage practitioners in the process of changing behavior, attitudes, and/or knowledge of practitioners by enhancing their readiness to change their clinical practice (Lal and Korner-Bitensky 2013).

The application of theories of behavior change to practitioner behaviors may also warrant some further exploration. This has already been explored in some contexts. For example, the theory of planned behavior variables (attitude toward the behavior, the subjective norm, and the perceived behavior control) have been found to be associated with healthcare professionals' intention to use clinical practice guidelines generally in their decisions on patient care (Kortteisto et al. 2010). For physicians, the factor associated most strongly with intention was their perceived behavioral control, but the subjective norm was most strongly associated for other professionals. The implications of these findings are that context- and guideline-based factors either encourage or hinder the intention for physicians to use clinical practice guidelines and that normative beliefs related to social pressures have a corresponding effect for allied health professionals (Kortteisto et al. 2010). Godin et al. (2008) published a systematic review to explore the extent to which health professional intentions and behaviors can be explained by social cognitive theories. They found the theory of planned behavior the most commonly used theory in research seeking to explain practitioner behavior with a frequency weighted mean R^2 of 0.31. They highlighted a range of methodological issues to be cognizant of, but cautiously proposed a hypothetical model of health professional intention and behavior, which is representative of their findings across studies. Key determinants of intentions included beliefs about consequences, social influences, model norm, role and identity, and characteristics of health professionals, while key determinants for behavior included intentions, beliefs about capabilities, and habit/past behavior. While the authors call for more research applying theories of behavior change to better understand health professional behavior, this model may provide a useful starting point for the development of theoretically informed interventions targeting change in practitioner behavior.

Another innovative approach that has been found to be useful in improving person-centered practice is the use of research-based theater (Kontos et al. 2012). The play appears to have evoked a strong degree of self-reflection from the therapists in the audience to the point where they felt compelled to change their practice. It is hypothesized that the dramatic performance was capable of engaging the imagination and fostered empathy of the therapists in the audience in a way that does not seem possible with other more traditional forms of dissemination. It is possible that the arts, more broadly, have a role to play in fostering deep reflection (Kontos and Poland 2009) that has largely been underutilized in health care.

In this section, we have provided a few examples of how we might actively intervene with practitioners to facilitate change in clinical behavior. Despite these examples, there is considerable scope for advancement in our knowledge and understanding of how to more effectively facilitate change in practitioner behavior. There are risks, for example, in relying only on existing behavior change theory to make sense of practitioner behavior, given the different populations and contexts that have underpinned those theoretical frameworks. Future research is needed to (a) better understand why we do what we do; (b) explore with practitioners what they perceive to be the key barriers and facilitators for change; (c) investigate the process of change in practitioner behavior in more depth (i.e., how do practitioners become ready to change, what are the tipping points for change, what drives practitioners to maintain status quo?); and (d) develop and test novel strategies for change to better understand what works, for how long, and for what outcomes.

Summary

In this chapter, we have drawn on empirical evidence to argue for a shift in practice and research from an almost singular focus on intervening with patients to a more active focus on the role that practitioners have in influencing rehabilitation outcome. Of all the variables that impact on rehabilitation outcome, there are many that we cannot change; but one that is within our control, which is subject to influence, is the role that *we* play— the way we think and the way we work with our patients. We have proposed a series of possible responses that range from a focused reflection on the way we think, making visible our assumptions, beliefs, and attitudes and considering their influence on practice; to a more active, targeted approach to intervening with practitioners and changing their behavior.

We suspect that we have only just begun to scratch the surface of what is possible if we turn our attention to this largely untapped resource, our human technology—the practitioner. The existing evidence base is relatively thin in comparison to what we know about patient-related factors, and so the scope for knowledge advance is immense. Even if we only take the evidence presented in this chapter, we would suggest that it is clear that a more explicit focus on the role of the practitioner in influencing rehabilitation outcomes has the potential to have a significant impact on practice and outcome. We urge rehabilitation practitioners and researchers to take up the challenge we have proposed and actively work to harness this largely untapped resource to optimize rehabilitation outcome.

References

Anderson, R. M., and Funnell, M. M. (2000). Compliance and adherence are dysfunctional concepts in diabetes care. *The Diabetes Educator, 26*(4), 597–604.

Atkinson, H. L., and Nixon-Cave, K. (2011). A tool for clinical reasoning and reflection using the international classification of functioning, disability and health (ICF) framework and patient management model. *Physical Therapy, 91*(3), 416–430.

Atwal, A., McIntyre, A., and Wiggett, C. (2012). Risks with older adults in acute care settings: UK occupational therapists' and physiotherapists' perceptions of risks associated with discharge and professional practice. *Scandinavian Journal of Caring Sciences, 26*(2), 381–393.

Benbassat, J., and Baumal, R. (2004). What is empathy, and how can it be promoted during clinical clerkships? *Academic Medicine, 79*, 832–839.

Bishop, A., Foster, N. E., Thomas, E., and Hay, E. M. (2008). How does the self-reported clinical management of patients with low back pain relate to the attitudes and beliefs of health care practitioners? A survey of UK general practitioners and physiotherapists. *Pain, 135*(1–2), 187–195.

Bordin, E. (1979). The generalizability of the psychoanalytic concept of the working alliance. *Psychotherapy, 16*(3), 252–260.

Bright, F., Boland, P., Rutherford, S., Kayes, N., and McPherson, K. (2012). Implementing a client-centred approach in rehabilitation: An autoethnography. *Disability and Rehabilitation, 34*(12), 997–1004.

Bright, F. A. S., Kayes, N. M., Worrall, L., and McPherson, K. M. (2014). A conceptual review of engagement in healthcare and rehabilitation. *Disability and Rehabilitation, 0*(0), 1–12.

Brown, C. A., and Richardson, C. (2006). Nurses' in the multi-professional pain team: A study of attitudes, beliefs and treatment endorsements. *European Journal of Pain, 10*(1), 13–22.

Brunner, E., De Herdt, A., Minguet, P., Baldew, S.-S., and Probst, M. (2013). Can cognitive behavioral therapy based strategies be integrated into physiotherapy for the prevention of chronic low back pain? A systematic review. *Disability and Rehabilitation, 35*(1), 1–10.

Buys, T., and Casteleijn, D. (2007). Preparing for work practice: Under- and postgraduate student training at the University of Pretoria (South Africa). *Work (Reading, Mass.), 29*(1), 25–29.

Cameron, L. D., and Leventhal, H. P. (2003). *The Self-Regulation of Health and Illness Behavior.* New York: Routledge.

Clay, D. L., and Hopps, J. A. (2003). Treatment adherence in rehabilitation: The role of treatment accommodation. *Rehabilitation Psychology, 48*(3), 215–219.

Cole, M. B., and McLean, V. (2003). Therapeutic relationships re-defined. *Occupational Therapy in Mental Health, 19*(2), 33–56.

Cooper, A. F., Jackson, G., Weinman, J., and Horne, R. (2002). Factors associated with cardiac rehabilitation attendance: A systematic review of the literature. *Clinical Rehabilitation, 16*(5), 541–552.

Damschroder, L. J., Aron, D. C., Keith, R. E., Kirsh, S. R., Alexander, J. A., and Lowery, J. C. (2009). Fostering implementation of health services research findings into practice: A consolidated framework for advancing implementation science. *Implementation Science, 4,* 50.

Darlow, B., Fullen, B. M., Dean, S., Hurley, D. A., Baxter, G. D., and Dowell, A. (2012). The association between health care professional attitudes and beliefs and the attitudes and beliefs, clinical management, and outcomes of patients with low back pain: A systematic review. *European Journal of Pain, 16,* 3–17.

Daykin, A. R., and Richardson, B. (2004). Physiotherapists' pain beliefs and their influence on the management of patients with chronic low back pain. *Spine, 29*(7), 783–795.

Demou, E., Gibson, I., and Macdonald, E. B. (2012). Identification of the factors associated with outcomes in a Condition Management Programme. *BMC Public Health, 12*(1), 927.

DiMatteo, M., Sherbourne, C., Hays, R., Ordway, L., Kravitz, R., McGlynn, E., Kaplan, S., and Rogers, W. (1993). Physicians' characteristics influence patients' adherence to medical treatment: Results from the Medical Outcomes Study. *Health Psychology, 12*(2), 93–102.

Duncan, P., Horner, R., Reker, D., Samsa, G., Hoenig, H., Hamilton, B., and Dudley, T. (2002). Adherence to postacute rehabilitation guidelines is associated with functional recovery in stroke. *Stroke, 33*(1), 167–177.

Evans, C. C., Sherer, M., Nakase-Richardson, R., Mani, T., and Irby, J. J. W. (2008). Evaluation of an interdisciplinary team intervention to improve therapeutic alliance in post-acute brain injury rehabilitation. *Journal of Head Trauma Rehabilitation, 23*(5), 329–338.

Ezrachi, O., Ben-Yishay, Y., Kay, T., Diller, L., and Rattok, J. (1991). Predicting employment in traumatic brain injury following neuropsychological rehabilitation. *Journal of Head Trauma Rehabilitation, 6*(3), 71–84.

Fadyl, J., McPherson, K. M., and Kayes, N. M. (2010). Perspectives on quality of care for people who experience disability. *Quality and Safety in Health Care, 20*(1), 87–95.

French, D. P., Cooper, A., and Weinman, J. (2006). Illness perceptions predict attendance at cardiac rehabilitation following acute myocardial infarction: A systematic review with meta-analysis. *Journal of Psychosomatic Research, 61*(6), 757–767.

Gelso, C. (2009a). The real relationship in a postmodern world: Theoretical and empirical explorations. *Psychotherapy Research, 19*(3), 253–264.

Gelso, C. (2009b). The time has come: The real relationship in psychotherapy research. *Psychotherapy Research, 19*(3), 278–282.

Gelso, C., and Carter, J. (1985). The relationship in counseling and psychotherapy: Components, consequences, and theoretical antecedents. *The Counseling Psychologist, 13*(2), 155–243.

Gibson, B. E., Nixon, S. A., and Nicholls, D. A. (2010). Critical reflections on the physiotherapy profession in Canada. *Physiotherapy Canada, 62*(2), 98–100, 101–103.

Godin, G., Bélanger-Gravel, A., Eccles, M., and Grimshaw, J. (2008). Healthcare professionals' intentions and behaviors: A systematic review of studies based on social cognitive theories. *Implementation Science, 3,* 36.

Graham, F., Robertson, L., and Anderson, J. (2013). New Zealand occupational therapists' views on evidence-based practice: A replicated survey of attitudes, confidence and behaviors. *Australian Occupational Therapy Journal, 60*(2), 120–128.

Guidetti, S., Asaba, E., and Tham, K. (2009). Meaning of context in recapturing self-care after stroke or spinal cord injury. *American Journal of Occupational Therapy, 63*(3), 323–332.

Hall, A. M., Ferreira, P. H., Maher, C. G., Latimer, J., and Ferreira, M. L. (2010). The influence of the therapist-patient relationship on treatment outcome in physical rehabilitation: A systematic review. *Physical Therapy, 90*(8), 1099–1110.

Hall, M., Meaden, A., Smith, J., and Jones, C. (2001). Brief report: The development and psychometric properties of an observer-rated measure of engagement with mental health services. *Journal of Mental Health, 10*(4), 457–465.

Hammell, K. (2006). *Perspectives on Disability and Rehabilitation: Contesting Assumptions, Challenging Practice*. Philadelphia, PA: Churchill Livingstone.

Hargreaves, S. (1982). The relevance of non-verbal skills in physiotherapy. *Australian Journal of Physiotherapy, 28*(4), 19–20.

Hendrick, P., Mani, R., Bishop, A., Milosavljevic, S., and Schneiders, A. G. (2013). Therapist knowledge, adherence and use of low back pain guidelines to inform clinical decisions—A national survey of manipulative and sports physiotherapists in New Zealand. *Manual Therapy, 18*(2), 136–142.

Hills, R., and Kitchen, S. (2007). Satisfaction with outpatient physiotherapy: Focus groups to explore the views of patients with acute and chronic musculoskeletal conditions. *Physiotherapy Theory and Practice, 23*(1), 1–20.

Hinckley, J. J. (2005). The piano lesson: An autoethnography about changing clinical paradigms in aphasia practice. *Aphasiology, 19*(8), 765–779.

Hoppes, S., Hamilton, T. B., and Robinson, C. (2007). A course in autoethnography: Fostering reflective practitioners in occupational therapy. *Occupational Therapy in Health Care, 21*(1–2), 133–143.

Houben, R. M. A., Ostelo, R. W. J. G., Vlaeyen, J. W. S., Wolters, P. M. J. C., Peters, M., and den Berg, S. G. M. S.-v. (2005). Health care providers' orientations toward common low back pain predict perceived harmfulness of physical activities and recommendations regarding return to normal activity. *European Journal of Pain, 9*(2), 173–183.

Hubbard, I. J., Harris, D., Kilkenny, M. F., Faux, S. G., Pollack, M. R., and Cadilhac, D. A. (2012). Adherence to clinical guidelines improves patient outcomes in Australian audit of stroke rehabilitation practice. *Archives of Physical Medicine and Rehabilitation, 93*(6), 965–971.

Jackson, B., Dimmock, J. A., Taylor, I. M., and Hagger, M. S. (2012). The tripartite efficacy framework in client-therapist rehabilitation interactions: Implications for relationship quality and client engagement. *Rehabilitation Psychology, 57*(4), 308–319.

Jensen, G. M., Gwyer, J., Shepard, K. F., and Hack, L. M. (2000). Expert practice in physical therapy. *Physical Therapy, 80*(1), 28–43.

Jones, F., Livingstone, E., and Hawkes, L. (2013). "Getting the balance between encouragement and taking over"—Reflections on using a new stroke self-management programme. *Physiotherapy Research International, 18*(2), 91–99.

Josephson, I., Hedberg, B., and Bülow, P. (2013). Problem-solving in physiotherapy— Physiotherapists' talk about encounters with patients with non-specific low back pain. *Disability and Rehabilitation, 35*(8), 668–677.

Kaplan, S., Greenfield, S., and Ware, J. (1989). Assessing the effects of physician-patient interactions on the outcomes of chronic disease. *Medical Care, 27*(3 Suppl), S110–S127.

Kayes, N., and McPherson, K. (2012). Human technologies in rehabilitation: "Who" and "How" we are with our clients. *Disability and Rehabilitation, 34*(22), 1907–1911.

Kenny, D. A., Veldhuijzen, W., Weijden, T., Leblanc, A., Lockyer, J., Légaré, F., and Campbell, C. (2010). Interpersonal perception in the context of doctor-patient relationships: A dyadic analysis of doctor-patient communication. *Social Science and Medicine, 70*(5), 763–768.

Kerns, R. D., Marcus, K. S., and Otis, J. (2001). Cognitive-behavioral approaches to pain management for older adults. *Topics in Geriatric Rehabilitation, 16*(3), 24–33.

King, G., Wright, V., and Russell, D. J. (2011). Understanding paediatric rehabilitation therapists' lack of use of outcome measures. *Disability and Rehabilitation, 33*(25/26), 2662–2671.

Klonoff, P., Lamb, D., and Henderson, S. (2001). Outcomes from milieu-based neurorehabilitation at up to 11 years post-discharge. *Brain Injury, 15*(5), 413–428.

Klonoff, P., Olson, K., Talley, M., Husk, K., Myles, S., Gehrels, J., and Dawson, L. (2010). The relationship of cognitive retraining to neurological patients' driving status: The role of process variables and compensation training. *Brain Injury, 24*(2), 63–73.

Klonoff, P., Talley, M., Dawson, L., Myles, S., Watt, L., Gehrels, J., and Henderson, S. (2007). The relationship of cognitive retraining to neurological patients' work and school status. *Brain Injury, 21*(11), 1097–1107.

Kolehmainen, N., Francis, J., and McKee, L. (2008). To provide or not to provide treatment? That is the question. *British Journal of Occupational Therapy, 71*(12), 510–523.

Kontos, P. C., Miller, K.-L., Gilbert, J. E., Mitchell, G. J., Colantonio, A., Keightley, M. L., and Cott, C. (2012). Improving client-centered brain injury rehabilitation through research-based theater. *Quality Health Research, 22*(12), 1612–1632.

Kontos, P. C., and Poland, B. D. (2009). Mapping new theoretical and methodological terrain for knowledge translation: Contributions from critical realism and the arts. *Implementation Science, 4*(1), 1.

Kortte, K., Falk, L., Castillo, R., Johnson-Greene, D., and Wegener, S. (2007). The Hopkins Rehabilitation Engagement Rating Scale: Development and psychometric properties. *Archives in Physical Medicine and Rehabilitation, 88*(7), 877–884.

Kortteisto, T., Kaila, M., Komulainen, J., Mäntyranta, T., and Rissanen, P. (2010). Healthcare professionals' intentions to use clinical guidelines: A survey using the theory of planned behavior. *Implementation Science, 5*(1), 51.

Lal, S., and Korner-Bitensky, N. (2013). Motivational interviewing: A novel intervention for translating rehabilitation research into practice. *Disability and Rehabilitation, 35*(11), 919–923.

Leach, M. J. (2005). Rapport: A key to treatment success. *Complementary Therapies in Clinical Practice, 11*, 262–265.

Lee, T. S., Sullivan, G., and Lansbury, G. (2006). Physiotherapists' perceptions of clients from culturally diverse backgrounds. *Physiotherapy, 92*(3), 166–170.

Lent, R. W., and Lopez, F. G. (2002). Cognitive ties that bind: A tripartite view of efficacy beliefs in growth-promoting relationships. *Journal of Social and Clinical Psychology, 21*(3), 256–286.

Lequerica, A. H., Rapport, L. J., Whitman, R. D., Millis, S. R., Vangel, Jr., S. J., Hanks, R. A., and Axelrod, B. N. (2006). Psychometric properties of the rehabilitation therapy engagement scale when used among individuals with acquired brain injury. *Rehabilitation Psychology, 51*(4), 331–337.

Lerner, B. H. (1997). From careless consumptives to recalcitrant patients: The historical construction of noncompliance. *Social Science and Medicine, 45*(9), 1423–1431.

Leventhal, H., Nerenz, D., and Steele, D. (1984). Illness representations and coping with health threats. In A. Baum, and J. Singer (Eds.), *A Handbook of Psychology and Health* (pp. 219–252). Hillsdale, NJ: Erlbaum.

Macgowan, M. (2006). The Group Engagement Measure: A review of its conceptual and empirical properties. *Journal of Groups in Addiction and Recovery, 1*(2), 33–52.

Maclean, N., and Pound, P. (2000). A critical review of the concept of patient motivation in the literature on physical rehabilitation. *Social Science and Medicine (1982), 50*(4), 495–506.

McGlynn, M., and Cott, C. A. (2007). Weighing the evidence: Clinical decision making in neurological physical therapy. *Physiotherapy Canada, 59*(4), 241–252.

McPherson, K. M., and Reid, D. A. (2007). New roles in health care: What are the key questions? *Medical Journal of Australia, 186*(12), 614–615.

Meaden, A., Hacker, D., Villiers, A., Carbourne, J., and Paget, A. (2012). Developing a measurement of engagement: The Residential Rehabilitation Engagement Scale for psychosis. *Journal of Mental Health, 21*(2), 183–192.

Medley, A., and Powell, T. (2010). Motivational interviewing to promote self-awareness and engagement in rehabilitation following acquired brain injury: A conceptual review. *Neuropsychological Rehabilitation, 20*(4), 481–508.

Michie, S., Abraham, C., Whittington, C., McAteer, J., and Gupta, S. (2009). Effective techniques in healthy eating and physical activity interventions: A meta-regression. *Health Psychology, 28*(6), 690–701.

Mikhail, B. (1981). The health belief model: A review and critical evaluation of the model, research, and practice. *Advances in Nursing Science, 4*, 65–82.

Miller, W. R., and Rollnick, S. (2002). *Motivational Interviewing: Preparing People for Change*. New York: London.

Mudge, S., Stretton, C., and Kayes, N. (2014). Are physiotherapists comfortable with person-centred practice? An autoethnographic insight. *Disability and Rehabilitation, 36*(6), 457–463.

Nicholls, D. A., and Holmes, D. (2012). Discipline, desire, and transgression in physiotherapy practice. *Physiotherapy Theory and Practice, 28*(6), 454–465.

Nilsagård, Y., and Lohse, G. (2010). Evidence-based physiotherapy: A survey of knowledge, behavior, attitudes and prerequisites. *Advances in Physiotherapy, 12*(4), 179–186.

Norris, M., and Kilbride, C. (2014). From dictatorship to a reluctant democracy: Stroke therapists talking about self-management. *Disability and Rehabilitation, 36*(1), 32–38.

O'Halloran, R., Grohn, B., and Worrall, L. (2012). Environmental factors that influence communication for patients with a communication disability in acute hospital stroke units: A qualitative metasynthesis. *Archives of Physical Medicine and Rehabilitation, 93*(1 Suppl), S77–S85.

Oakley, E. T., Katz, G., Sauer, K., Dent, B., and Millar, A. L. (2010). Physical therapists' perception of spirituality and patient care: Beliefs, practices, and perceived barriers. *Journal of Physical Therapy Education, 24*(2), 45–52.

Parry, R. H., and Brown, K. (2009). Teaching and learning communication skills in physiotherapy: What is done and how should it be done? *Physiotherapy, 95*(4), 294–301.

Persch, A. C., and Page, S. J. (2013). Protocol development, treatment fidelity, adherence to treatment, and quality control. *The American Journal of Occupational Therapy, 67*(2), 146–153.

Pinsof, W. M. (1994). An integrative systems perspective on the therapeutic alliance: Theoretical, clinical and research implications. In A. O. Horvath, & L. S. Greenburg (Eds.), *The Working Alliance: Theory, Research and Practice* (pp. 173–195). New York: John Wiley & Sons.

Potter, M., Gordon, S., and Hamer, P. (2003a). The difficult patient in private practice physiotherapy: A qualitative study. *Australian Journal of Physiotherapy, 49*(1), 53–61.

Potter, M., Gordon, S., and Hamer, P. (2003b). The physiotherapy experience in private practice: The patients' perspective. *Australian Journal of Physiotherapy, 49*, 195–202.

Prigatano, G., Klonoff, P., O'Brien, K., Altman, I., Amin, K., Chiapello, D., Shepherd, J., Cunningham, M., and Mora, M. (1994). Productivity after neuropsychologically oriented milieu rehabilitation. *The Journal of Head Trauma Rehabilitation, 9*(1), 91–102.

Punwar, J., and Peloquin, M. (2000). *Occupation Therapy: Principles and Practice.* Philadelphia, PA: Lippincott, Williams & Wilkins.

Resnik, L., and Hart, D. L. (2003). Using clinical outcomes to identify expert physical therapists. *Physical Therapy, 83*(11), 990–1002.

Resnik, L., and Jensen, G. M. (2003). Using clinical outcomes to explore the theory of expert practice in physical therapy. *Physical Therapy, 83*(12), 1090–1106.

Roberts, L., and Bucksey, S. J. (2007). Communicating with patients: What happens in practice? *Physical Therapy, 87*(5), 586–594.

Rose, J. (2013). A preliminary investigation into the influence of therapist experience on the outcome of individual anger interventions for people with intellectual disabilities. *Behavioral and Cognitive Psychotherapy, 41*(4), 470–478.

Ruppe, M. D., Feudtner, C., Hexem, K. R., and Morrison, W. E. (2013). Family factors affect clinician attitudes in pediatric end-of-life decision making: A randomized vignette study. *Journal of Pain and Symptom Management, 45*(5), 832–840.

Salbach, N. M., Guilcher, S. J., Jaglal, S. B., and Davis, D. A. (2010). Determinants of research use in clinical decision making among physical therapists providing services post-stroke: A cross-sectional study. *Implementation Science, 5*(1), 77.

Salbach, N. M., Jaglal, S. B., Korner-Bitensky, N., Rappolt, S., and Davis, D. (2007). Practitioner and organizational barriers to evidence-based practice of physical therapists for people with stroke. *Physical Therapy, 87*(10), 1284–1303.

Schönberger, M., Humle, F., and Teasdale, T. W. (2006a). The development of the thera-peutic working alliance, patients' awareness and their compliance during the pro-cess of brain injury rehabilitation. *Brain Injury, 20*(4), 445–454.

Schönberger, M., Humle, F., and Teasdale, T. W. (2006b). Subjective outcome of brain injury rehabilitation in relation to the therapeutic working alliance, client compli-ance and awareness. *Brain Injury, 20*(12), 1271–1282.

Schönberger, M., Humle, F., Zeeman, P., and Teasdale, T. W. (2006c). Working alliance and patient compliance in brain injury rehabilitation and their relation to psychosocial outcome. *Neuropsychological Rehabilitation, 16*(3), 298–314.

Schutz, S. (2007). Reflection and reflective practice. *Community Practitioner, 80*(9), 26–29.

Sela-Kaufman, M., Rassovsky, Y., Agranov, E., Levi, Y., and Vakil, E. (2013). Premorbid personality characteristics and attachment style moderate the effect of injury sever-ity on occupational outcome in traumatic brain injury: Another aspect of reserve. *Journal of Clinical and Experimental Neuropsychology, 35*(6), 584–595.

Shaw, J. A., and DeForge, R. T. (2012). Physiotherapy as bricolage: Theorizing expert prac-tice. *Physiotherapy Theory and Practice, 28*(6), 420–427.

Shields, N., Bruder, A., Taylor, N. F., and Angelo, T. (2013). Getting fit for practice: An innovative paediatric clinical placement provided physiotherapy students opportu-nities for skill development. *Physiotherapy, 99*(2), 159–164.

Shontz, F., and Fink, S. (1957). The significance of patient-staff rapport in the rehabilita-tion of individuals with chronic physical illness. *Journal of Consulting Psychology, 21,* 327–334.

Sran, M. M., and Murphy, S. (2009). Postgraduate physiotherapy training: Interest and perceived barriers to participation in a clinical master's degree programme. *Physiotherapy Canada, 61*(4), 234–243.

Stenmar, L., and Nordholm, L. A. (1994). Swedish physical therapists' beliefs on what makes therapy work. *Physical Therapy, 74*(11), 1034–1039.

Stevenson, T. J., Barclay-Goddard, R., and Ripat, J. (2005). Influences on treatment choices in stroke rehabilitation: Survey of Canadian physical therapists. *Physiotherapy Canada, 57*(2), 135.

Strauser, D. R., Lustig, D. C., Chan, F., and O'Sullivan, D. (2010). Working alliance and vocational outcomes for cancer survivors: An initial analysis. *International Journal of Rehabilitation Research, 33*(3), 271–274.

Thomas, P. W., Thomas, S., Hillier, C., Galvin, K., and Baker, R. (2006). Psychological interventions for multiple sclerosis. *Cochrane Database of Systematic Reviews* (1).

Trede, F. V. (2000). Physiotherapists' approaches to low back pain education. *Physiotherapy, 86*(8), 427–433.

Turner, P., and Whitfield, T. W. (1997). Physiotherapists' use of evidence based practice: A cross-national study. *Physiotherapy Research International, 2*(1), 17–29.

van Almenkerk, S., Smalbrugge, M., Depla, M. F. I. A., Eefsting, J. A., and Hertogh, C. M. P. M. (2013). What predicts a poor outcome in older stroke survivors? A systematic review of the literature. *Disability and Rehabilitation, 35*(21), 1774–1782.

van der Hulst, M., Vollenbroek-Hutten, M. M. R., and Ijzerman, M. J. (2005). A systematic review of sociodemographic, physical, and psychological predictors of multidisciplinary rehabilitation—Or, back school treatment outcome in patients with chronic low back pain. *Spine, 30*(7), 813–825.

van der Ploeg, H. P., Streppel, K. R., van der Beek, A. J., van der Woude, L. H., Vollenbroek-Hutten, M. M., van Harten, W. H., and van Mechelen, W. (2007). Successfully improving physical activity behavior after rehabilitation. *American Journal of Health Promotion, 21*(3), 153–159.

van Hal, L., Meershoek, A., Nijhuis, F., and Horstman, K. (2013). A sociological perspective on 'the unmotivated client': Public accountability and professional work methods in vocational rehabilitation. *Disability and Rehabilitation, 35*(10), 809–818.

Van Vuuren, S., and Nel, M. (2010). Clinical Skills Centre: Training of South African occupational therapists. *South African Journal of Occupational Therapy, 40*(2), 3–5.

Vieira, E. R., Freund-Heritage, R., and da Costa, B. R. (2011). Risk factors for geriatric patient falls in rehabilitation hospital settings: A systematic review. *Clinical Rehabilitation, 25*(9), 788–799.

Waldron, B., Casserly, L. M., and O'Sullivan, C. (2013). Cognitive behavioral therapy for depression and anxiety in adults with acquired brain injury. What works for whom? *Neuropsychological Rehabilitation, 23*(1), 64–101.

Wall, C. L., Ogloff, J. R. P., and Morrissey, S. A. (2006). The psychology of injured workers: Health and cost of vocational rehabilitation. *Journal of Occupational Rehabilitation, 16*(4), 513–528.

Walters, S. J. (2010). Therapist effects in randomised controlled trials: What to do about them. *Journal of Clinical Nursing, 19*(7–8), 1102–1112.

Whitman, J. M., Fritz, J. M., and Childs, J. D. (2004). The influence of experience and specialty certifications on clinical outcomes for patients with low back pain treated within a standardized physical therapy management program. *The Journal of Orthopaedic and Sports Physical Therapy, 34*(11), 662–672; discussion 672–675.

Wilding, P. M. (2008). Reflective practice: A learning tool for student nurses. *British Journal of Nursing, 17*(11), 720–724.

Wright, B. J., Galtieri, N. J., and Fell, M. (2014). Non-adherence to prescribed home rehabilitation exercises for musculoskeletal injuries: The role of the patient practitioner relationship. *Journal of Rehabilitation Medicine, 46*, 153–158.

14

Rehab as an Existential, Social Learning Process: A Thought Experiment

David A. Stone

Christina
Papadimitriou

Prologue to the Thought Experiment

In this chapter, we ask you, the reader, to suspend your attachments to the current approach to inpatient rehabilitation; we also ask that you suspend your commitment to incremental change as the only realistic opportunity for improving the care and treatment of persons with traumatic injuries that require inpatient rehabilitation. These requests respect the attachments that providers, advocates, and researchers have to the current approach to traumatic injuries rehabilitation and recognize the value of deliberate, evidence-based change in the ways in which rehabilitation is understood, conducted, organized, and financed. Nevertheless, we ask you to take a leap with us into a completely different way of approaching this type of rehabilitation in the hopes that fresh perspective and the articulation of large-scale change will foster productive dialogue about its future and the best ways to meet the needs of those who require its services.

273

To help you take the leap, we organized this chapter as a thought experiment, a grand "what-if" through which we will explore new approaches to our fundamental understanding of traumatic spinal cord injury (TSCI), how treatment is conducted, how it is organized institutionally, and how it is recognized and financed within our larger political and economic systems. To get us started, consider the following what-if:

How would our world be different if: you are standing on a dock along a slow-moving river. You are a white male. You dive headfirst into the water, strike your head, lose consciousness, and awaken to discover that you are now a black female. You live in a world where this happens hundreds of thousands of times a year, but you never thought it would happen to you. And to your good fortune, you also live in a world that recognizes that this kind of immediate transformation requires a number of services that assist individuals in adjusting to their new reality. In the blink of an eye, you have left the world you know as a man, as white, as a member of your culture's dominant categories (politically, economically, socially, culturally, linguistically, etc.). You now need to learn how to function in the world from your new status. Your body is different, your voice is different, you look different, the world looks at you differently, myriad possibilities that existed for you in your future as a white male no longer exist for you at all, and many that do only do so in a very different way. You now find yourself a member of new identity groups, though as a special case, not as one born into these groups, but one who becomes a member only through trauma. It slowly dawns on you that these changes will also affect your love life, your relationships with friends and colleagues, your work life. You will feel *everything* has changed!

But not to worry, in this scenario, society has provided for the needs it anticipates you will have transitioning from male and white to female and black.* It conceptualizes this transition as an existential and social learning process—learning to *become* female and black—and it supports services to assist you in your transition. Some of the services would need to address aspects of your physical situation. There would be medical attention for the bump on your head; there would be attention paid to other new medical needs related to hormone and endocrine changes, changes in bodily strength, bone density, and the like. But there would also need to be assistance in how to do things as a female—you would be taught how to get your point across as someone whose position and standpoint are devalued by virtue of who you are, not what you are saying. But the point is, it would be apparent to everyone, and be reflected in the nature and types of services provided, that the transitions with which people in this situation would need the most help would involve learning *how to be*, *how to embody being*, female and black, and how these differ from what it is to be, to live, from the position of being male and white.

Using this example as the basis for our thought experiment, this chapter will argue that, in fact, the core transitions involved in TSCI rehabilitation are better understood and addressed as an existential, social learning process involving moving from one way of being in the world to another (moving from nondisabled to disabled to newly abled) than as a medical transition from well (preinjury) to ill (disabled) to recovering (coping with a disability). In TSCI, individuals move from the dominant status of nondisabled,

---

* While this scenario assumes a US context, the point in this and any context is to imagine a situation in which one goes from a dominant social and cultural position to a minority position for which there are structural obstacles that do not exist for those in the dominant group.

a status that comes with physical, social, economic, and political worlds arranged in their favor, to a minority status within which the physical, social, economic, and political worlds are structurally arranged to result in disadvantage (Oliver 1990, 1993, 1996; Mitchell and Snyder 1997; Linton 1998; Hammell 2006, 2007a,b; Shakespeare 2006; Siebers 2008; Davis 2010).

For reasons that pertain largely to the history of how disability was perceived (DeJong 1979; Davis 1995; Wilson 2010; see also Chapters 1, 2 and 10), rehabilitation has been located within and focused around medical practices and approaches to care. It has only dealt with these social, existential considerations at the margin (see Chapters 3 and 9). In this chapter, we argue that advances in medical treatment for TSCI have for a long time now made it possible for rehabilitation to be reframed along the lines suggested by our hypothetical diver, and that, indeed, in many ways this process has already begun (Hoff 2010; Fredericks et al. 2012). In what follows, we will flesh out this thought experiment by making the case for *reconceptualizing* TSCI rehabilitation as a form of existential, social learning and will create basic frameworks for further discussion on how the various levels of rehabilitation (practice, organization, and institution) could be rethought in light of this reconceived understanding of the nature of rehabilitation itself.

The Current State of Affairs

In this section, we briefly outline historical, conceptual, and political dimensions of the current state of rehabilitation services.

Over the last fifty years, understanding of disability and the rights of so-called disabled persons in the developed world has changed dramatically. The Western world has moved from one in which disability is understood as deficit (as handicap)—and for which pity, charity, and care were the overt responses—to one in which disability is coming to be viewed as just one form of difference among the many that create a diversity of abilities, experiences, and perspectives, and for which the overt response is recognition of and respect for those differences. In the United States, these changes are most visible in political, legal, and social advances such as the Rehabilitation Act of 1973, its amendments, and the Americans with Disabilities Act of 1990, which affords persons with disabilities the same protections against job discrimination as women and minorities received in the Civil Rights Act of 1964 (Albrecht 1992, p. 109). In terms of accommodations, the 2008 amendments* move away from a medical model approach determining and defining disability in the workplace to a more social approach.

Along with this change has come a new and better understanding of the human and social experiences of disability (Hoff 2010) and of what it means to become a "newly abled" person (Papadimitriou 2008a,b; Papadimitriou and Stone 2011; Gibson et al. 2012;

---

* In the United States, the Amendments Act of 2008 (Public Law 110-325, ADAAA) requires that courts interpreting the ADA and other federal disability nondiscrimination laws focus on whether the covered entity has discriminated, rather than whether the individual seeking the law's protection has an impairment that fits within the technical definition of the term *disability*. Similar changes have occurred in other countries. For an example of Canadian legal developments, see Pothier (2006).

Kavanagh 2012). As we mentioned earlier, in the case of a nondisabled person with a TSCI who becomes paralyzed, the social transformation that can occur is from nondisabled to injured and disabled, to newly abled. To become newly abled means to re-have, to become able once more. We explain this becoming in the section "Carrying Out the Thought Experiment." Researchers, clinicians, and advocates now have a richer understanding of the ways in which becoming a newly abled person has existential, social, political, practical, and identity-based dimensions. Currently, the professions and the institutional settings within which rehabilitation is undertaken have come to recognize the varied dimensions involved in the "becoming" process. However, we argue that health-care institutions and practices have not adopted a full appreciation of the importance of these dimensions and the role that inpatient rehabilitation can play in facilitating the process of becoming newly abled and adopting a social justice approach to the experience of disability.

Inpatient rehabilitation is rooted in and has developed through medical contexts, thereby creating medically oriented thinking, practices, institutions, and funding models. In practical terms, the professions of occupational therapy (OT), physical therapy (PT), and rehabilitation nursing view themselves and are viewed externally as allied *health* professions (DeJong et al. 2002; Magasi 2008a,b; Shakespeare et al. 2009; Siegert et al. 2010). They operate necessarily, and ever more explicitly, within the larger context of medical practices, institutions, and insurance systems (whether public or private). Training in these professions involves intensive study of biology, chemistry, mathematics, musculoskeletal disorders, human anatomy, physiology, and the neurological basis for human movement, and includes hands-on training in hospital or other medically oriented clinical settings. Clinical practices are focused on health-related and functional outcomes (Whyte and Hart 2003; Kayes and McPherson 2010; Verkaaik et al. 2010; Whiteneck and Gassaway 2010; Levack et al. 2011; see also Chapter 13 for a reflection on how practice might be different). Inpatient spinal cord rehabilitation units are almost always located in or adjacent to hospitals. Patient records are protected under health-related regulations and protocols.

Conceptually, protestations of holism to the contrary, the human body is still viewed primarily in mechanistic terms (Pawloski 2001; Nicholls and Gibson 2010). Human bodies are anatomical structures potentially capable of certain functions (movements, actions, activities). And while the goals of these functional movements may require understanding of daily living needs or social requirements, or psychological matters relating to motivation or lack thereof, the basis for these understandings remains the physical body as mechanism. Tying this conceptual level back to the practical, we can cite the Functional Independence Measure (FIM™) score as one of the dominant measures (for both clinical decision making about progress and discharge) and payment (U.S. insurance requirements for progress are based on the FIM™ score).

This brings us more specifically to the issue of the *financing of care* and how ensconced it is within the orbits of medicine and health in the United States. Currently, in the United States, funding for inpatient rehabilitation facilities (IRFs) under Medicare and Medicaid is based on the prospective payment system, which means that IRFs are paid per discharge using a methodology that calculates the average cost per discharge for a given condition, which is then adjusted for geography, case-mix, and other factors. Core elements of the case-mix are cognitive and functional impairment. Functional impairment is determined by a score on the FIM™, which measures 13 motor (e.g., bathing, grooming, transferring, locomotion) and 5 cognitive tasks. Rehabilitation thus operates alongside other forms of inpatient treatment in terms of the ways in which it is assessed and paid for and is driven by these payment methods (Murer 2008).

While there have been clear benefits to these approaches, such as increased survival, it is less clear whether medical model–based rehabilitation best meets the needs of persons who have been disabled or is the best use of societal resources in the effort to facilitate transition to life as a newly abled individual (O'Day et al. 2002). This concern has come into sharper focus in part in response to the ever-growing cost and complexity of healthcare and rehabilitation (DeJong et al. 2002, 2011). For our purposes, the concern comes from the intersection of two important literatures. The first one is from the steady growth in the academic literature, as well as in writing emanating from the disability advocacy community, that speaks directly from the experiences of disabled people and newly abled individuals. These literatures, which explore in detail and depth the experiences of those whom rehabilitation seeks to serve, go some way in replacing the third person, theoretical, and clinician-based accounts of the rehabilitation process and permit the possibility of thinking with some informed sophistication about how these practices and institutional settings are experienced (Toombs 1992, 1995, 2001; Wendell 1996; Seymour 1998; Beckett and Wrighton 2000; Sparkes and Smith 2002; Kielhofner 2005; Smith and Sparkes 2007, 2008; Magasi et al. 2008; Charlton 2010 [1998]; Price et al. 2011; Magasi 2012). And second, there is a growing literature on efforts to make inpatient rehabilitation more client-centered (Cain 2002; Hagenow 2003; Cott 2004; Donnelly et al. 2004; Cott et al. 2006; Hammell 2006; Cohen and Schemm 2007; Gzil et al. 2007; Leplege et al. 2007; McLeod and McPherson 2007; Papadimitriou and Cott 2012; Papadimitriou and Carpenter 2013). Much of this literature lays bare the struggle of practitioners to work in client-centered ways in conflict with medically based conceptual approaches, therapeutic regimens, institutional and organizational policies, and payment systems that augur against such efforts (see Fredericks et al. 2012 for a similar point).

The question we are taking up with the thought experiment is whether more of the same, refinements at the margins of what is essentially a medically oriented approach to TSCI rehabilitation, is likely to successfully address the needs of persons with SCI.

Carrying Out the Thought Experiment

In this section, we introduce and describe an existential, social approach to the process of facilitating someone's transition from disabled to newly abled as an advance over the traditional mechanistic medical-model approach to rehabilitation. The point of the story at the beginning was to ask you to think analogically about how the shift from nondisabled to disabled to newly abled is, like the shift from white male to black female, a shift that is fundamentally existential and social and so to imagine along with us how understanding of and our social response to TSCI could be rethought and restructured to meet the needs of that kind of life change.

Conceptually, we conceive of this approach as a modification of the meaning of the term *rehabilitation*. In its present, mechanistic medical meaning, the term rehabilitation traces its roots to the verb *habile*, to be able, and thus rehabilitation is the process by which the individual is re-enabled to do x or y. We propose instead that the term be rooted in the verbs *habe* or *haber*—to have. The roots *habe* and *haber* also give us the terms habituate, habitus, and habitual; hence we recast rehabilitation as re-having and re-habituating: re-having oneself, that is, reconnecting with one's existential self, one's possibilities for existing as a human being, in both a temporal and embodied way.

To elucidate this, we provide two exemplars of what it means to rethink the therapeutic process as an existential learning process of re-having: (1) developing capability in wheelchair use post traumatic injury and (2) what it takes for inpatients to make sense of the work they are doing in therapy. Both have been described elsewhere in greater methodological and epistemological detail (Papadimitriou 2008b; Stone and Papadimitriou 2010; Papadimitriou and Stone 2011). We summarize them here and use them to move the thought experiment forward. As you read these exemplars, consider again our white male who suddenly finds himself a black female, and the ways in which those kinds of transformations are mirrored in the case of individuals with TSCI who find themselves in an inpatient rehabilitation unit.

Learning to *Live from* a Wheelchair: Enwheeling

Enwheeling is a concept coined after listening to wheelchair users talk about their experiences of re-learning to live through the wheelchair both within and outside inpatient rehab after spinal cord injury (Papadimitriou 2008b). It is not a term that they used, but one that we believe captures lived experience as not merely mechanical (physical/functional) or practical (skill acquisition) but also existential and embodied. Enwheeling is the process of learning to make the chair a "natural" part of one's body. Inpatient rehabilitation health providers (esp. PT and OT) focus on this process by developing exercises or tasks that can work on or increase skill acquisition for the patient, such as learning to do wheelies (balancing on back two wheels and "bumping" curbs), all the while the chair slowly becomes a more habitual part of one's movement. Enwheeling is a process that points to this learning not only at the functional and skill acquisition levels but also, coevally, at the existential and embodied levels. This means that as a person is learning to do wheelies, she is also learning to use her body via the chair, thus transforming her body to include possibilities and limitations afforded by the chair (ways it can move, ways it cannot move, its physical dimensions, the height it provides, its weight, its sounds, the obstacles it can traverse, the obstacles it cannot traverse, etc.). In this re-embodiment process, her body is the chair, the chair is her body; she doesn't "have" a chair—rather she is enwheeled. As the chair is becoming part of her habitual and tacit ways of acting (hurrying, waiting, strolling, watching, cooking, working, dancing, toileting, etc.)—she is inhabiting the world anew.

Whether we are enwheeled or not, we all inhabit the world through our bodies. Our focus here is the *re-learning cum new learning* that takes place for someone transforming from walking to enwheeled. This learning transcends the functional or task-oriented approaches that skills such as wheelies are commonly presented as, to engulf an emotional, existential, and social new beginning. Many disability activists and scholars would also include political aspects to becoming enwheeled (Shakespeare 2006) and thus newly abled (Papadimitriou 2000, 2008a).

This means that in becoming enwheeled, and thus newly abled, one's habitual ways of doing things incorporate the wheelchair as an extension of the physical body—as integral to how one is, physically, in the world. In that sense, successful wheelies signify not only the competence to maneuver the wheelchair and move that indicates the patient is closer to discharge goals, but they provide evidence of living *from* an enwheeled body. In "living from,"* the person no longer has to think about how to get themselves and their chairs out of the way of someone approaching in a narrow hallway; they just get out of the way (not just learning how to walk in high heels, but learning how to make those heels *work for you*). Such persons are living from a new habitual, kinesthetic dimension of embodiment: they are enwheeled, that is, they are in the process of re-embodiment as they live with from and through the body–chair.

We argue, that this way of seeing the development of wheelchair skills and abilities is distinguishable from the received view, wherein the focus is on wheelies as mere movement, or a task and a PT/OT goal, which misses that it is also about an existential and embodied moment in a person's re-habilitation—a re-habituation.† The existential details do not need to be explained to patients, but practitioners need to recognize the significance of the existential for wheelchair users and how, in enwheeling, a person now lives through a transformed embodiment. As we re-think what rehabilitation might look like from an existential or non-medical approach, we argue that the existential and embodiment implications of tasks, skills, and exercises (the functional, mechanical, physical aspects of body and action) that OT and PT focus on can *be contextualized as meaningful life experiences that point to ability, competence, and re-embodiment*. Clinicians can be made (more) aware of the need to make therapy meaningful to patients. Once clinicians can see the existential aspects of their care, they may be able to address it both directly and indirectly during therapy (Cott 2004; Hamell 2006; Shattell et al. 2008; Nichols and Gibson 2010; Levack et al. 2011; Bright et al. 2013; Papadimitriou and Carpenter 2013). We recognize that this is not an easy task, and that providers' are ill prepared to adopt the suggestions we are making with confidence, flexibility, and expertise (Higgs and Titchen 2001; Carpenter 2004; Parry 2004; MacLeod and McPherson 2007). We hope that this thought experiment will trigger interest in fostering discussion and learning modules that can assist providers in adopting an attitude that accepts and promotes that an existential understanding of and approach to rehabilitation care means that providers focus beyond the mechanical body and its assessment to a more encompassing/holistic one that starts with what is meaningful to the patient.

---

* The locution living "from" derives from the phenomenological tradition of Merleau-Ponty (1962 [1947]). See Papadimitriou (2008c) for a fuller phenomenological exploration of re-embodiment and living from the wheelchair. This is a similar idea to Gibson et al. (2007) characterization as "living through."

† Equally, there is a process of social learning here, which we will not cover in detail. However, one can readily see how one's capacity to function socially is altered, and needs to be readjusted to, when one moves from standing and walking to sitting and wheeling. This dimension of the need to rehabituate would also need to be addressed in the therapeutic process.

Making Sense of Rehabilitation Therapy: The Case of Human Temporality

Our second exemplar introduces the role that human temporality plays in how people make meaning, that is, how they explicitly make sense of the world around them, and how this process is disrupted in cases of TSCI.

We have argued elsewhere (Stone and Papadimitriou 2010; Papadimitriou and Stone 2011)* that the way human beings live in and through time and how that process is central to human meaning-making (that is, the process of understanding) are disrupted after spinal cord injury. Viewed existentially, understanding is not something that is primarily cognitive or mental; understanding is putting possibilities into play—it is action, not knowledge. And as action it is intentional, it is movement with a goal (even if that goal is unclear), and therefore it is more than simple movement or embodied function. The sum of those actions is human life. Human life is the process of becoming who we already are, the process of fulfilling what it is to be me. Our ability as human beings to make meaning, that is, to give meaning to or accept the meaning of the things around us and the activities that we are engaged in, arises existentially from our ability to project ourselves futurally into possibilities that exist for us. These possibilities, which arise for us individually both from our cultures, societies, and histories, and from our personal experiences (our past—what we bring with us from our own experience to the meaning-making process), are projected out in front of us on the basis of our plans, goals, intentions, and our sense of who it is we are trying to be (to become). This rock *is* now (means) a hammer because I need it to pound in this tent stake so I can set my tent up before it rains, so that I can have a nice vacation in the forest and teach my children the value of nature, and so be (or become) the kind of parent I want to be. Together, our past and our future come together to allow what we are doing now and the things we are engaged with to be meaningful for us. (Literally, this is the structure through which the world arises for us as intelligible at all and through which it makes sense to us in the way it does in each individual instance.)

People who have been recently disabled in some significant way (like a spinal cord injury) lack a rich and concrete sense of their own future possibilities as disabled because they lack a past as disabled. Without this ability to project themselves into their own future in real and concrete ways (not generally by, for example, thinking of disabled people in movies), they struggle to make more than the most superficial sense of the work they are being asked to engage in in rehab. That is, they struggle to connect the apparatus they are being asked to use and the exercises and activities they are being asked to perform, to any real sense of how their lives will be structured and operate once they leave the rehab unit. Without being able to fully make sense of the narrowly proscribed exercises, activities, and tasks they are being ask to perform, tasks that are often difficult, painful, and embarrassing, patients often struggle to find the energy and the motivation to complete them. Rehabilitation providers' received, mechanistic understanding of this experience see this struggle, but are forced to account for it in psychological terms

---

* These works draw on the hermeneutic phenomenology of Heidegger (1962).

(depression, loss, lack of motivation) because they fail to be able to take into account the existential dimension of human existence and so wind up treating the wrong cause (see Chapter 3 for a related discussion).

What these two exemplars demonstrate is a basis for shifting our conceptual approach to TSCI from a narrow, mechanistic, medically oriented view of the body and of disability to an existential, social learning understanding persons with TSCI that reveals the transition from nondisabled to disabled to newly abled as a process of re-having, of being shown how to reappropriate dimensions of existence that allow us to live fully as human beings. Think here again of our white male diver and the work it would take to live from a female body and to make sense of the work required to learn how to operate in the word as a black female as you just begin to develop a concrete sense of your new future.

Rethinking Practice

Certainly, persons with TSCI arrive at inpatient rehab units with both acute medical issues that require attention and in some cases with chronic medical issues that will need to be addressed throughout rehab and perhaps also once the individual returns to the community (e.g., pressure ulcers). But what the two exemplars reveal are the ways in which the kind of strict mechanistic, medical conceptualization of re-enabling fails to properly account for and address essential human characteristics like embodiment and temporality. If, therefore, we were to shift our conceptual understanding from the mechanistic medically oriented approach to one based on existential and social learning processes, this would then require a shift in our approach to therapeutic practices.

TSCI rehabilitation is the process that promotes and fosters relearning how to live from one's embodiment—becoming enwheeled—and how one reestablishes the capacity to make sense of one's existence by connecting patients to a rich and concrete sense of their life possibilities as newly abled persons. Enwheeling, for example, would require that practitioners reorient work with the wheelchair to more directly focus on how it is one lives from the wheelchair and how one becomes habituated in their bodily reactions to operate through the chair. This would mean more than training to meet purely functional goals as we mentioned above; it would mean connecting the body with the meaningful and embodied aspects of learning to maneuver a wheelchair.

In our prior work on the role of temporality and how SCI patients develop a concrete sense of their futures, the value of engaging peer mentors early (and at critical junctures) during the inpatient stay was evident (Stone and Papadimitriou 2010; Papadimitriou and Stone 2011). The presence of peer mentors allowed patients to see first-hand the kinds of possibilities (both life possibilities like relationships and practical possibilities like driving) out of which they can begin to fashion their own future possibilities. There are a number of ways that access to peer mentors provides a more concrete connection between the incremental, physical work that PTs and OTs demand of their patients and the long-term outcomes that patients can only dimly view without a richer sense of their own futures. Identifying and addressing the existential dimension of injury and rehabilitation directly (in this case, for example, by employing peer mentors to give patients a more concrete sense of the real possibilities of a newly abled life) is what we mean by re-having—giving the patients back their human existence.

A rethinking of rehabilitation services asks researchers, clinicians, and advocates to focus on these kinds of *re-having* strategies, teachings, and learning and identify additional ways in which the practices of OT, PT, and rehab nursing could be reestablished as means for fostering the existential social learning process of re-having.

Rethinking Organization and Institution

The section "Rethinking Practice" described how SCI can be reconceptualized as a human experience that is primarily existential and social in character rather than physical and medical. It showed how this reconceptualization can lead to a reorientation of existing professional and lay practices (OT, PT, and peer mentoring) that emphasizes the existential and social dimensions of care over the functional and mechanical. To continue the thought experiment, the next question is, How would the organizational and institutional dimensions of such rehabilitation be structured? By organizational we mean what an inpatient rehab unit would look like. By institutional we mean, Where would rehabilitation be located relative to other social institutions (e.g., healthcare, social services, public or private sector, etc.) and how would it be financed?

Inpatient care of SCI is provided in settings designed primarily to address the medical needs of patients (nursing, medicine), their physical and functional development (PT, SLP), and their ability to carry out basic activities of daily living (OT). In these ways, inpatient facilities are physically and structurally organized like a cross between a hospital and a gymnasium. Functionally, inpatient treatment is most commonly provided by teams of health professionals who view themselves as applying their expert knowledge to develop strength, range of motion, and a range of discrete activities that the patient will need to function independently. Operationally, these teams tend to groups of individual specialists who divide the required labor and coordinate the dimension of care they provide with the other members of the team. Thought another way, they divide up the patient and ply their expertise on that dimension of the patient or the patient's needs. As has been pointed out repeatedly, this combination of structure, function, and operation tends to work against the collective desire of the profession, the professionals, the patients, and their families that care be provided in a client-centered fashion (Hagenow 2003; Donnelly et al. 2004; Gzil et al. 2007; Patson 2007). And, as we might expect, it hardly seems the proper organization to support activities intended primarily to address the existential and social learning process involved in becoming newly abled.

What we are calling for as part of this thought experiment is research and development in the area of organization and operation of rehabilitation that would redesign the inpatient experience to meet the needs of an existential, social learning process. There is already ample literature in the area of "learning enterprises" on how to design spaces and organize functions in ways that promote learning (Randeree 2006; Moultrie et al. 2007; Taylor 2009). This literature could serve as a starting point for developing learning environments that promote re-having. Additionally, further studies are necessary to better capture the nature of the work involved in the existential and social learning processes that are involved in TSCI rehabilitation. For example, in Papadimitriou (2008a), we explored the "work" of physical therapy, and in that context, we noted a range of invisible activities (i.e., activities that the PT does but that are not commonly discussed explicitly as part of the role of the PT). For example, the kind of coaching that a PT is required to perform in order to move a patient from one level of ability to the next (at least to do it well) requires that the PT has already developed a significant level of trust with their patient. This kind of trust building would be essential to an existential, social approach to rehabilitation. Therefore, further research is required to understand the range of these invisible activities and the kinds of physical spaces, organizational structures, and operational models that would best facilitate those kinds of activities (Cameron 2010; Park 2012; Magasi 2012).

As discussed in the section "Rethinking Practice," the guide for rethinking the organizational dimensions of rehabilitation involves shifting the focus from the deficit model of well to ill to recovering to the difference model of nondisabled to disabled to newly abled. In the difference model, the focus is on what it is to "be" newly abled, and so the process of rehabilitation becomes the process of rehabituating, of facilitating the process of the person's re-having their ability to become who they are, to live again habitually through their bodies and to develop for themselves a concrete sense of future that permits them to make (their own) sense of their lives. Our thought experiment asks the community of researchers and practitioners of organizations: what organizational forms or types are required (or are optimal) for meeting those needs.

Rethinking Rehabilitation as an Institution

In the United States, as it is in many countries, rehabilitation is at present securely ensconced in the health-care system, embedded as an institution in both the private and public insurance systems, and financed for the most part through those means. But it did not always hold this institutional position, nor this means of financial support. Like care services for people with developmental disabilities in the United States, rehabilitation began as a social service, financed originally through public and private charity arrangements and later through social-welfare funds. The history of the institution and finance of services for people with developmental disabilities shows how the shift from a social benefit to a healthcare need has left millions of individuals and families struggling to obtain resources that were once guaranteed (Braddock and Hemp 2008).* The case of people with developmental disabilities is a cautionary tale for those who support the view that TSCI rehabilitation is best served institutionally and financially by remaining within the health-care orbit. This, of course, is the U.S. perspective, and though we surmise that the U.S. experience has not been unique, any effort to rethink the institutional place of SCI rehabilitation should also consider the history and experiences of other countries, both their successes and their failures.

Again, our question is, What institutional and financial arrangements would be most appropriate to organizations and professional practices dedicated to the existential and social learning of persons with SCI?

---

* It should, however, be noted that in the United States this phenomenon was complicated by the move to deinstitutionalization in the twentieth century and by the unique role played by the states in Medicaid funding.

As we described in the section "Prologue to the Thought Experiment," rehabilitation in the United States is presently financed as part of the web of public and private healthcare insurance schemes, though relying heavily on Medicare and Medicaid. And, as we noted above, placing rehabilitation within the healthcare institutionally brings with it a medically oriented approach to assessment and justification and so has promoted the use of scores to measure discrete bodily functions and ability to perform ADLs, as evidence of progress for "recovery." These scores, though ostensibly useful to clinicians to evaluate progress and assist in goal setting, appear to be driving (or heavily influencing) treatment decisions and recording practices, without concern to client-centered philosophy and goals. In short, the present situation in the United States sees the medically oriented, healthcare financed approach to rehabilitation driving practitioners away from best practices, away from lengths of stay that best serve their patients, and away from client-centered practices (i.e., that recognize the autonomy and decision-making authority that should reside with patients and their families).

On the issue of finance, the question the thought experiment asks is, What model for financing care would best serve an approach to rehabilitation that was less driven by achieving medically oriented functional and physical goals and more driven by the needs of individuals working to become newly abled to live happily and productively from their embodied selves and into futures that they can appropriate for themselves? Thinking back to our hypothetical, how would a society finance the process of a white male becoming a black female? Would this be a medical cost? Assuming a society saw the value in ensuring that such changed lives remained productive and fulfilling, would it leave such costs to private insurance thereby depriving many who could not afford it to manage on their own? Presumably not. And if society thus recognized the need for equitable treatment of those individuals, the question of finance would move to the question of institution. If this transformation is not medical, and is not best driven by medically oriented measures of progress and success, then what national-level institution should serve as the conduit for society's intention to see these people assisted?

Education is one option. Clearly, we have suggested that the process of becoming newly abled is one of existential and social learning. If this is a learning process, is there a role for the Department (or Ministry) of Education to play? As part of our thought experiment, one could certainly imagine the Department of Education providing funding and expertise to assist in research on the process of becoming newly abled. One could also see them supporting the training of rehabilitation practitioners. Indeed, the U.S. Department of Education already does provide some level of support for these two efforts. Beyond that, the Department has promoted the intention to see persons recovering from disability return to active, productive, and self-sustaining lives. They have also actively promoted the shift from expert-driven care to client-centered care as the appropriate philosophy attendant to rehabilitation. As such, one can imagine, in the United States at least, the Department of Education as the locus for an existential approach to rehabilitation and as taking up the role of primary funding source for rehabilitation rethought as promoting and facilitating the process of re-having for persons becoming newly abled.

Conclusion

In this chapter, we have provided a starting place for rethinking rehabilitation. We have asked you, the reader, to reimagine the process of becoming newly abled as akin to what it would be like to transition from a white male to a black female and how professionals and society might best support that transition. We have provided evidence for a way of reconceiving disability as a process that is less primarily medical, physical, and functional, and more a matter of the transformations that must take place at an existential level for a person to move from disabled to newly abled. We have shown, if only very preliminarily, how these processes are and can be conceived as existential, one of living-bodily from and of becoming, rather than one that is primarily a matter of regaining physical function. We have characterized the process of rehabilitation as one better understood as a re-having, or one's rehabituated sense of how one lives from his or her bodily existence and out into his or her projected future possibilities, than as a process of regaining abilities. We then provided a guide for beginning to rethink key dimensions of rehabilitation: professional practice, financing, and social policy institutional stewardship.

Our hope is that you, the reader, will take up any one or more of these dimensions and help carry them forward. Clearly there is more research, writing, and conceptual clarification required to firmly secure our understanding of rehabilitation as a process of existential and social learning. There are professional practitioners and researchers interested in rehabilitation professions and practices who, like us, would like to see rehabilitation is rethought in this way—as a process of promoting and facilitating re-having, and developing approaches and techniques that would foster existential and social learning in their clients. Similarly there are those who manage rehabilitation units and those who study organizational practices and how they align with the purposes of the work being undertaken. We look to those leaders and scholars to take up exploration of organization forms that promote the process of becoming newly abled and better balance the physical and medical needs of patients with their existential and social learning needs.

Finally, we look to those who develop social policy and those who study it to explore variations in current practices that would promote a more existentially oriented approach to the financing and institutional support for rehabilitation.

In the end, this is neither a purely academic matter nor should rethinking be a purely academic exercise. Every year, tens of thousands of individuals in the United States alone find themselves in an inpatient SCI unit struggling to go through the process of re-having their lives, of learning to live from their newly abled bodies and to carve out a future for themselves that allows them to continue the fragile human process of becoming who they already are. As populations around the world become older, these individuals will live longer; as social resources become more limited, these individuals will need to remain productive.

The arc of understanding of disability has shifted. Disabled people are no longer "other"; it is no longer acceptable that they be treated differently. They are no longer (dis)abled; they are newly abled, equally but differently abled. Our current practice of assisting with re-abling them, as though all that has changed for them is their physical capacities, is out of date and out of step with the recognition that they have every right to be reassimilated into society. This is our goal for rethinking rehabilitation.

Acknowledgments

The authors thank the participants of the Rethinking Rehabilitation Symposium (Toronto, Canada, July 2012) for their feedback and support. We especially thank the editors, Kath McPherson and Barbara Gibson, for their critical readings of earlier drafts, and D. Nicholls and G. DeJong for keeping us conceptually honest and grounded. We also thank our children who played nicely together as we worked on this manuscript during nights and weekends.

References

Albrecht, G. (1992). *The Disability Business: Rehabilitation in America*. Newbury Park, CA: Sage Publications.

Beckett, C., and Wrighton, E. (2000). "What matters to me is not what you're talking about": Maintaining the social model of disability in "public and private" negotiations. *Disability and Society, 15*, 991–999.

Braddock, D. L., and Hemp, R. (2008). *Services and Funding for People with Developmental Disabilities in Illinois: A Multi-State Comparative Analysis*. Denver, CO: Department of Psychiatry, University of Colorado, Denver School of Medicine.

Bright, F. A. S., Kayes, N. M., McCann, C. M., and McPherson, K. M. (2013). Hope in people with aphasia. *Aphasiology, 27*(1), 41–58.

Cain, P. (2002). "Partnership" is not enough: Professional-client relations revisited. In K. W. M. Fulford, D. L. Dickenson, and T. H. Murray (Eds.), *Healthcare Ethics and Human Values* (pp. 278–281). Oxford: Blackwell.

Cameron, I. (2010). Models of rehabilitation—Commonalities of interventions that work and of those that do not. *Disability and Rehabilitation, 32*(12), 1051–1058.

Carpenter, C. (2004). Dilemmas of practice as experienced by physical therapists in rehabilitation settings. *Physiotherapy Canada, 57*(1), 63–74.

Charlton, J. (2010). The dimensions of disability oppression. In L. J. Davis (Ed.), *The Disability Studies Reader* (3rd ed., pp. 147–159). New York: Routledge.

Cohen, M. E., and Schemm, R. L. (2007). Client-centered occupational therapy for individuals with spinal cord injury. *Occupational Therapy in Health Care, 21*, 1–15.

Cott, C. (2004). Client-centred rehabilitation: The client's perspective. *Disability and Rehabilitation, 26*, 1411–1422.

Cott, C. A., Teare, G., McGilton, K. S., and Lineker, S. (2006). Reliability and construct validity of the client-centred rehabilitation questionnaire. *Disability and Rehabilitation, 28*(22), 1387–1397.

Davis, L. (1995). *Enforcing Normalcy: Disability, Deafness, and the Body*. New York: Verso.

Davis, L. J. (Ed.). (2010). *The Disability Studies Reader* (3rd ed.). New York: Routledge.

DeJong, G. (1979). Independent living: From social movement to analytic paradigm. *Archives of Physical Medicine and Rehabilitation, 60*, 435–446.

DeJong, G., Hoffman, J., Meade, M. A., Bombardier, C., Deutsch, A., Nemunaitis, G., and Forchheimer, M. (2011). Postrehabilitative health care for individuals with SCI: Extending health care into the community. *Topics in Spinal Cord Injury Rehabilitation, 17*(2), 46–58.

DeJong, G., Palsbo, S., Beatty, P., Jones, G., Knoll, T., and Neri, M. (2002). The organization and financing of health services for persons with disabilities. *Milbank Quarterly, 80*(2), 261–301.

Donnelly, C., Eng, J. J., Hall, J., Alford, L., Giachino, R., Norton, K., and Kerr, D. S. (2004). Client-centered assessment and the identification of meaningful treatment goals for individuals with a spinal cord injury. *Spinal Cord, 42*, 302–307.

Fredericks, S., Lapum, J., Schwind, J., Beanlands, H., Romaniuk, D., and McCay, E. (2012). Discussion of patient-centered care in health care organizations. *Quarterly Management and Health Care, 21*(3), 127–134.

Gibson, B. E., Carnevale, F. A., and King, G. (2012). "This is my way": Reimagining disability, in/dependence and interconnectedness of persons and assistive technologies. *Disability and Rehabilitation, 34*(22), 1894–1899.

Gibson, B. E., Young, N. L., Upshur, R. E. G., and McKeever, P. (2007). Men on the margin: A Bourdieusian examination of living into adulthood with muscular dystrophy. *Social Science & Medicine, 65*(3), 505–517.

Gzil, F., Lefeve, C., Cammelli, M., Pachoud, B., and Ravaud, J. F. (2007). Why is rehabilitation not yet fully person-centered and should it be more person-centered? *Disability and Rehabilitation, 29*, 1616–1624.

Hagenow, N. R. (2003). Why not person-centered care? The challenges of implementation. *Nursing Administration Quarterly, 27*(3), 203–207.

Hammell, K. W. (2006). *Perspectives on Disability and Rehabilitation: Contesting Assumptions, Challenging Practice.* Edinburgh: Churchill Livingstone.

Hammell, K. W. (2007a). Client-centred practice: Ethical obligation or professional obfuscation? *The British Journal of Occupational Therapy, 70*(6), 264–266.

Hammell, K. W. (2007b). Reflections on … a disability methodology for the client-centred practice of occupational therapy research. *Canadian Journal of Occupational Therapy, 74*(5), 365–369.

Heidegger, M. (1962). *Being and Time* (J. Macquarrie and E. Robinson, Trans.). New York: Harper and Row.

Higgs, J., and Titchen, A. (2001). Rethinking the practice-knowledge interface in an uncertain world: A model for practice development. *The British Journal of Occupational Therapy, 64*(11), 526–533.

Hoff, T. (2010). Managing the negatives of experience in physician teams. *Health Care Management Review, 35*(1), 65–76.

Kavanagh, E. (2012). Affirmation through disability: One athlete's personal journey to the London Paralympic Games. *Perspectives in Public Health, 132*, 68–74.

Kayes, N. M., and McPherson, K. M. (2010). Measuring what matters: Does "objectivity" mean good science? *Disability Rehabilitation, 32*, 1018–1026.

Kielhofner, G. (2005). Rethinking disability and what to do about it: Disability studies and its implications for occupational therapy. *American Journal of Occupational Therapy, 59*, 487–496.

Leplege, P. A., Gzil, F., Cammelli, M., Lefeve, C., Pachoud, B., and Ville, I. (2007). Person centeredness: Conceptual and historical perspective. *Disability and Rehabilitation, 29*(20–21), 1555–1565.

Levack, W. M., Dean, S. G., Siegert, R. J., and McPherson, K. M. (2011). Navigating patient-centered goal setting in inpatient stroke rehabilitation: How clinicians control the process to meet perceived professional responsibilities. *Patient Education and Counseling, 85*(2), 206–213.

Linton, S. (1998). *Claiming Disability: Knowledge and Identity.* New York: New York University Press.

Macleod, R., and McPherson, K. M. (2007). Care and compassion: Part of person-centred rehabilitation, inappropriate response or a forgotten art? *Disability and Rehabilitation, 29*(20–21), 1589–1595.

Magasi, S. (2008a). Disability studies in practice: A work in progress. *Topics of Stroke Rehabilitation, 15*(6), 611–617.

Magasi, S. (2008b). Infusing disability studies into the rehabilitation sciences. *Topics of Stroke Rehabilitation, 15*(3), 283–287.

Magasi, S. (2012). Negotiating the social service systems: A vital yet frequently invisible occupation. *OTJR-Occupation, Participation, and Health, 32*(1), S25–S33.

Magasi, S., Heinemann, A., Whiteneck, G., Bogner, J., and Rodriguez, E. (2008). What does participation mean? An insider perspective from people with disabilities. *Disability and Rehabilitation, 30*(19), 1445–1460.

Merleau-Ponty, M. (1962). *Phenomenology of Perception* (C. Smith, Trans.). London: Routledge and Kegan Paul.

Mitchell, D. T., and Snyder, S. L. (1997). *The Body and Physical Difference: Discourses of Disability.* Ann Arbor, MI: University of Michigan Press.

Moultrie, J., Nilsson, M., Dissel, M., Haner, U.-E., Janssen, S., and Van der Lugt, R. (2007). Innovation spaces: Towards a framework for understanding the role of the physical environment in innovation. *Creativity and Innovation Management, 16*(1), 53–65.

Murer, C. G. (2008). Inpatient rehab overview. *Rehab Management, 21*(2), 38–39.

Nicholls, D. A., and Gibson, B. E. (2010). The body and physiotherapy. *Physiotherapy Theory and Practice, 26*(8), 497–509.

O'Day, B., Palsbo, S., Dhont, K., and Scheer, J. (2002). Health plan selection criteria by people with impaired mobility. *Medical Care, 40*(9), 732–742.

Oliver, M. (1990). *The Politics of Disablement: A Sociological Approach.* New York: St. Martin's Press.

Oliver, M. (1993). Disability and dependency: A creation of industrial societies. In J. Swain, V. Finkelstein, S. French, and M. Oliver (Eds.), *Disabling Barriers—Enabling Environments.* London: Sage.

Oliver, M. (1996). *Understanding Disability: From Theory to Practice.* New York: St. Martin's Press.

Papadimitriou, C. (2008a). "It was hard but you did it": Work in a clinical context among physical therapists and spinal cord injured adults. *Disability and Rehabilitation, 30*(5), 365–774.

Papadimitriou, C. (2008b). Becoming en-wheeled: Re-embodiment as a wheelchair user after spinal cord injury. *Disability and Society, 23*(7), 691–704.

Papadimitriou, C. (2008c). The "i" of the beholder: Phenomenological "seeing" in the context of disability. *Sports, Ethics and Philosophy, 2*(2), 216–233.

Papadimitriou, C., and Carpenter, C. (2013). Client-centered practice in spinal cord injury: A field guide. Retrieved from http://www.carf.org/manuals.

Papadimitriou, C., and Cott, C. (2012). Team functioning and client-centered care in inpatient rehabilitation. *8th International Organizational Behavior in Healthcare Conference Proceedings*, Trinity College, Dublin, Ireland.

Papadimitriou, C., and Stone, D. A. (2011). Addressing existential disruption in traumatic spinal cord injury: A new approach to human temporality in inpatient rehabilitation. *Disability and Rehabilitation 33*(21–22), 2121–2133.

Park, M. (2012). Pleasure, throwing breaches, and embodied metaphors: Tracing transformations-in-participation for a child with Autism to a sensory integration-based therapy session. *OTJR-Occupation, Participation, and Health, 32*(1), S34–S47.

Parry, R. H. (2004). Communication during goal-setting in physiotherapy treatment sessions. *Clinical Rehabilitation, 18*(6), 668–682.

Patson, P. (2007). Constructive functional diversity: A new paradigm beyond disability and impairment. *Disability and Rehabilitation, 29*(20–21), 1625–1633.

Pawloski, D. (2001). Work of staff with disabilities in an urban medical rehabilitation hospital. *Disability Studies Quarterly, 21*(3), 67–75.

Pothier, D. (2006). Legal developments in the supreme court of Canada regarding disability. In D. Pothier, & R. Devlin (Eds.), *Critical Disability Theory: Essays in Philosophy, Politics, Policy, and Law* (pp. 305–316). Vancouver, Toronto: UBC Press.

Price, P., Stephenson, S., Krantz, L., and Ward, K. (2011). Beyond my front door: The occupational and social participation of adults with spinal cord injury. *OTJR: Occupation, Participation and Health, 31*(2), 81–88.

Randeree, E. (2006). Structural barriers: Redesigning schools to create learning organizations. *International Journal of Educational Management, 20*(5), 397–404.

Seymour, W. (1998). *Remaking the Body: Rehabilitation and Change.* London: Routledge.

Shakespeare, T. (2006). *Disability Rights and Wrongs.* New York: Routledge.

Shakespeare, T., Iezzoni, L. I., and Groce, N. E. (2009). Disability and the training of health professionals. *The Lancet, 374*(9704), 1815–1816.

Shattell, M. M., Bartlett, R., and Rowe, T. (2008). "I have always felt different": The experience of attention-deficit/hyperactivity disorder in childhood. *Journal of Pediatric Nursing, 23*(1), 49–57.

Siebers, T. (2008). *Disability Theory.* Ann Arbor, MI: University of Michigan Press.

Siegert, R. J., Ward, T., and Playford, D. E. (2010). Human rights and rehabilitation outcomes. *Disability and Rehabilitation, 32*(12), 965–971.

Smith, B., and Sparkes, A. C. (2007). Sport, spinal cord injury, and body narratives: A qualitative project. *Health Psychology Update, 16*, 26–32.

Smith, B., and Sparkes, A. C. (2008). Changing bodies, changing narratives and the consequence of tellability: A case study of becoming disabled through sport. *Sociology of Health and Illness, 30*, 217–236.

Sparkes, A. C., and Smith, B. (2002). Sport, spinal cord injury, embodied masculinities, and the dilemmas of narrative identity. *Men and Masculinities, 4*(3), 258–285.

Stone, D. A., and Papadimitriou, C. (2010). Exploring Heidegger's ecstatic temporality in the context of embodied breakdown. *Schutzian Research, 2,* 137–154.

Taylor, A. (2009). *Linking Architecture and Education: Sustainable Design for Learning Environments.* Albuquerque, NM: University of New Mexico Press.

Toombs, S. K. (1992). *The Meaning of Illness: A Phenomenological Account of the Different Perspectives of Physician and Patient.* Boston: Kluwer Academic Publishers.

Toombs, S. K. (1995). The lived experience of disability. *Human Studies, 18*(1), 9–23.

Toombs, S. K. (ed.) (2001). *Handbook of Phenomenology and Medicine.* Boston: Kluwer Academic Publishers.

Verkaaik, J., Sinnott, K. A., Cassidy, B., Freeman, C., and Kunowski, T. (2010). The productive partnerships framework: Harnessing health consumer knowledge and autonomy to create and predict successful rehabilitation outcomes. *Disability and Rehabilitation, 32*(12), 978–985.

Wendell, S. (1996). *The Rejected Body: Feminist Philosophical Reflections on Disability.* New York: Routledge.

Whiteneck, G., and Gassaway, J. (2010). SCIRehab: A model for rehabilitation research using comprehensive person, process, and outcome data. *Disability Rehabilitation, 32*(12), 1042–1049.

Whyte, J., and Hart, T. (2003). It's more than a black box; it's a Russian doll: Defining rehabilitation treatments. *American Journal of Physical Medicine and Rehabilitation, 82,* 639–652.

Wilson, J. C. (2010). Disability and the human genome. In L. J. Davis (Ed.), *The Disability Studies Reader* (3rd ed., pp. 52–62). New York: Routledge.

Index

Page numbers followed by f, t and b indicate figures, tables and boxes, respectively.